Nutrient Use Efficiency in Plants

Plant Ecophysiology

Volume 10

Series Editors:

Luit J. De Kok

University of Groningen, The Netherlands

Malcolm J. Hawkesford

Rothamsted Research, United Kingdom

Aims & Scope:

The Springer Series in *Plant Ecophysiology* comprises a series of volumes that deals with the impact of biotic and abiotic factors on plant functioning and physiological adaptation to the environment. The aim of the *Plant Ecophysiology* series is to review and integrate the present knowledge on the impact of the environment on plant functioning and adaptation at various levels: from the molecular, biochemical and physiological to a whole plant level. This series is of interest to scientists who like to be informed of new developments and insights in plant ecophysiology, and can be used as advanced textbooks for biology students.

More information about this series at http://www.springer.com/series/6193

Malcolm J. Hawkesford • Stanislav Kopriva
Luit J. De Kok
Editors

Nutrient Use Efficiency in Plants

Concepts and Approaches

Springer

Editors
Malcolm J. Hawkesford
Rothamsted Research
Harpenden, Hertfordshire
United Kingdom

Stanislav Kopriva
Botanical Institute
University of Cologne
Cologne, Germany

Luit J. De Kok
Laboratory of Plant Physiology
University of Groningen
Groningen, The Netherlands

ISSN 1572-5561 ISSN 2405-4321 (electronic)
ISBN 978-3-319-38217-3 ISBN 978-3-319-10635-9 (eBook)
DOI 10.1007/978-3-319-10635-9
Springer Cham Heidelberg New York Dordrecht London

Softcover reprint of the hardcover 1st edition 2014

Printed on acid-free paper

Springer is part of Springer Science+Business Media (www.springer.com)

Preface

Nutrient Use Efficiency in Plants: Concepts and Approaches is the ninth volume in the *Plant Ecophysiology* series. It presents a broad overview of topics related to improvement of nutrient use efficiency of crops.

Nutrient use efficiency (NUE) is a measure of how well plants use the available mineral nutrients. It can be defined as yield (biomass) per unit input (fertiliser, nutrient content). NUE is a complex trait: it depends on the ability to take up the nutrients from the soil, but also on transport, storage, mobilization, usage within the plant, and even on the environment. NUE is an important contributor to growth control and yield; the same levels of nutrients may cause growth and yield penalty in one species or variety but not in another one. NUE is of particular interest as a major target for crop improvement. Improvement of NUE is an essential pre-requisite for expansion of crop production into marginal lands with low nutrient availability but also a way to reduce use of inorganic fertiliser. Aspects of NUE have been covered in detail within the *Plant Ecophysiology* series, in the volumes on nitrogen, phosphorus, sulfur, or root physiology. In this volume, the expanding field of nutrient use efficiency is comprehensively discussed, with the aim to present the current approaches, concepts, and ideas on how to better understand the genetic control of NUE and how to use this knowledge for its improvement.

This volume, however, is special not only because of the new topic. It is also a presentation of a Marie Curie Initial Training Network BIONUT-ITN (BIOchemical and genetic dissection of control of plant NUTrition). BIONUT-ITN is a network of eight research institutions and one company who came together to provide state-of-the-art research training in different aspects of plant nutrient use efficiency. The individual student's projects have been linked to ensure that a fully integrated approach is taken to get the whole picture of plant nutrition. This integration is a key feature of the network, as it advances the science beyond focusing on one mineral nutrient, such as nitrogen or sulfur, to look at the combined nutritional needs of the plant using models and crops, in the laboratory as well as in the field. BIONUT is also a hub for activities in plant nutrition field, organizing conferences and fostering new collaborations. This volume is evidence of such integration. The contributions of the eight students span a broad range of themes within NUE. There are detailed reviews focused on single nutrients – sulfur, phosphorus, iron, and boron. Thus, macronutrients are discussed alongside micronutrients needed in small quantities, but still essential. But these reviews stress also the importance of interaction between different nutrients and the need for integrative view on plant nutrition. Other chapters bring overviews of large and complex topics or approaches – natural variation, autophagy, or effects of elevated CO_2. Included are contributions targeting nutrient deficiency on a molecular level as well as its monitoring in the field. Together with a thorough introduction into the NUE topic, the book presents ten chapters that wrap up the theme of NUE and the potential for crop improvement.

We hope that this book will find broad audience. It is not only an overview of an interesting and important research area; it is also a snapshot of current activities in the field and introduction of new generation of scientists from the BIONUT-ITN. We believe that it will be of interest to graduate students and researchers in a wide range of disciplines including plant nutrition, plant physiology, plant biochemistry, and agriculture.

Harpenden, UK — Malcolm J. Hawkesford
Cologne, Germany — Stanislav Kopriva
Groningen, The Netherlands — Luit J. De Kok

Acknowledgement

The editors thank Helen Jenkins for grammatical and technical editing of the chapters.

Contents

Chapter 1
Physiological Basis of Plant Nutrient Use Efficiency – Concepts, Opportunities and Challenges for Its Improvement

Martin Reich, Tahereh Aghajanzadeh, and Luit J. De Kok

Abstract Knowledge on the underlying physiological processes and variables which bias their contribution to nutrient use efficiency (NUE) is crucial to develop strategies for improvement in agroecosystems. This chapter aims to contribute to the understanding of the physiological basis of NUE to develop strategies for improvement by modern breeding, but also conceive the challenges and current limits to do so. General concepts will be summarized briefly and broken down to the main components before, in the main part of this chapter, the involved physiological processes are reviewed and discussed in their relation to NUE. This is followed by an identification of the factors that make the individual contributions of these processes to NUE so variable and impede one general concept for all crops, environmental conditions and nutrients. The last part of the chapter is dedicated to a critical analysis of the opportunities and challenges to improve NUE, which arise from physiological interactions and trade-offs on a whole plant level.

Keywords NUE (nitrogen use efficiency) • Nitrogen • Agroecosystem • MRT (mean residence time) • Crop yield • Oscillations • Acquisition efficiency • Utilization efficiency

Introduction

Plants are principally, as are all living organisms, chemical compartments, which are in thermodynamic disequilibrium with their environment. This is actively maintained by the utilization of solar energy for driving selective chemical exchange with the environment. In addition to carbon dioxide and water, which provide the structural and metabolic backbone elements C, O and H, the complex functioning of plants requires the uptake of at least 13 additional essential nutrients from the soil. Nutrients can be classified into two very distinct groups depending on

M. Reich • T. Aghajanzadeh • L.J. De Kok (✉)
Laboratory of Plant Physiology, University of Groningen, P.O. Box 11103, 9700 CC, Groningen, The Netherlands
e-mail: l.j.de.kok@rug.nl

M.J. Hawkesford et al. (eds.), *Nutrient Use Efficiency in Plants*,
Plant Ecophysiology 10, DOI 10.1007/978-3-319-10635-9_1

their concentration in plant tissues (macro- and micronutrients), and the specific roles they fulfil in the plant's metabolism are as diverse as their physiochemical properties (Marschner 2012).

Plant NUE is a term, which describes a highly complex, multigenic trait with various interconnected physiological processes involved and modified by numerous factors. Consequently, there are numerous approaches to define, analyse and possibly improve NUE. For a long time it has been known that the ability of plants to utilize nutrients can differ substantially between species and cultivars, and that this could be the basis for further improvement through breeding (Gerloff 1963; Shea et al. 1968; Siddiqi and Glass 1981). In order to develop a common framework for NUE, scientists started to formulate concepts and definitions that should serve as a basis for comparison and discussion of research. Since then countless studies in various scientific disciplines dealing with different plant species, in different contexts, under different conditions and focusing on different nutrients failed to find one definition of NUE that describes all cases satisfactory but rather revealed that the issue is too complex to do so.

In this chapter the definition of NUE will be discussed from a whole plant perspective. It starts with the transfer of NUE from an ecological to an agronomical context and the different levels of organization on which it can be discussed. This is followed by a brief introduction into the conceptual framework of NUE, especially for nitrogen (N) and equipped with these theoretical concepts the attention of the reader will be drawn to the concrete physiological basis of NUE. These processes are the targets of potential improvement of NUE for agricultural production by modern breeding. However, how relevant a certain physiological process is in a particular cropping system depends on three main variables: environment, plant and nutrient, which influence the physiological basis of NUE. The last parts of this chapter are dedicated to the need to improve NUE in modern agriculture and ends with a critical review of the chances and challenges to improve plant NUE from a whole plant perspective.

Nutrient Use Efficiency – Contexts and Concepts

A general definition of "efficiency" is: *The achievement of an intended outcome with a lowest possible input of costs.* While the input in the concept of NUE obviously is nutrients, the *intended outcome* needs to be further specified. This can happen in different ways, which leads to many different versions of what NUE actually means and how it can be improved. Very fundamental is the difference between an ecological and agronomical context. Understanding this difference is crucial to develop strategies of improving a plant with its complex ecophysiological background in the straightforward input-output system of agriculture.

The environment of a plant is far from being a stationary equilibrium. Arising from the way our planet turns around its own axis and follows its orbit around the sun, all abiotic and biotic factors on its surface underlie oscillations over time and all forms of living organisms are forced to adapt to the local oscillations in their respective habitat in order to reach a lifespan of hours, days, weeks, months, years or decades and finally produce successful offspring. This is especially true of plants which being sessile, have evolved strategies to synchronize their internal processes with the external oscillations of their environment to their best advantage. This synchronization includes life cycle, developmental program, morphology, diurnal physiological rhythms (Somers et al. 1998; McClung 2006) and also uptake and assimilation of nutrients (Zhang et al. 1991; Bot and Kirkby 1992; Delhon et al. 1995; Haydon et al. 2011) on different timescales. A well-adjusted synchronization with the environment will increase the performance of a plant and increase competiveness. From this ecological and evolutionary point of view plants can be called "nutrient efficient", if they use the temporal and spatial availability of nutrients for an optimal and balanced vegetative and reproductive growth, which is most suitable to survive and compete in their respective habitat and niche.

In an agricultural context, however, the quality of the *intended outcome* shifts. Instead of offspring the plant produces a desired yield product, which can be utilized for food production and other economically relevant purposes. With agricultural practice and plant breeding to increase the production of this agronomic *intended outcome* the plant is detached from its ecological and evolutionary context. No longer exposed to the natural selection pressure but the artificial selection by man, plants are reshaped for agriculture: development, morphology and fluxes of resources are rerouted towards increased production of whatever yield is desired. Even after thousands of years of breeding, plants still bear their ecological heritage, which may conflict with agricultural interests and may limit the potential for traditional plant breeding to improve NUE. Bringing these two contexts together is one of the main tasks for plant scientists to understand the functioning of a plant in the semi-natural system of agriculture. In this way, ecophysiological potentials of plants might be further exploited for agricultural production and the limits for improving plants with traditional breeding might be identified and overcome. A profound understanding of the physiological background of NUE is the basis for modern plant breeding using molecular techniques.

In an ecological context, NUE can be examined at the level of individuals, populations, species, communities or entire ecosystems (Nardoto et al. 2006). NUE in agronomy can also be discussed on several levels (Fig. 1.1). On each level, input and output differ in kind, and different components have to be considered to adequately calculate the NUE of the respective system. In scientific discussions it is important to consider the same level to avoid confusion and misunderstanding. During the past few decades, scientists have become increasingly aware that agricultural systems can be regarded as ecosystems in which the role of soil composition and fertility, the influences of biotic interactions as well as abiotic environmental factors should not be underestimated. This is brought together in the concept of agroecology (Gliessman 1990; Schnug and Haneklaus 1998; Francis et al. 2003). In this holistic approach, not only the *intended outcome* but also the *input of costs* becomes very complex, as negative impacts of fertilisation, pesticides etc. have to be considered. In modern approaches many different benefits that an intact ecosystem delivers to society are assessed. These "ecosystem services"

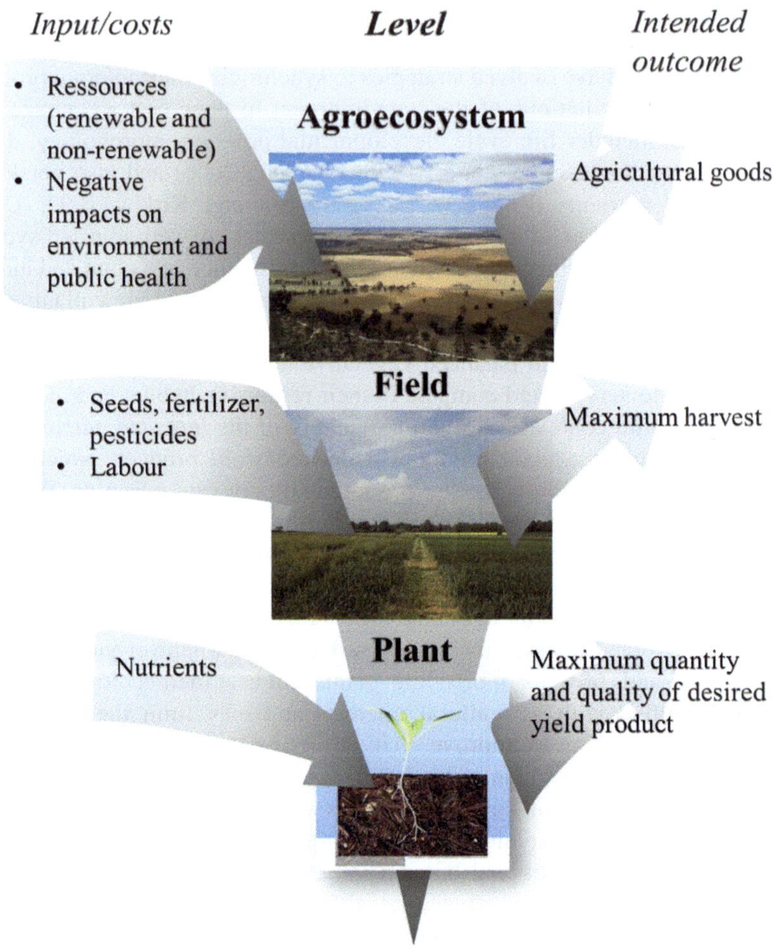

Fig. 1.1 NUE can be analysed and discussed on different levels of organization. On each level input and intended outcome differ in quality and a different terminology is used. Agroecologists regard the agricultural system with all associated ecosystems and society as the entity with a particular NUE, whereas agronomists put the field into focus. Plant physiologists deal with the plant as a complex input-output system with an inherent NUE. To avoid confusion in comparative research it helps to clarify on which level NUE is assessed

(Daily 1997) include delivery of agricultural goods as well as indirect beneficial properties such as the protective role of an intact forest against flooding or its capacity to purify water and bind carbon dioxide (Costanza et al. 1997; Daily et al. 2000; Tilman et al. 2002). Although it is difficult to assess the actual monetary value of all the components of an ecosystem, this concept is the only one that adequately expresses the efficiency of an agricultural system for a society in the long term by including all detrimental effects to the environment and to public health in the *input of costs* side of the calculation.

From the less holistic perspective of agronomy, the field is usually the level of choice for calculations of NUE. For farmers, the field represents an economic entity, where the inputs are the costs for labour and materials while the output is the harvest, usually measured by criteria such as "harvest index" (Donald and Hamblin 1976). In addition to NUE the term "fertiliser efficiency" is often used to describe how efficiently the applied fertiliser is used by the agronomic system: how much yield is produced, the quantity of nutrients which remains in the soil and how much is lost from the system by leaching and emission (Saurbeck and Helal 1990; Oenema et al. 2009). However, NUE can also be discussed at the plant level where a single plant instead of a whole field is regarded as an input-output system and in the present chapter we will deal with the individual plant and its NUE to discuss all the other dimensions and factors in relation to it. Improvement of NUE at the plant level also has the potential to improve NUE at higher levels and therefore has strong agronomic and environmental implications.

A common conceptual framework ensures a consistent use of terminology and definitions. As was noted above, NUE can be discussed in an ecological as well as an agronomical context. One context again can be divided into a sub-set of different levels of organization. By coming down to the level of an individual plant in an agronomical context, much of the universality of the term has been reduced. As already mentioned, the term *efficiency* implies the achievement of an intended outcome with a lowest possible input of costs. Therefore a very simplified definition for the efficiency of a given system can be expressed in the equation:

$$Efficiency = Output/Input$$

If values for input and output are competitive, the maximum value of efficiency is 1, in an ideal case where input equals output. Either decreasing the input or increasing the output might achieve higher efficiency. Every economical concept, which has to generate profit, in principle follows this simplified equation and agronomy is no exception. In all sectors of an industrialized economy it is desirable to make working processes less costly while maintaining or increasing the output. Technical innovation leads to more sophisticated techniques and methods, which also revolutionized the efficiency of agricultural practice at the field level. There is, however, an essential difference in improving an inanimate machine or process, which has been planned and constructed by man and improving a living organism such as a plant whose functioning is still far from being fully understood.

From a whole plant perspective NUE consists of several components and by regarding a plant as an input-output system, physiologists have established equations that put these components into context in relation to NUE. In the most universal approach, NUE at a plant level can be divided into two main components: the efficiency of nutrient acquisition (NAcE) and the efficiency with which the nutrient is utilized to produce the desired yield (nutrient utilization efficiency, NUtE):

$$NUE = NAcE * NUtE$$

While Chapin (1980) defined NUE simply as the inverse of the tissue nutrient concentration, NUtE can be further sub-divided into nutrient productivity (NP) and mean residence time, the period in which a certain nutrient can be used for production (MRT; Berendse and Aerts 1987). In the 1980s, Vitousek and co-workers (Vitousek 1982; Birk and Vitousek 1986) defined the nitrogen use efficiency (NitUE) of perennials as the amount of organic matter, which is lost from a plant or permanently stored in wood, divided by the amount of N lost or permanently stored. It was shown that NitUE of *Pinus taeda* L. stands decreased with increasing N availability. A more general definition was suggested by Berendse and Aerts (1987), who identified MRT and nitrogen productivity (NitP) as the main components of NitUE:

$$NitUE = NitP * MRT$$

According to Berendse and Aerts (1987), NitP describes the instantaneous rate of carbon fixation or biomass production per unit N present in the plant while MRT is a measure for the period in which N can be used for carbon fixation. This concept of MRT can theoretically be extended to other nutrients and plant species. In fertilisation models NP can be used to calculate the nutrient flux density that is necessary to maintain an optimal nutrient concentration in the plant (Ingestad 1988) but this again refers to the field and not to NUE at the plant level.

It is well known that plant species and ecotypes, which naturally grow in nutrient-poor soils possess mechanisms to increase the MRT of nutrients *e.g.* slow growth, high accumulation of nutrients and efficient remobilization of such storage capacities or a reduction of nutrient loss (Vázquez de Aldana and Berendse 1997). In soils where nutrients are available in excess or at least where nutrient availability is not the limiting factor, there is less selective pressure on developing such mechanisms. It is more important to have a high NP to grow fast and compete with neighbouring individuals for space and light and one way to reach this might be having a high nutrient throughput rather than a long MRT. Studies under controlled conditions with plants from both soil types showed that in the short term fast growing species were the better competitors in both optimal and limiting N conditions while in the longer term, plants from nutrient-poor soils outcompeted fast growing species under limiting conditions (Chapin 1980; Wedin and Tilman 1990; Berendse et al. 1992). It is considered that these differences in NUE between plants, which originated from soils with different nutrient concentrations, are due to differences in the underlying physiology, morphology and development. Van der Werf et al. (1993) showed how important morphological traits are for adapting NUE to the respective nutrient concentration in the soil. For instance, a high investment in root mass served for the high NP of fast growing species, though it should be noted that the majority of these studies on NUE dealt with wild species and N.

Expression of NP as unit biomass produced per unit nutrient may not always be the most suitable measure. The desired product in an agricultural system is not always biomass, consisting of structural or non-structural carbohydrates, but more often seeds that are rich in proteins or oil. It is thereby not only important how a nutrient contributes to growth but also how it improves the yield and quality of the desired product. Therefore, the respective nutrient can itself be a substrate for the production (*e.g.* as N and sulfur (S) for proteins) or a facilitator of the production (*e.g.* by being a component of an enzyme involved). Consequently NP on the basis of biomass may not always be the best measure and a more general indicator may be yield productivity (YP), which includes quantity and quality of the desired yield product per unit nutrient in the plant tissue. However, in agriculture and in general for all nutrients, NP (or YP) and MRT can be seen as sub-components of NUtE.

Physiological Processes Involved in NUE

After the derivation of a conceptual framework the key physiological processes involved in the complex trait NUE will be briefly summarized (Fig. 1.2). As described above, NAcE is one main component of NUE and consequently nutrient uptake is one of the key processes involved. Although some nutrients can be derived from the atmosphere (*viz.* N and S; Faller 1972; Stulen et al. 1998; De Kok et al. 2007), the plant largely depends on mineral nutrients taken up from the soil (Mengel and Kirkby 1987; Marschner 2012). These are either derived from weathering of parental rock material or biological breakdown of organic matter and the chemical availability to the plant depends on soil-specific properties which in turn determine the proportion of nutrients dissolved in the soil water (usually less than 0.2 %), bound to organic detritus (around 98 %) or adsorbed by soil colloids (Larcher 1995).

There has been much discussion on the significance of NAcE in explaining differences in NUE between plants. Most studies were focussed exclusively on N and came to different conclusions. For corn (*Zea mays L.*) it has been concluded from a study with different hybrids that NAcE is only relevant for differences in NitUE if the outside N concentrations are high, while NUtE of accumulated N was the driving variable if the supply was low (Moll et al. 1982). Whereas in pumpkin NAcE was not a possible target to improve NUE at either high or low N concentrations (Swiader et al. 1994). However, recent studies have suggested that an increased acidification capacity of the rhizosphere could be targeted to increase nitrate uptake and improve NUE (Paez-Valencia et al. 2013). In addition, the NUE of an agricultural system may be improved if plants could maintain internal nutrient concentrations and optimal growth with a lower outside concentration in the soil. Therefore understanding the response mechanisms of NAcE to nutrient deficiencies may improve the ability of crops to tolerate lower nutrient concentrations in the soil and thereby save fertiliser and reduce potential pollution.

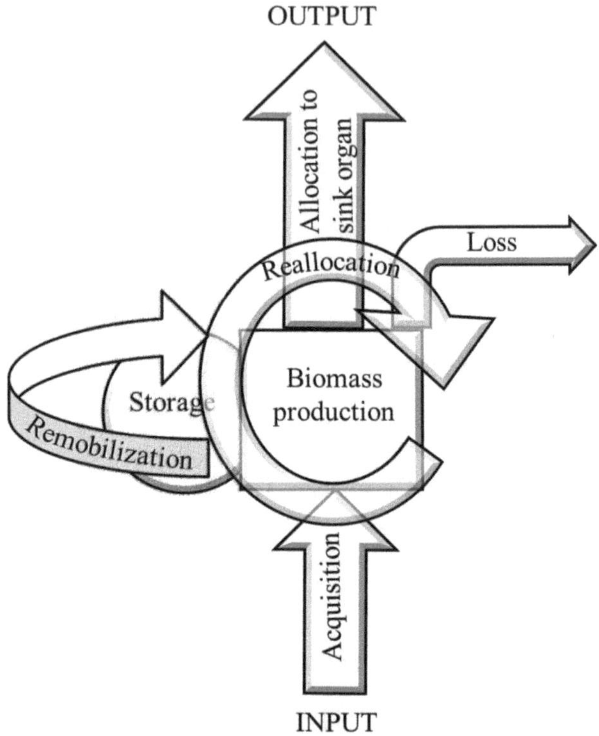

Fig. 1.2 Plant NUE has a complex physiological basis with interacting cellular and whole plant processes. After the acquisition of a nutrient it contributes directly or indirectly to the production of biomass and the final yield. Storage and remobilization are important processes that buffer asynchronies in nutrient demand and availability and the efficient reallocation of nutrients between different plant organs is a crucial process during plant development. Especially in cereals the translocation of nutrients to the finally harvested sink organ, the grains, is of particular importance. Nutrient loss can happen in several ways and displays a general constraint for NUE

Nutrient storage is another process of importance and can be functionally sub-divided into accumulation, reserve formation and recycling (Chapin et al. 1990). Accumulation summarizes the increase of compounds that are not directly related to growth. They accumulate simply because the availability exceeds the demand of the plant metabolism for these compounds. Reserve formation in contrast describes metabolically controlled storage in designated storage compounds. In this way compounds that otherwise would promote growth are stored in a form that does not. The formation of these storage compounds directly competes with growth and other processes that would use the compound in its original form as a substrate. In the process of recycling, compounds that originally contributed to growth promotion or other physiological functions but which would be lost are actively broken down to be used for future growth (Chapin et al. 1990). The significance of nutrient storage and remobilization for NUE depends on nutrient availability. A study with a number of hybrids of corn (*Zea mays L.*) revealed that under low N supply differences in NitUE between hybrids are related to variation in the utilization of stored N. However if N supply was high, acquisition efficiency became more important (Moll et al. 1982). A low ability to remobilize N leads to a lowered N harvest index in *Brassica napus* (Rossato et al. 2001). Similarly for S the limits of storage capacity and remobilization efficiency of sulfate are regarded as a constraint to NUE and a possible target for its improvement (Hawkesford 2000).

For biomass production plants convert inorganic carbon dioxide from the atmosphere to organic carbohydrates via photosynthesis. Fuelled by the energy of the sun, this process is the primary generator of all biomass on earth. Although N is most directly linked to photosynthesis, all the essential nutrients contribute in some form to growth promotion *i.e.* biomass production. This can happen directly if the nutrient is part of the carbon-assimilating apparatus or indirectly if it plays a role in energy transfer, defence, homeostasis, tolerance and other processes that facilitate optimal plant functioning. Consequently for every nutrient a respective NP can be assigned which is a measure for the biomass produced per unit of the nutrient in the plant. However, the mechanisms underlying this component of NUE and how the NP (and consequently the NUtE) of a certain nutrient can be improved are manifold and depend on the specific role that a nutrient plays in plant metabolism.

Again N is studied most intensively, and due to its direct link to photosynthesis and biomass production, there is a clear cut correlation with the NP of N and (i) the amount of total N invested in the photosynthetic tissue, (ii) the N efficiency of photosynthesis and (iii) relatively low loss of carbon due to respiration (Ågren 1985; Poorter et al. 1990). The carbon-assimilating enzyme Rubisco is currently one of the most prominent targets for possible genetic improvement of photosynthesis (Loomis and Amthor 1999; Parry et al. 2011), largely due to its apparent catalytic inefficiency in carboxylation and its consequent high abundance. The idea is that a higher efficiency would lead to less Rubisco being needed to maintain the same rate of photosynthesis and consequently, as this enzyme contains high amounts of N, a higher NUE of N. One intriguing approach is the attempt to express the Rubisco of some non-green algae, which have a greater specificity for CO_2, into higher plants (Whitney et al. 2001).

However, is the biochemical inefficiency of photosynthesis really the bottleneck that hinders higher biomass production and NUE? Although photosynthetic efficiency is in theory one of the key limiting factors for increasing biomass and crop yields (Long et al. 2006; Parry et al. 2011), supportive correlations in practice are not easy to assess and studies come to different conclusions. Studies on closely related germplasm of wheat showed a correlation of photosynthetic rate and yield (Watanabe et al. 1994), while comparisons of cultivated crops with their wild ancestors showed that the latter have a higher photosynthetic rate (Evans and Dunstone 1970). The potential limiting role of photosynthesis apparently depends to a greater extent on other processes with negative feedback on photosynthesis. If the capacity of the sink declines and the flux of photosynthates into sink products stagnates, this results in a compensating down-regulation of photosynthesis. Consequently the strength of the sink is just as important if not more so for yield as the efficiency of the source (Zelitch 1982; Borrás et al. 2004; Reynolds et al. 2005). According to these studies an increased sink capacity is required to increase photosynthesis and not the other way around. However, field studies with C_3 plant species under exposure to elevated levels of carbon dioxide (eCO_2) suggested that an increase in yield is, indeed source-limited or at least that sink capacity is stimulated by the increased source activity (higher net photosynthetic rate under eCO_2). These studies suggested that sink capacity is not necessarily a constraint to increase yield production by means of improving photosynthesis (Kimball et al. 2002; Ainsworth et al. 2004).

Apart from the question of how source and sink, or in other words supply and demand, determine and influence each other, the duality of photosynthesis and photorespiration makes the issue more complex. Up to one third of C fixed by the carboxylation activity of Rubisco is again lost by photorespiration. While some authors propose that photorespiration is of vital importance for plant functioning (Kozaki and Takeba 1996), others see these functions as at least partially redundant and suggest that a reduction of the C lost by this process would improve the efficiency of photosynthesis and biomass production (Long et al. 2006; Peterhansel and Maurino 2011). However, knowledge about the different roles of photorespiration in plant metabolism and NUE is still limited.

Another important process involved in NUE is the reallocation of nutrients. Re-use of nutrients from senescing leaves reduces nutrient loss and thereby increases NUE. Once more N is tightly coupled to C gain and its efficient allocation from one leaf to another contributes to optimal C fixation (Field 1983). In this process older leaves with declining photosynthetic N efficiency are exploited as a source for N, which is reallocated to young leaves to promote their growth. In this way N is used efficiently for photosynthesis at a whole plant level (Westoby et al. 2002; Escudero and Mediavilla 2003) and also NUE is increased, as loss is reduced. Resorption of nutrients from senescing leaves has also been studied for P (Lajtha 1987; Chapin and Moilanen 1991; Killingbeck 1996). It is generally assumed that the costs of this process are very low for the plant (Givnish 2002), which further supports reallocation of nutrients as a key process for the improvement of NUE. However, nutrients, which are efficiently recycled within the plant and thereby are not lost during senescence, will also not end up in the decomposition cycle in the soil. Whether this has negative feedback consequences for the plant and NUE is not fully understood and much will depend on the particular system. However, there are speculations about a general trade-off between efficient nutrient re-sorption in plants and the decomposability of litter (Aerts 1997).

The translocation of nutrients to the harvestable yield organ follows the same principles as the allocation to other plant organs. For obvious reasons it is, however, the most crucial allocation process for yield production and therefore regarded as a special case that is worthy of additional attention. Plant breeding has resulted in a wide diversity of crops in which virtually any part of a plant might serve as a yield organ: roots, stems, leaves, seeds, fruits. However, the six most important crops in terms of worldwide food and feed production are all grain crops (corn, rice, wheat, soybean, barley and sorghum) with seeds being the plant organ of interest and grain filling as a crucial step for yield production (Borrás et al. 2004; Foulkes et al. 2011). The reallocation of N from senescing leaves to the developing seeds is of particular importance in determining the quality of the crop and thus increasing the efficiency of reallocating N from leaves to grains is a potential target for improving NUE

(Barbottin et al. 2005). While N translocation to grain determines its quality, on improving NUE of P could be achieved by decreasing its translocation to the grain (Rose et al. 2010). Additionally it should be noticed that the efficient translocation of nutrients (as well as C) to yield organs is not only determined by the efficiency of the exploitation of the sink organs but also by the sink strength (*e.g.* for wheat: Reynolds et al. 2005; for rice: Ntanos and Koutroubas 2002).

The loss of nutrients to the outside may be the most obvious constraint to NUE. There are several paths by which nutrients can be lost by plants to their environment. Leaves lose nutrients by leaching, in some cases as gases or other volatile compounds (Eichert and Fernández 2012) and finally by litter fall, *i.e.* senescence. These processes of nutrient loss can either be a way for the plant to balance its nutritional status or may be unavoidable, for example due to wash-off by rain or evaporation due to a trade-off with stomatal conductance. In the latter case, a reduction of nutrient loss from the plant to its environment before harvest may be a target for the improvement of NUE, particularly in crop systems where litter and its nutritional status play a minor role. Further research is required to more fully understand the physiological significance of these losses in order to distinguish between avoidable leaks and metabolic valves, which assure internal nutrient balance and thereby optimal plant functioning.

Factors Affecting NUE

The complex physiological basis of NUE becomes even more complex in reality, as the contribution of these processes is modified by numerous factors, which can be categorized into plant, environment and nutrient (Fig. 1.3). Numerous studies and reviews have pointed out that concrete definitions of NUE depend to a great extent on plant species and growth type. Again it should be stated that NUE is an artificial term based on a hypothetical input-output concept. The diversity of NUE in nature, however, reflects the diversity of plant strategies to survive and produce successful offspring in their very different niches. How they perform if we apply the agricultural standard of NUE does not reflect their ecological and evolutionary fitness. Despite breeding the strong influence of the respective phylogenetic background of a cultivated plant on its performance and peculiarities in the field adds another degree of difficulty when defining one general concept for NUE. Fundamental differences between crops can be metabolic in nature, *e.g.* between C_3 and C_4 plants (Brown 1978) or arise from different growth forms such as trees which are harvested after decades, and herbaceous crop species which produce their yield within months. Processes such as nutrient storage and reallocation function very differently and have different significance for NUE in annual and perennial species (Aerts and Chapin 1999), as well as in deciduous and evergreens (Chapin and Kedrowski 1983; Aerts 1990; Franklin et al. 2009). This variability in plant growth strategy makes it hard to derive one concept and set of definitions for NUE and its

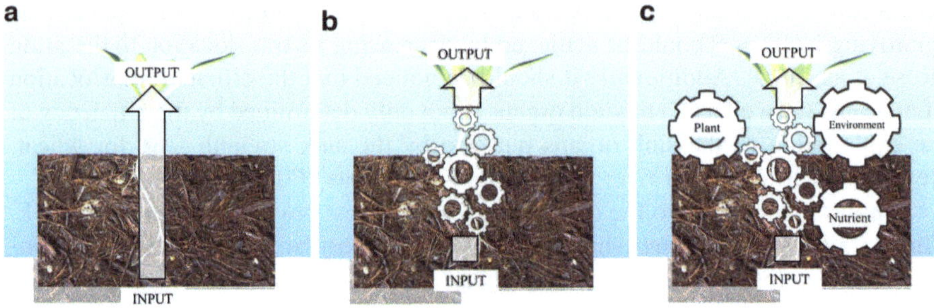

Fig. 1.3 The complexity of NUE – In an agronomical context the plant is regarded as an input-output system with an inherent efficiency that shall be improved (**a**). Numerous physiological processes are determining the NUE of a plant (**b**) and the actual contribution of each process to the NUE of the plant is biased by many factors of the three variables plant, environment and nutrient (**c**). To develop strategies for the improvement of NUE in an agricultural system both the physiological processes involved and the factors that influence their contribution to NUE have to be considered

improvement for all species (and even cultivars). To identify the physiological processes whose modification could increase the NUE of a respective crop, the particular characteristics should be carefully taken into account and adequate comparisons of NUE are often only possible within cultivars or strains of the same species. However, the repertoire of physiological strategies in nature can also serve as a pool of mechanistic possibilities to improve NUE, *e.g.* by transferring beneficial traits within distantly related species via transgenic methods.

Another plant-specific variable is the kind of yield the crop will produce. As described in the introduction the *intended outcome* in agriculture is a maximum quantity of yield. However, the specific NUE and the way to improve it will differ fundamentally depending on the desired yield *quality*. The relevance of the physiological processes described above for NUE shifts completely if the desired yield is starch or sugar and not proteins or oil. While for the former efficient biomass production and C storage will be important, for the latter allocation to the seeds increases in relevance. Furthermore, virtually all morphological parts can be the yield organ into which the desired compounds are allocated before it is finally harvested.

Environment is the second variable that has an important impact on NUE. According to Evans and Fischer (1999) yield potential (Yp) can be defined as 'the yield of a cultivar when grown in environments to which it is adapted, with nutrients and water non-limiting, and with pests, diseases, weeds, lodging and other stresses effectively controlled'. The yield of a crop depends largely on the environmental conditions during the growing period. Nutrients are not always the limiting factor for plant growth and crop yield. Environmental factors such as temperature, light and rain or soil-specific factors such as soil composition, pH or pollution with salts or heavy metals may also be of great significance. If this is the case, there are more urgent steps to be taken to increase productivity than increasing NUE (Boyer 1982). Even if nutrient availability is the limiting factor, the physiological processes that are involved in NUE may be modified by environmental conditions. For instance, drought has numerous implications for the mineral nutrition of plants. As the uptake of many nutrients is depending on the mass flow of water there is a direct relationship between water availability and the uptake of mobile nutrients such as nitrate (Smika et al. 1965; Buljovcic and Engels 2001). In turn, an optimal N supply was shown to alleviate detrimental effects of drought in *Zea mays* (Zhang et al. 2007). However, not only the uptake but also the translocation to the shoot via the xylem is impaired during drought stress, which may affect all nutrients but which has been shown for P (Rasnick 1970). Even moderate drought stress might cause P deficiency in crops, which is an explanation for the often-observed positive effect of increased P fertilisation under dry conditions (Turner 1985; Garg et al. 2004).

It is known that growth rate and the uptake of certain nutrients (*e.g.* P but not N) may show a strong decline under low light compared to high light conditions, if the nutrient supply is adequate (Bloom 1985; Chapin 1991). The same is true for both low and high temperatures (Tindall et al. 1990). This dependency of growth and nutrient uptake on temperature and light leads to a different relevance of these processes for NUE in different patches of a field, during different seasons and in different climatic regions of the world.

The mineralization and cycles of essential nutrients such as N are mainly driven by the properties of the soil. Litter decomposition plays a crucial role in overall nutrient cycling in an ecosystem as well as in an agroecosystem. Globally the extent of litter decomposition depends mainly on temperature, while on a regional scale the chemical composition of the litter becomes most important (Aerts and Chapin 1999). In this way the soil not only plays a crucial role in the NUE of a field or a whole agroecosystem but also at plant level where NUE is partly governed by soil specific factors. Nutrient uptake is the process, which is affected most obviously. In natural ecosystems nutrient uptake is ultimately dependent on the nutrient supply rate by the parental rock material. Its mineral composition, age and weathering rate determine the nutritional status of a soil (Lambers et al. 2008). In an agricultural system the nutrient composition of the soil is mainly controlled by fertilisation. However, there are other factors with an impact on nutrient uptake, which are usually under much less control. One is the pH of the rhizosphere, which has a tremendous impact on the ability of the plant root to acquire different nutrients. In addition the soil is much more than just part of the plant's *abiotic* environment. It hosts a still widely unknown diversity of microbial life. Interactions with arbuscular mycorrhiza and rhizobacteria that increase the uptake surface of plant roots and provide nutrients to it may alter the NUE of crops, increase yield significantly and improve fertiliser management on a field scale (Smith et al. 1992; Adesemoye et al. 2008; Adesemoye and Kloepper 2009).

While the below ground part of the plant is surrounded by soil, the above-ground part is exposed to the atmosphere and although almost 100 % of its volume is equal in composition around the globe, there are traces of gases which fluctuate in their local concentrations (Kraus 2006). Some of these gases can have a significant impact on plant metabolism. Most of these N- and S-containing gases are usually referred to as pollutants, as they originate to a great extent from anthropogenic activities but also from volcanic activities. It has, however, been shown that some of these gases have an ambivalent mode of action on plants, as they are not only toxic if too high in concentration but can also serve as nutrients (De Kok et al. 2002, 2007). This can either happen by wet deposition, *i.e.* a deposition to the soil by rain, or by dry deposition via the stomata of plants. This additional nutrition from the atmosphere and predictions for changes in the concentration of these gases should be included in future calculations of NUE and fertilisation regimes.

The third variable that complicates efforts to define one general concept for NUE is the nutrient itself, as nutrients differ not only in their physiochemical properties but also in their uptake by and function in the plant. For each of them, the relevance of the underlying physiological processes for NUE differs and so do the necessary strategies for the improvement of NUE. For one nutrient the capacity of storage might be crucial, while for another the uptake or the allocation to the sink organ limits production. For this reason, studies on NUE in which different nutrients are used are often difficult to compare. More specific definitions and components have to be developed to accommodate this diversity. A recent study suggested three components for NitUE: N uptake efficiency, grain-specific N efficiency and grain N concentration (Weih et al. 2011). S for example is involved in plant defence and thereby using less S while maintaining the same defence status should increase S use efficiency. Some micronutrients are co-factors of particular enzymes so consequently a higher efficiency and decreased amount of these enzymes could increase the NUE of these micronutrients. Furthermore, some nutrients are of nutritional value, which again increases the relevance of translocation to the yield organ, *e.g.* the grain, while other nutrients are not desired to be part of the yield but play a role in its production. Conclusively the NUE of a given nutrient depends on its importance for yield production and its value for human nutrition, *i.e.* its concentration in the yield. Each case needs a different set of strategies for an improvement of NUE.

Improvement of NUE

Although controlled by man, agriculture remains a semi-artificial environment, which is still subject to oscillations, particularly from outside the agroecosystem. These oscillations may be entirely different from those in the ancient, natural habitat of the plant species and thus transferring it to an agricultural system may lead to sub-optimal growth and yield. Since the beginning of agricultural practice (ca. 11,000–13,000 years ago; Allard 1999) farmers have tried to solve these problems with two different approaches:

(i) By reducing the amplitude of environmental oscillations either by growing plants only in a certain season to avoid extremes in weather (*e.g.* cold winter, dry summers or rainy seasons) or by creating more stable conditions in the field (*e.g.* by using hedges as wind protection and digging moats to avoid flooding or, in modern times, building partially closed and controlled environments such as greenhouses).

Another important factor is the nutrient content of the soil, which was controlled by deploying manure before the revolutionary discovery of Liebig's law of the minimum and the advent of modern fertilisers. (ii) By reducing all kinds of unfavourable dynamics and traits in the plant's phenotype, which are relics of adaptation to its ancient habitat and/or part of its developmental program but no longer needed in cultivation. This is done by breeding, which makes use of the same mechanisms as evolution (variation and selection). Plants are selected for traits of agricultural interest with a main focus on bigger sink organs and higher concentrations of the compounds of interest at the expense of traits and adaptations that are no longer necessary.

In theory a combination of (i) and (ii) could result in a stable farming system in which environment and plant are under full control and no unfavourable oscillations and dynamics should occur anymore. Nutrient loss from the system would be at a minimum resulting in an optimal input-output ratio. The result would be to equal out all the variables that make NUE such a complicated trait: plant, environment and nutrient specific factors (see above). In reality, however, this is an ideal scenario and is presently far from being achievable. First, the technological effort to control all environmental factors and their respective oscillations is uneconomical and second the complex ways in which plant functioning is still largely unknown. Consequently the compromise that has developed over thousands of years of plant domestication is the attempt to synchronize the oscillations of environment and plant as well as possible. This is especially true for fertilisation because the discrepancy between demand of the plant and availability of nutrients in the soil is an important factor that can hinder optimal growth. Nutrient storage can buffer this discrepancy only to some extent and matching nutrient supply by fertiliser application to plant demand is regarded as one of the most promising ways to reach higher fertiliser use efficiency (Cassman et al. 1993; Frink et al. 1999; Tilman et al. 2002).

Without a deeper understanding of the biochemical and physiological processes involved, traditional breeding managed for more than 10,000 years of agriculture to develop plants with massive yield organs containing high protein, starch or oil content, compared to their ancestors (Mazoyer and Roudart 2006). At the same time major steps towards increasing output-input-ratios have been made by the progress of agricultural practice, technology and science (Russell 1966; Thompson 2011).

With Liebig's postulation of the "law of the minimum" (see review Browne 1942) and its resolution in terms of N supply by industrial production of relatively cheap nitrogenous fertilisers (mainly by the Haber-Bosch process in which N and hydrogen are directly converted to ammonia, see *e.g.* Tour 1920), the modern age of agriculture began and led to a historical change of paradigms, also named the "Green Revolution" (Borlaug 1972). Instead of trying to have an optimal input-output ratio, as it is necessary if fertiliser is a rare commodity, the highest possible output became the primary aim and remains so in present agricultural practice. The increased growth of cereals due to super-optimal N supply led to another factor becoming a major constraint for yield, namely the damage caused by lodging. The solution was the breeding of "dwarf cultivars" of wheat and rice in the 1960s by deployment of dwarfing genes. This in turn was only possible with better weed control through the development of herbicides so that the smaller cereals were not overgrown by wild weeds. This innovative trinity of the Green Revolution made it possible to neglect the input-output ratio and exclusively focus on maximum output (Evans 1998). As a result, yield per hectare increased tremendously during the last century and enabled an explosion of human population counting billions instead of millions. However, it has become more and more apparent that this practice cannot continue in the future. Parallel to a linear increase of global yields since the 1960s, the NUE of agricultural system (measured as unit yield per unit fertiliser applied) continuously declined. This "law of diminishing returns" implies that further increases in fertiliser application will not lead to higher yields in the same proportion as in the past (Tilman et al. 2002). High levels of nutrient input have resulted in pollution of the environment on the one hand and anticipated shortages of non-renewable resources such as inorganic P on the other. Furthermore, the benefits of these high outputs are very unevenly distributed over the world and with some production wasted in Europe and North America, spikes in food prices and hunger crises are expected to occur more frequently in developing countries. The awareness of this alarming trend led policymakers to put "food security" on the top of political agendas and consequently also into scientific focus (Rosegrant and Cline 2003; Vitousek et al. 2009; Godfray et al. 2010; Hawkesford et al. 2013).

To some extent the agriculture of the future has to come back to the old paradigm of requiring a more optimal input-output ratio. However, maintenance of the ongoing trend of increasing output in form of yield is an imperative due to the still-growing world population and no concept, which reduces the output, can be realistically considered (Evans 1998). Consequently, modifications of the input-output ratio have to concentrate on the input side of the equation. At field level (Fig. 1.1) the approaches to improve the input-output ratio may be categorized in two dimensions: time and space. This is again based on the general conception of plant and environment as two non-stationary, heterogeneous and oscillating systems. The supply with nutrients (fertiliser) can be adapted to the need of the plant for growth over time to lose fewer nutrients from the system in times of a low crop demand and to avoid concentrations being too low if the demand is high. In addition, nutrient supply and its demand in the field are not homogenously distributed in space but instead graduated or patchy. A homogenous application of nutrients thereby leads to excess supply in some parts of the field, resulting in nutrient loss, and to a sub-optimal supply in other parts, resulting in a non-optimal growth and yield. For both dimensions technological innovations have been developed, commonly summarized with the term "precision agriculture" (Pierce and Nowak 1999; Stafford 2000; Zhang et al. 2002). However, in addition there might be ways to breed or genetically manipulate the plant in a way that it uses nutrients more efficient and consequently produces the desired yield with less nutrient input. While transgenic and genetic techniques offer the possibility for accelerated improvement through genetic modifications and marker assisted breeding, whole plant physiology provides the knowledge to use these tools in an effective way. To practically improve the NUE of a plant in a particular agricultural system, the relevance of the underlying physiological processes (Fig. 1.2) has to be analysed in relation to the variables that modify it (Fig. 1.3). Once the limiting processes are identified, breeding or transgenic methods might lead to further improvement of NUE. For success, the physical and physiological trade-offs that limit an improvement in NUE have to be identified. This will be the topic of the last part of this chapter.

NUE – Challenges from a Whole Plant Perspective

As stated previously, NP and MRT can be seen as sub-components of NUtE. Consequently both could be targeted to improve NUE in cropping systems. Berendse and Aerts (1987) have pointed out the apparent ecological trade-off of NitP and MRT. While in nutrient-poor soils a long MRT is favourable, a high NitP gives advantages in nutrient-rich soils. Plant species in nutrient-rich soils generally possess a larger photosynthetic apparatus and can thereby rapidly make use of higher N availability, while species in nutrient-poor soils are able to use spare nutrients more economically. Theoretically, generalist species should combine both traits, but evidence suggests that they are competitive (Berendse et al. 1987). A later study on species with different life forms in a sub-arctic environment confirmed this negative relationship of NitP and MRT (Eckstein and Karlsson 1997), and others followed (Yasumura et al. 2002; Silla and Escudero 2004). Within the same species, this trade-off also appears to be consistent, with higher N supply leading to higher NitP but lower MRT (Yuan et al. 2005) or the other way around (Yuan et al. 2008). High nutrient supply, however, may even lead to a decline of both components in certain circumstances. A decline in NP can be caused by cross- and self-shading and an accompanied decrease in photosynthetic activity and a lower MRT can be the result of enhanced litter production or leaching (Meuleman et al. 2002).

As studies have almost exclusively focused on natural systems, perennials and N, it is difficult to assess if a strict trade-off between NP and MRT exists in agricultural crops and for all nutrients. As litter decomposability is not an issue in annual crop production, a high NP and long MRT at the same time are desired to maximize yield output and minimize nutrient input. The fact that traits, which lead to either a high NP or a long MRT, are not co-occurring in nature does not necessarily mean that breeding could not combine them. More integrated research on the described underlying physiological processes and their possible trade-offs are needed, including research with nutrients other than N.

In general, NUE is only studied for one single nutrient. There are few studies, which address the question whether improving the NUE for one nutrient will affect the NUE of others. Given the interactive and competitive nature of nutrients in many physiological processes, it seems likely that increasing NUE for one nutrient will also alter the NUE of others in a positive or negative manner.

On the level of nutrient uptake from the soil there are manifold co-influences known, but their whole extent is still far from being understood. Many of these are caused by the charged nature of ions when dissolved in water. If, for example, N is taken up predominantly as positive charged ammonium, the uptake of other cations such as magnesium and calcium is impaired, probably in order to maintain a balance of charge (Haynes and Goh 1978). On the other hand, plants that are supplied with nitrate absorb less phosphate than plants supplied with ammonium (Riley and Barber 1971). Similarly, do sulfate or nitrate show a higher accumulation in the plant if the other is missing in the root medium (Steingröver et al. 1986; Koralewska et al. 2009), either due to a replacement as an osmolyte in the vacuole or to a balance of anion-cation uptake, *i.e.* a balance of charge. This is a hint that increasing the storage capacity of the vacuole for one nutrient could decrease the capacity for others of the same charge. Interactions of the uptake of nutrients with different charge have been observed, for example for sulfate and iron (Paolacci et al. 2013). Here a direct interaction on the level of uptake does not seem likely. In contrast a higher uptake of sulfate under iron deficiency is suggested to serve for the production of S-containing defence compounds and a coupling of both nutrients could be due to their combination in Fe-S clusters (Forieri et al. 2013). Another example of the interactive effects of the uptake of different nutrients has been shown for sulfate, nitrate and ammonium (Clarkson et al. 1989) and there are many more reported. The direct linkage between the uptake of different nutrients, however, is under critical discussion. Studies in which plants were exposed to an atmospheric S source while deprivation occurred in the rhizosphere show an apparent uncoupling of sulfate and nitrate uptake (Westerman et al. 2000, 2001; Stulen and De Kok 2012).

Another potential conflict in the uptake of different nutrients arises if they use the same uptake system. This may be true, for instance, for the transport of phosphate and sulfate across the chloroplast membrane, which was reported to be competitive (Gross et al. 1990). As well as Liebig's "law of the minimum" states that identifying and increasing the amount of the most limiting factor can increase plant production, there is the far less popular but equally important "law of the optimum" formulated by Liebscher (see review Browne 1942). It states that the increase of such a limiting factor contributes more to the productivity of the system, the closer all other factors are to their optimum. Liebscher studied N, P and K nutrition of crops and was one of the first researchers to demonstrate the strong interactive component of different nutrients and their contributions to yield. It does not contradict Liebig's "law of the minimum" but shows that reality is more complex. Improvement of NUtE for any of the three nutrients N, P and K requires a balanced supply with the other two (Janssen 1998).

An important question for the future improvement of plant NUE is whether interferences with other efficiencies exist, especially if those turn out to be real physical trade-offs. Such trade-offs usually arise from the involvement of physiological mechanisms in several efficiencies. Studies on different tree species showed intraspecific inverse relationships between water use efficiency (WUE) and NitUE (Field et al. 1983; Reich et al. 1989), which appear to explain the spatial distribution of species on either N-poor or water-deficient soils (Patterson et al. 1997) but also have implications for the improvement of NUE in agriculture. This trade-off has stomatal as well as non-stomatal components (Reich et al. 1989) depending on the water and N status of the soil. How relevant it is for agricultural practice and crop breeding towards an increased NUE has still to be clarified. It might help the case to distinguish between the levels of NAcE and NUtE. If water is a limiting factor during crop growth, NAcE usually increases in its relevance, as nutrient uptake is physically coupled to water uptake. Climate change may mean seasonal droughts appear more frequently in many regions of the world which further increases the need for crops with a high WUE (Reynolds et al. 2011). For wheat, for example, it has been shown that under water limiting conditions genotypes with a greater root biomass produce more grains, probably due to both a high WUE and NAcE during early growth due to enhanced ability to capture water and reduce nitrate leaching (Ehdaie et al. 2010). Root traits are considered as a selection criterion especially under drought conditions (Ren et al. 2012) but there are strong interactions with NAcE that have to be considered. For example, the responses of root architecture to P and N deficiency are very different. While P deficiency leads to a shallow root system, foraging for the immobile phosphate, N deficiency causes a deep, scarcely branched root system. The latter is also associated with an efficient capture of water during periodical drought (White et al. 2013) implying that the NAcE for N and an efficient water uptake are not in competition but share a similar morphological trait. How the shallow root system that leads to a higher NAcE of P displays a constraint to WUE needs to be further evaluated. However, there are also reports that show an apparent negative interaction of N supply with WUE. High N supply leads to an inhibition of the typical growth characteristics that lead to a higher WUE in *Sophora davidii* while appropriate or low supply alleviated drought stress (Wu et al. 2008).

NUtE, NitUE and WUE show strong interactions due to their correlative effects on stomatal conductance, gas exchange and photosynthesis. In wheat, a higher N supply is reported to lead to an increase in WUE but at the same time to a decreased NitUE (Shangguan et al. 2000; Cabrera-Bosquet et al. 2007). This trade-off between NUtE and WUE is particularly noticeable if the N-to-grain price ratio in an agricultural system is high (Sadras and Rodriguez 2010). How the NUE of nutrients other than N interferes with WUE is still an open question.

Other studies have revealed a trade-off between N and light use efficiency (LUE) in canopies (Niinemets and Tenhunen 1997; Hirose and Bazzaz 1998). This can be explained by the fact that a high concentration of N in leaves leads to a high LUE but low NitUE. The impact of this trade-off again depends on the factors described above. It becomes more relevant under shading conditions and for trees because the build-up of the desired timber strongly depends on exploiting light efficiently during the growing season.

Again, it should be noted that inverse relationships between certain plant traits observed in nature do not necessarily display real physical trade-offs that cannot be overcome in plant breeding. They might just reflect genetic adaptations to different habitats or an ecological disadvantage of the combination of both traits (Veneklaas et al. 2012). If this is the case traits, which never occur together in nature may still be combined in one crop.

Finding plant traits, which improve Yp and NUE is a major aim of modern breeding programs. The genetic variability of crops, as well as of the model plant *Arabidopsis thaliana*, may serve as sources of diversity and promising traits may then be incorporated into transgenic crop plants. Additionally future techniques might enable the transfer of physiological traits such as C_4 metabolism (Leegood 2002) or N fixation (Charpentier and Oldroyd 2010; Beatty and Good 2011) between distantly related species. In this way, ecological but non-physical trade-offs could theoretically be overcome faster, and crop plants could be complemented with physiological properties that they would never gain through breeding. The diversity present in nature is potentially a rich source of traits that could improve the NUE of crops. There may be plant species that evolved mechanisms to use nutrients much more efficiently than crops but which science has yet to exploit. In particular species from nutrient-poor habitats are very promising candidates. Thus transgenics could be a useful tool to engineer crops whilst ecophysiology delivers the ideas of how these modifications will look.

Conclusions

Plant NUE is as complex and multi-dimensional as the plant itself. Consequently the greatest challenge, but also the greatest opportunity for modern plant nutrition research, is the integration of various disciplines and approaches. Use of state-of-the-art methods in transcriptomics, metabolomics, and proteomics give researchers unprecedented opportunities to obtain huge amounts of information with high resolution (see the other chapters of this book), not least enabled by the exponential increase of computing performance since its emergence (Moore 1975; Kurzweil 2001). The growing understanding of complex molecular networks and their holistic responses to alterations in nutrient availability contributes to uncovering the genetic basis of NUE and shows how complex metabolic pathways are interacting on a molecular level. However, the real power of molecular techniques for future crop breeding can only unfold if the complexity of the underlying physiological processes and their contribution to NUE is to be further understood.

Additionally there are physiological trade-offs that challenge the improvement of NUE. It might be possible to overcome some of them by modern molecular breeding techniques; others might be physically determined and display a real limit for the improvement of NUE. Further investigation is necessary to determine the physiological basis of these trade-offs before tools such as genetic manipulation may be used to overcome them. As studies on NUE usually focus on only one nutrient (and usually only N or P), there is little data available on possible trade-offs between the NUE of different

(continued)

nutrients. However, the compensatory and interactive nature of the uptake and storage of different nutrients suggests that such inter-nutritional trade-offs could exist. Adding to this complexity is the knowledge that all these possible trade-offs are strongly influenced by environment, plant and nutrient specific variables.

In conclusion, the physiological basis of NUE still displays a wide field for future research and the complexity and plasticity of plant metabolism may yet have many surprises in store in our attempts to "improve" it. More integrative studies, connecting several scientific disciplines are required to understand the complexity of all the variables that influence NUE resulting in physiologically relevant strategies for the improvement of plant NUE in agricultural production.

References

Adesemoye AO, Kloepper JW (2009) Plant–microbes interactions in enhanced fertilizer-use efficiency. Appl Microbiol Biotechnol 85:1–12

Adesemoye AO, Torbert HA, Kloepper JW (2008) Enhanced plant nutrient use efficiency with PGPR and AMF in an integrated nutrient management system. Can J Microbiol 54:876–886

Aerts R (1990) Nutrient use efficiency in evergreen and deciduous species from heathlands. Oecologia 84:391–397

Aerts R (1997) Nitrogen partitioning between resorption and decomposition pathways: a trade-off between nitrogen use efficiency and litter decomposibility? Oikos 80:603–606

Aerts R, Chapin FS III (1999) The mineral nutrition of wild plants revisited: a re-evaluation of processes and patterns. Adv Ecol Res 30:1–67

Ågren GI (1985) Theory for growth of plants derived from the nitrogen productivity concept. Physiol Plant 64:17–28

Ainsworth EA, Rogers A, Nelson R, Long SP (2004) Testing the "source–sink" hypothesis of down-regulation of photosynthesis in elevated [CO_2] in the field with single gene substitutions in *Glycine max*. Agric Forest Meteorol 122:85–94

Allard RW (1999) Principles of plant breeding. Wiley, New York

Barbottin A, Lecomte C, Bouchard C, Jeuffroy MH (2005) Nitrogen remobilization during grain filling in wheat. Crop Sci 45:1141–1150

Beatty PH, Good AG (2011) Future prospects for cereals that fix nitrogen. Science 333:416–417

Berendse F, Aerts R (1987) Nitrogen-use-efficiency: a biologically meaningful definition? Funct Ecol 1:293–296

Berendse F, Oudhof H, Bol J (1987) A comparative study on nutrient cycling in wet heathland ecosystems. Oecologia 74:174–184

Berendse F, Elberse WT, Geerts RHME (1992) Competition and nitrogen loss from plants in grassland ecosystems. Ecology 73:46–53

Birk EM, Vitousek PM (1986) Nitrogen availability and nitrogen use efficiency in loblolly pine stands. Ecology 67:69–79

Bloom AJ (1985) Wild and cultivated barleys show similar affinities for mineral nitrogen. Oecologia 65:555–557

Borlaug NE (1972) The green revolution, peace, and humanity. Speech delivered upon receipt of the 1970 Nobel Peace Prize. CIMMYT reprint and translation series No. 3. Centro Internacional de Mejoramiento de Maiz y Trigo, El Batan, Mexico

Borrás L, Slafer GA, Otegui ME (2004) Seed dry weight response to source–sink manipulations in wheat, maize and soybean: a quantitative reappraisal. Field Crop Res 86:131–146

Bot JL, Kirkby EA (1992) Diurnal uptake of nitrate and potassium during the vegetative growth of tomato plants. J Plant Nutr 15:247–264

Boyer JS (1982) Plant productivity and environment. Science 218:443–448

Brown RH (1978) A difference in N use efficiency in C_3 and C_4 plants and its implications in adaptation and evolution. Crop Sci 18:93–98

Browne CA (1942) Liebig and the law of the minimum. In: Moulton FA (ed) Liebig and after Liebig. A century of progress in agricultural chemistry. American Association for the Advancement of Science, Washington, DC, pp 71–82

Buljovcic Z, Engels C (2001) Nitrate uptake ability by maize roots during and after drought stress. Plant Soil 229:125–135

Cabrera-Bosquet L, Molero G, Bort J, Nogués S, Araus JL (2007) The combined effect of constant water deficit and nitrogen supply on WUE, NUE and $\Delta^{13}C$ in durum wheat potted plants. Ann Appl Biol 151:277–289

Cassman KG, Kropff MJ, Gaunt J, Peng S (1993) Nitrogen use efficiency of rice reconsidered: what are the key constraints? Plant Soil 155:359–362

Chapin FS III (1980) The mineral nutrition of wild plants. Annu Rev Ecol Syst 11:233–260

Chapin FS III (1991) Integrated responses of plants to stress. Bioscience 41:29–36

Chapin FS III, Kedrowski RA (1983) Seasonal changes in nitrogen and phosphorus fractions and autumn retranslocation in evergreen and deciduous taiga trees. Ecology 64:376–391

Chapin FS III, Moilanen L (1991) Nutritional controls over nitrogen and phosphorus resorption from Alaskan birch leaves. Ecology 72:709–715

Chapin FS III, Schulze ED, Mooney HA (1990) The ecology and economics of storage in plants. Annu Rev Ecol Syst 21:423–447

Charpentier M, Oldroyd G (2010) How close are we to nitrogen-fixing cereals? Curr Opin Plant Biol 13:556–564

Clarkson DT, Saker LR, Purves JV (1989) Depression of nitrate and ammonium transport in barley plants with diminished sulphate status. Evidence of co-regulation of nitrogen and sulphate intake. J Exp Bot 40:953–963

Costanza R, d'Arge R, De Groot R, Farber S, Grasso M, Hannon B, Limburg K, Naeem S, O'Neill RV, Paruelo J, Raskin RG, Sutton P, Van den Belt M (1997) The value of the world's ecosystem services and natural capital. Nature 387:253–260

Daily GC (1997) Nature's services: societal dependence on natural ecosystems. Island, Washington, DC

Daily GC, Söderqvist T, Aniyar S, Arrow K, Dasgupta P, Ehrlich PR, Folke C, Jansson A, Jansson B-O, Kautsky N, Levin S, Lubchenco J, Mäler K-G, Simpson D, Starrett D, Tilman D, Walker B (2000) Ecology. The value of nature and the nature of value. Science 289:395

De Kok LJ, Stuiver CEE, Westerman S, Stulen I (2002) Elevated levels of hydrogen sulfide in the plant environment: nutrient or toxin. In: Omasa K, Saji H, Youssefian S, Kondo N (eds) Air pollution and biotechnology in plants. Springer, Tokyo, pp 201–213

De Kok LJ, Durenkamp M, Yang L, Stulen I (2007) Atmospheric sulfur. In: Hawkesford MJ, De Kok LJ (eds) Sulfur in plants – an ecological perspective. Springer, Dordrecht, pp 91–106

Delhon P, Gojon A, Tillard P, Passama L (1995) Diurnal regulation of NO_3^- uptake in soybean plants I. Changes in NO_3^- influx, efflux, and N utilization in the plant during the day/night cycle. J Exp Bot 46:1585–1594

Donald CM, Hamblin J (1976) The biological yield and harvest index of cereals as agronomic and plant breeding criteria. Adv Agron 28:361–405

Eckstein RL, Karlsson PS (1997) Above-ground growth and nutrient use by plants in a subarctic environment: effects of habitat, life-form and species. Oikos 79:311–324

Ehdaie B, Merhaut DJ, Ahmadian S, Hoops AC, Khuong T, Layne AP, Waines JG (2010) Root system size influences water-nutrient uptake and nitrate leaching potential in wheat. J Agron Crop Sci 196:455–466

Eichert T, Fernández V (2012) Uptake and release of elements by leaves and other aerial plant parts. In: Marschner P (ed) Marschner's mineral nutrition of higher plants. Academic, Oxford, UK, pp 71–84

Escudero A, Mediavilla S (2003) Decline in photosynthetic nitrogen use efficiency with leaf age and nitrogen resorption as determinants of leaf life span. J Ecol 91:880–889

Evans LT (1998) Feeding the ten billion: plants and population growth. Cambridge University Press, Cambridge, UK

Evans LT, Dunstone RL (1970) Some physiological aspects of evolution in wheat. Aust J Biol Sci 23:725–742

Evans LT, Fischer RA (1999) Yield potential: its definition, measurement, and significance. Crop Sci 39:1544–1551

Faller N (1972) Schwefeldioxid, Schwefelwasserstoff, nitrose Gase und Ammoniak als ausschließliche S- bzw. N-Quellen der höheren Pflanze. J Plant Nutr Soil Sci 131:120–130

Field C (1983) Allocating leaf nitrogen for the maximization of carbon gain: leaf age as a control on the allocation program. Oecologia 56:341–347

Field C, Merino J, Mooney HA (1983) Compromises between water-use efficiency and nitrogen-use efficiency in five species of California evergreens. Oecologia 60:384–389

Forieri I, Wirtz M, Hell R (2013) Towards new perspectives on the interaction of iron and sulfur metabolism in *Arabidopsis thaliana*. Front Plant Sci 4:357

Foulkes MJ, Slafer GA, Davies WJ, Berry PM, Sylvester-Bradley R, Martre P, Calderini DF, Griffiths R, Reynolds MP (2011) Raising yield potential of wheat. III. Optimizing partitioning to grain while maintaining lodging resistance. J Exp Bot 62:469–486

Francis C, Lieblein G, Gliessman S, Breland TA, Creamer N, Harwood R, Salomonsson L, Helenius J, Rickerl D, Salvador R, Wiedenhoeft M, Simmons S, Allen P, Altieri M, Flora C, Poincelot R (2003) Agroecology: the ecology of food systems. J Sustain Agric 22:99–118

Franklin O, McMurtie ROSS, Iversen CM, Crous KY, Finzi AC, Tissue DT, Ellsworth DS, Oren R, Norby RJ (2009) Forest fine-root production and nitrogen use under elevated CO_2: contrasting responses in evergreen and deciduous trees explained by a common principle. Glob Change Biol 15:132–144

Frink CR, Waggoner PE, Ausubel JH (1999) Nitrogen fertilizer: retrospect and prospect. Proc Natl Acad Sci U S A 96:1175–1180

Garg BK, Burman U, Kathju S (2004) The influence of phosphorus nutrition on the physiological response of moth bean genotypes to drought. J Plant Nutr Soil Sci 167:503–508

Gerloff GC (1963) Comparative mineral nutrition of plants. Annu Rev Plant Physiol 14:107–124

Givnish TJ (2002) Adaptive significance of evergreen vs. deciduous leaves: solving the triple paradox. Silva Fenn 36:703–743

Gliessman SR (1990) Agroecology: researching the ecological basis for sustainable agriculture. Springer, New York

Godfray HCJ, Beddington JR, Crute IR, Haddad L, Lawrence D, Muir JF, Pretty J, Robinson S, Thomas SM, Toulmin C (2010) Food security: the challenge of feeding 9 billion people. Science 327:812–818

Gross A, Brückner G, Heldt HW, Flügge UI (1990) Comparison of the kinetic properties, inhibition and labelling of the phosphate translocators from maize and spinach mesophyll chloroplasts. Planta 180:262–271

Hawkesford MJ (2000) Plant responses to sulphur deficiency and the genetic manipulation of sulphate transporters to improve S-utilization efficiency. J Exp Bot 51:131–138

Hawkesford MJ, Araus JL, Park R, Calderini D, Miralles D, Shen T, Zhang J, Parry MA (2013) Prospects of doubling global wheat yields. Food Energy Secur 2:34–48

Haydon MJ, Bell LJ, Webb AA (2011) Interactions between plant circadian clocks and solute transport. J Exp Bot 62:2333–2348

Haynes RJ, Goh KM (1978) Ammonium and nitrate nutrition of plants. Biol Rev 53:465–510

Hirose T, Bazzaz FA (1998) Trade-off between light-and nitrogen-use efficiency in canopy photosynthesis. Ann Bot 82:195–202

Ingestad T (1988) A fertilization model based on the concepts of nutrient flux density and nutrient productivity. Scand J Forest Res 3:157–173
Janssen BH (1998) Efficient use of nutrients: an art of balancing. Field Crop Res 56:197–201
Killingbeck KT (1996) Nutrients in senesced leaves: keys to the search for potential resorption and resorption proficiency. Ecology 77:1716–1727
Kimball BA, Kobayashi K, Bindi M (2002) Responses of agricultural crops to free-air CO_2 enrichment. Adv Agron 77:293–368
Koralewska A, Buchner P, Stuiver CEE, Posthumus FS, Kopriva S, Hawkesford MJ, De Kok LJ (2009) Expression and activity of sulfate transporters and APS reductase in curly kale in response to sulfate deprivation and re-supply. J Plant Physiol 166:168–179
Kozaki A, Takeba G (1996) Photorespiration protects C_3 plants from photooxidation. Nature 384:557–560
Kraus H (2006) Die Atmosphäre der Erde: Eine Einführung in die Meteorologie. (The atmosphere of the earth: an introduction to meteorology). Springer, Berlin/Heidelberg, p 23
Kurzweil R (2001) The law of accelerating returns. Retrieved from www.kurzweilai.net in 2013
Lajtha K (1987) Nutrient reabsorption efficiency and the response to phosphorus fertilization in the desert shrub *Larrea tridentata* (DC.) Cov. Biogeochemistry 4:265–276
Lambers H, Chapin FS III, Pons TL (2008) Plant physiological ecology, 2nd edn. Springer, Berlin
Larcher W (1995) Plant physiological ecology, 3rd edn. Springer, Berlin
Leegood RC (2002) C_4 photosynthesis: principles of CO_2 concentration and prospects for its introduction into C_3 plants. J Exp Bot 53:581–590
Long SP, Zhu XG, Naidu SL, Ort DR (2006) Can improvement in photosynthesis increase crop yields? Plant Cell Environ 29:315–330
Loomis RS, Amthor JS (1999) Yield potential, plant assimilatory capacity, and metabolic efficiencies. Crop Sci 39:1584–1596
Marschner H (2012) In: Marschner P (ed) Marschner's mineral nutrition of higher plants. Academic, Oxford, UK
Mazoyer M, Roudart L (2006) A history of world agriculture: from the neolithic age to the current crisis. Monthly Review Press, New York
McClung CR (2006) Plant circadian rhythms. Plant Cell 18:792–803
Mengel K, Kirkby EA (1987) Principles of plant nutrition, 4th edn. International Potash Institute, Bern
Meuleman AF, Beekman JHP, Verhoeven JT (2002) Nutrient retention and nutrient-use efficiency in *Phragmites australis* stands after waster water application. Wetlands 22:712–721
Moll RH, Kamprath EJ, Jackson WA (1982) Analysis and interpretation of factors which contribute to efficiency of nitrogen utilization. Agron J 74:562–564
Moore GE (1975) Progress in digital integrated electronics. IEEE, international electron devices meeting, IEDM Tech. Digest 1975, pp 11–13
Nardoto GB, da Cunha Bustamante MM, Pinto AS, Klink CA (2006) Nutrient use efficiency at ecosystem and species level in savanna areas of Central Brazil and impacts of fire. J Trop Ecol 22:191–201
Niinemets Ü, Tenhunen JD (1997) A model separating leaf structural and physiological effects on carbon gain along light gradients for the shade-tolerant species *Acer saccharum*. Plant Cell Environ 20:845–866
Ntanos DA, Koutroubas SD (2002) Dry matter and N accumulation and translocation for Indica and Japonica rice under Mediterranean conditions. Field Crop Res 74:93–101
Oenema O, Witzke HP, Klimont Z, Lesschen JP, Velthof GL (2009) Integrated assessment of promising measures to decrease nitrogen losses from agriculture in EU-27. Agric Ecosyst Environ 133:280–288
Paez-Valencia J, Sanchez-Lares J, Marsh E, Dorneles LT, Santos MP, Sanchez D, Winter A, Murphy S, Cox J, Trzaska M, Metler J, Kozic A, Facanha AR, Schachtman D, Sanchez C, Gaxiola RA (2013) Enhanced proton translocating pyrophosphatase activity improves nitrogen use efficiency in romaine lettuce. Plant Physiol 161:1557–1569

Paolacci AR, Celletti S, Catarcione G, Hawkesford MJ, Astolfi S, Ciaffi M (2013) Iron deprivation results in a rapid but not sustained increase of the expression of genes involved in iron metabolism and sulfate uptake in tomato (*Solanum lycopersicum* L.) seedlings. J Integr Plant Biol 56. Early View Online Version

Parry MA, Reynolds M, Salvucci ME, Raines C, Andralojc PJ, Zhu XG, Price GD, Condon AG, Furbank RT (2011) Raising yield potential of wheat. II. Increasing photosynthetic capacity and efficiency. J Exp Bot 62:453–467

Patterson TB, Guy RD, Dang QL (1997) Whole-plant nitrogen-and water-relations traits, and their associated trade-offs, in adjacent muskeg and upland boreal spruce species. Oecologia 110:160–168

Peterhansel C, Maurino VG (2011) Photorespiration redesigned. Plant Physiol 155:49–55

Pierce FJ, Nowak P (1999) Aspects of precision agriculture. Adv Agron 67:1–85

Poorter H, Remkes C, Lambers H (1990) Carbon and nitrogen economy of 24 wild species differing in relative growth rate. Plant Physiol 94:621–627

Rasnick M (1970) Effect of mannitol and polyethylene glycol on phosphorus uptake by maize plants. Ann Bot 34:497–502

Reich PB, Walters MB, Tabone TJ (1989) Response of *Ulmus americana* seedlings to varying nitrogen and water status. 2 Water and nitrogen use efficiency in photosynthesis. Tree Physiol 5:173–184

Ren Y, He X, Liu D, Li J, Zhao X, Li B, Tong Y, Zhang A, Li Z (2012) Major quantitative trait loci for seminal root morphology of wheat seedlings. Mol Breed 30:139–148

Reynolds MP, Pellegrineschi A, Skovmand B (2005) Sink limitation to yield and biomass: a summary of some investigations in spring wheat. Ann Appl Biol 146:39–49

Reynolds MP, Manes Y, Izanloo A, Langridge P (2011) Raising yield potential of wheat. I. Overview of a consortium approach and breeding strategies. J Exp Bot 62:439–452

Riley D, Barber SA (1971) Effect of ammonium and nitrate fertilization on phosphorus uptake as related to root-induced pH changes at the root-soil interface. Soil Sci Soc Am J 35:301–306

Rose TJ, Pariasca-Tanaka J, Rose MT, Fukuta Y, Wissuwa M (2010) Genotypic variation in grain phosphorus concentration, and opportunities to improve P-use efficiency in rice. Field Crop Res 119:154–160

Rosegrant MW, Cline SA (2003) Global food security: challenges and policies. Science 302:1917–1919

Rossato L, Laine P, Ourry A (2001) Nitrogen storage and remobilization in *Brassica napus* L. during the growth cycle: nitrogen fluxes within the plant and changes in soluble protein patterns. J Exp Bot 52:1655–1663

Russell EJ (1966) A history of agricultural science in great Britain 1620–1954. George Allen & Unwin, London

Sadras VO, Rodriguez D (2010) Modelling the nitrogen-driven trade-off between nitrogen utilisation efficiency and water use efficiency of wheat in eastern Australia. Field Crop Res 118:297–305

Saurbeck DC, Helal HM (1990) Factors affecting the nutritional efficiency of plants. In: Bassam NEL, Dambroth M, Loughman BC (eds) Genetic aspects of plant mineral nutrition. Martinus Nijhoff, Dordrecht, pp 361–372

Schnug E, Haneklaus S (1998) Diagnosis of sulphur nutrition. In: Schnug E (ed) Sulphur in agroecosystems. Springer, Dordrecht, pp 1–38

Shangguan ZP, Shao MA, Dyckmans J (2000) Nitrogen nutrition and water stress effects on leaf photosynthetic gas exchange and water use efficiency in winter wheat. Environ Exp Bot 44:141–149

Shea PF, Gerloff GC, Gabelman WH (1968) Differing efficiencies of potassium utilization in strains of Snapbeans, *Phaseolus vulgaris* L. Plant Soil 28:337–346

Siddiqi MY, Glass AD (1981) Utilization index: a modified approach to the estimation and comparison of nutrient utilization efficiency in plants. J Plant Nutr 4:289–302

Silla F, Escudero A (2004) Nitrogen-use efficiency: trade-offs between N productivity and mean residence time at organ, plant and population levels. Funct Ecol 18:511–521

Smika D, Haas H, Power W (1965) Effects of moisture and nitrogen fertilizer on growth and water use by native grass. Agron J 57:483–486

Smith SE, Robson AD, Abbott LK (1992) The involvement of mycorrhizas in assessment of genetically dependent efficiency of nutrient uptake and use. Plant Soil 146:169–179

Somers DE, Devlin PF, Kay SA (1998) Phytochromes and cryptochromes in the entrainment of the Arabidopsis circadian clock. Science 282:1488–1490

Stafford JV (2000) Implementing precision agriculture in the 21st century. J Agric Eng Res 76:267–275

Steingröver E, Woldendorp J, Sijtsma L (1986) Nitrate accumulation and its relation to leaf elongation in spinach leaves. J Exp Bot 181:1093–1102

Stulen I, De Kok LJ (2012) Exploring interactions between sulfate and nitrate uptake at a whole plant level. In: De Kok LJ, Tausz M, Hawkesford MJ, Hoefgen R, McManus MT, Norton RM, Rennenberg H, Saito K, Schnug E, Tabe L (eds) Sulfur metabolism in plants: mechanisms and application to food security, and responses to climate change. Springer, Dordrecht, pp 1–8

Stulen I, Perez-Soba M, De Kok LJ, Van der Eerden L (1998) Impact of gaseous nitrogen deposition on plant functioning. New Phytol 139:61–70

Swiader JM, Chyan Y, Freiji FG (1994) Genotypic differences in nitrate uptake and utilization efficiency in pumpkin hybrids. J Plant Nutr 17:1687–1699

Thompson RP (2011) Agro-technology: a philosophical introduction. Cambridge University Press, Cambridge, UK

Tilman D, Cassman KG, Matson PA, Naylor R, Polasky S (2002) Agricultural sustainability and intensive production practices. Nature 418:671–677

Tindall JA, Mills HA, Radcliffe DE (1990) The effect of root zone temperature on nutrient uptake of tomato. J Plant Nutr 13:939–956

Tour RS (1920) The direct synthetic ammonia process. Ind Eng Chem Res 12:844–852

Turner LB (1985) Changes in the phosphorus content of *Capsicum annuum* leaves during water-stress. J Plant Physiol 121:429–439

Van der Werf A, Van Nuenen M, Visser AJ, Lambers H (1993) Contribution of physiological and morphological plant traits to a species' competitive ability at high and low nitrogen supply. Oecologia 94:434–440

Vázquez de Aldana BR, Berendse F (1997) Nitrogen-use efficiency in six perennial grasses from contrasting habitats. Funct Ecol 11:619–626

Veneklaas EJ, Lambers H, Bragg J, Finnegan PM, Lovelock CE, Plaxton WC, Price CA, Scheible W-A, Shane MW, White PJ, Raven JA (2012) Opportunities for improving phosphorus-use efficiency in crop plants. New Phytol 195:306–320

Vitousek P (1982) Nutrient cycling and nutrient use efficiency. Am Nat 119:553–572

Vitousek PM, Naylor R, Crews T, David MB, Drinkwater LE, Holland E, Johnes PJ, Katzenberger J, Martinelli LA, Matson PA, Nziguheba G, Ojima D, Palm CA, Robertson GP, Sanchez PA, Townsend AR, Zhang FS (2009) Nutrient imbalances in agricultural development. Science 324:1519

Watanabe N, Evans JR, Chow WS (1994) Changes in the photosynthetic properties of Australian wheat cultivars over the last century. Funct Plant Biol 21:169–183

Wedin DA, Tilman D (1990) Species effects on nitrogen cycling: a test with perennial grasses. Oecologia 84:433–441

Weih M, Asplund L, Bergkvist G (2011) Assessment of nutrient use in annual and perennial crops: a functional concept for analyzing nitrogen use efficiency. Plant Soil 339:513–520

Westerman S, De Kok LJ, Stuiver CEE, Stulen I (2000) Interaction between metabolism of atmospheric H_2S in the shoot and sulfate uptake by the roots of curly kale (*Brassica oleracea*). Physiol Plant 109:443–449

Westerman S, Stulen I, Suter M, Brunold C, De Kok LJ (2001) Atmospheric H_2S as sulphur source for *Brassica oleracea*: consequences for the activity of the enzymes of the assimilatory sulphate reduction pathway. Plant Physiol Biochem 39:425–432

Westoby M, Falster DS, Moles AT, Vesk PA, Wright IJ (2002) Plant ecological strategies: some leading dimensions of variation between species. Annu Rev Ecol Syst 33:125–159

White PJ, George TS, Gregory PJ, Bengough AG, Hallett PD, McKenzie BM (2013) Matching roots to their environment. Ann Bot 112:207–222

Wu F, Bao W, Li F, Wu N (2008) Effects of drought stress and N supply on the growth, biomass partitioning and water-use efficiency of *Sophora davidii* seedlings. Environ Exp Bot 63:248–255

Yasumura Y, Hikosaka K, Matsui K, Hirose T (2002) Leaf-level nitrogen-use efficiency of canopy and understorey species in a beech forest. Funct Ecol 16:826–834

Yuan ZY, Li LH, Huang JH, Han XG, Wam SQ (2005) Effect of nitrogen supply on the nitrogen use efficiency of an annual herb, *Helianthus annuus* L. J Integr Plant Biol 47:539–548

Yuan ZY, Chen HY, Li LH (2008) Nitrogen use efficiency: does a trade-off exist between the N productivity and the mean residence time within Species? Aust J Bot 56:272–277

Zelitch I (1982) The close relationship between net photosynthesis and crop yield. Bioscience 32:796–802

Zhang FS, Römheld V, Marschner H (1991) Diurnal rhythm of release of phytosiderophores and uptake rate of zinc in iron-deficient wheat. Soil Sci Plant Nutr 37:671–678

Zhang N, Wang M, Wang N (2002) Precision agriculture – a worldwide overview. Comput Electron Agric 36:113–132

Zhang LX, Li SX, Zhang H, Liang ZS (2007) Nitrogen rates and water stress effects on production, lipid peroxidation and antioxidative enzyme activities in two maize (*Zea mays* L.) genotypes. J Agron Crop Sci 193:387–397

Chapter 2
Natural Variation as a Tool to Investigate Nutrient Use Efficiency in Plants

Giorgiana Chietera and Fabien Chardon

Abstract A huge natural variation exists between individuals within a given plant species. Most of the responses of growth-related traits to different environmental scenarios are genotype dependent. Hence, natural variation in plants provides an interesting and valuable source of genetic diversity to study plant responses to environmental factors. The identification of genes that underlie phenotypic variation has an enormous practical implication by providing a means to improve crop yield and quality. The approach based on natural variation aims to use naturally occurring differences to improve our knowledge about complex physiological responses of plants to their environment, including nutrition efficiency. An overview of different approaches currently used in plant research aimed at dissecting complex quantitative traits is presented here, with a special focus on those related to Nutrient Use Efficiency, to explain strategies based on QTL mapping in segregating populations and association mapping in wild populations. Some case studies regarding each of the investigative strategies described are detailed.

Keywords NUE (nitrogen use efficiency) • Nitrogen • Assimilation efficiency • Remobilization efficiency • QTL • MAGIC populations • GWAS • Nutrient limitation • Candidate genes • Arabidopsis • Natural variation

Introduction

Plants are considered adapted to variable and sub-optimal environments when they show the ability to successfully grow and reproduce in them. In order to deal with changing and challenging environmental conditions, plants exhibit a wide range of integrated responses, which usually display complex quantitative variation. Plant adaptation interests a large community of scientists, from ecologists and molecular geneticists working on fundamental mechanisms of adaptation, to crop breeders looking for natural variants which could optimise environmental resources and provide targets for breeding programs (Trontin et al. 2011). The identification of genes that underlie phenotypic variation can have enormous practical implications

G. Chietera • F. Chardon (✉)
Institute Jean-Pierre Bourgin INRA, Versailles, France
e-mail: Fabien.Chardon@versailles.inra.fr

M.J. Hawkesford et al. (eds.), *Nutrient Use Efficiency in Plants*,
Plant Ecophysiology 10, DOI 10.1007/978-3-319-10635-9_2

by providing a means to increase crop yield and quality in an agricultural context (Bergelson and Roux 2010). Many elements in the soil serve as mineral nutrients for plants. Among them, nitrogen (N), potassium (K), phosphorus (P) and sulfur (S) are required in relatively large amounts for plant growth. Therefore, deficiency in any one of these four elements in the field affects plant metabolism, and shows a dramatic impact on yield, nutritional quality and taste as well as pathogen and pest resistance in crops (Laegreid et al. 1999). In developed countries, plentiful amounts of fertilisers are applied, resulting in abundant inorganic nutrients in the ecosystem that may cause disturbance in normal bio-geochemical cycles of nutrients. However, this trend is being challenged by the current emphasis on developing more efficient cultivars for sustainable, low-input agriculture, fuelled by increasing cost of fertilisers, restrictions to minimise environmental impact and the increased use of poor quality land (Rengel and Damon 2008). Some plant species and genotypes within species have a capacity to grow and yield well on soils with a low level of available nutrients; these species and genotypes are considered as tolerant to nutrient deficiency and with high nutrient use efficiency (NUE; Good et al. 2004). The NUE can be described as the proportion of potential yield that can be achieved under that mineral deficiency availability. It is the product of nutrient uptake efficiency (NUpE) and nutrient utilisation efficiency (NUtE), which is the optimal combination between nutrient assimilation efficiency (NAE) and nutrient remobilisation efficiency (NRE) (Masclaux-Daubresse et al. 2010). Improving each crop individually requires a global knowledge of the different mechanisms that control all the steps involved in nutrient management in plants, and also a good knowledge of the specificities of each plant species in terms of metabolism (Chardon 2012).

Investigation of Natural Variation in Plants Reveals Different Strategies of Response to Nutrient Limitation

A huge natural variation exists between individuals within a given plant species, affecting for example the colour or the shape of leaves, grain composition, seed dormancy, flowering date or maturity date. These variations could have an important impact on yield or quality of their products in crops, fruit trees and forestry management species (Saisho et al. 2011; Arikita et al. 2013; Eduardo et al. 2010; Łata et al. 2005; Robinson et al. 2012). Some developmental traits, such as flowering time or seed dormancy, have drawn particular attention, partly because they are of applied interest to crop breeding, and partly because they are easy to investigate (Shindo et al. 2007). Natural variation involves not only the morphology of plants but also their behaviour when facing contrasted environments. The responses or growth-related traits to different environmental scenarios are genotype dependent. Hence, natural variation in plants provides an interesting and valuable source of genetic diversity to study plant responses to environmental factors. The plant's capacity to adapt to environmental constraints is called plant plasticity. It reflects both plant age and adaptation to environment (*e.g.* light intensity, photoperiod, temperature and nutrient availability). The approach based on natural variation aims to use naturally occurring differences to improve our knowledge about complex physiological responses of plants to their environment, including nutrition efficiency.

Plant morphology and physiology are complex quantitative traits, implying that they are genetically controlled but also influenced by the environment. The measurements of such traits show differences of averages between genotypes as well as variations due to chance and technical inaccuracies. Therefore, the investigation of such complex traits needs statistical tools for taking into account fluctuations due to chance in order to correctly understand the genetics behind them.

Investigation of natural variation is of interest from two general points of view. First, analysing this variation makes it possible to identify the function of individual genes. Despite the fact that mutant approaches have been very powerful for functional analysis, the small number of genetic backgrounds analysed limits the definition of gene functions using these procedures. Ultimately, the sort of mutant phenotypes that can be identified depends on the wild type genotype (Koornneef et al. 2004). There are specific alleles in nature that would not be easily recognised in mutant screens because they require very specific amino acid changes and therefore appear at an extremely low frequency (El-Din El-Assal et al. 2001). Second, analysis of natural variation has an increasing interest from an ecological and evolutionary perspective since diversity of physiologies and evolved responses in nature result from millions of generations of evolution (Ungerer et al. 2008). While much has been learned from bringing organisms into the laboratory to study elements of their biology in isolation, ignoring the ecological context in which these elements arose and persist runs the risk of a suboptimal understanding of particular biological responses and processes. Thus, the patterns of phenotypic and molecular variation observed are analysed to elucidate the mechanisms generating and maintaining this variation, and to identify which allelic variants are adaptive under specific environmental conditions.

Several recent papers demonstrated that natural variation exists for the different steps of plant nutrition: nutrient uptake and roots to shoot translocation, their assimilation in leaves, as well as their recycling and remobilisation for seed filling.

The uptake efficiency is dependent on the root system architecture (Dunbabin et al. 2004), and specific software have been developed for analysis (Armengaud et al. 2009; Ristova et al. 2013; Galkovskyi et al. 2012). It has long been known that root architecture and plasticity reveal a response of plants to scarce nutrients, and natural variation exists for these traits in different species, related to potassium (Kellermeier et al. 2013; Jia et al. 2008), nitrogen (De Pessemier et al. 2013), and phosphorus availability (Wang et al. 2010). Natural variation for NUpE has been shown directly by measuring a specific mineral content in different genotypes, as Burns et al. demonstrated for nitrate in different varieties of lettuce. Otherwise, a complementary approach aimed at studying the activity of enzymes involved in mineral assimilation in roots can be used. For example, Blair et al. (2010) identified important differences between varieties of beans for their ability to reduce iron when grown at various hydroponic iron concentrations, ranging from 0 to 20 μM Fe. Interestingly, these differences were more evident in plants grown at low Fe concentrations (iron limiting conditions) than at high iron concentrations (sufficiency conditions), revealing a genotype x environment interaction. Finally, the use of isotope labelling has been used to evaluate nutrient uptake capacity. In maize, nitrogen-15 labelling was used to study NUpE (Coque et al. 2008), showing that 28.3 % of whole-plant nitrogen was taken up after silking, and 93 % of this post-silking nitrogen uptake was allocated to kernels.

A similar approach could be used to analyse the roots to shoot translocation of nutrients. As an example, in order to understand why *Noccaea caerulescens* has good properties for phytoremediation, Xing et al. (2008) investigated the root-to-shoot translocation of Cd and Zn in different genotypes. The percentages of Cd and Zn transported to shoots within 24 h exposure varied widely among the 11 accessions analysed. Interestingly, the translocation efficiency did not correlate with the uptake for either metal, suggesting independent variation in uptake and translocation among different accessions of *Noccaea.*

Post-genomic studies integrating all "omics" sciences can depict precise pictures of nutrient assimilation in plants (Hirai et al. 2004). Sulpice et al. (2013) studied the response of 97 accessions of *Arabidopis* to different nitrogen and carbon conditions, an important number of traits showed significant natural variation between accessions. For example, biomass differed between the 97 accessions by 3.1-fold and 2.8-fold in high and low nitrogen, respectively, relative to the accession with the lowest biomass in that growth regime. The impact of low nitrogen and low carbon differed between accessions, with some accessions showing a greater than 70 % decrease in biomass and others showing no decrease. Moreover, accessions that maintained a relatively high biomass in low nitrogen tended to show only a small increase in biomass in high nitrogen, whereas accessions that showed a relatively small biomass in low nitrogen showed a large (greater than 3-fold) increase in biomass in high nitrogen. The total nitrogen concentration in the rosette was unrelated to the biomass difference between low and high nitrogen The nitrogen content (mg/N per rosette) was strongly related to the response of an accession to nitrogen; accessions that maintained biomass in low nitrogen contained more nitrogen in the rosette than accessions that showed a large gain in biomass in high nitrogen. These results imply that accessions differ in the extent to which they can acquire nitrogen from low nitrogen soil and that this is far more important for the response of biomass to nitrogen supply than changes in the nitrogen content of the rosette.

As for uptake efficiency, natural variation for remobilisation efficiency can be directly investigated by measuring nutrient content in the seeds. For instance, Khan et al. (2012) revealed differences in oil content in the seeds of various *Acacia* species, revealing some species as a novel source of edible vegetable fat. As for the studies conducted on N uptake, an isotope labelling technique is suitable to investigate the remobilisation efficiency. This was shown in the work of Coque et al. (2008), who investigated N remobilisation in maize with the aim of mapping and characterising loci involved in the variation of quantitative traits (QTL) related to NUpE, grain N yield, N remobilisation and post-silking N uptake. They stated that QTLs for remobilisation mainly coincided in clusters with loci for leaf senescence, underlying the role of a "stay-green" phenotype in favoring N uptake capacity, and thus grain yield and N grain yield. Similarly, in *Arabidopsis* a range of variation for Harvest Index (HI) and Nitrogen Harvest Index (NHI) has been found when 20 accessions were cultivated with a limited or ample supply of nitrogen (Masclaux-Daubresse and Chardon 2011). It was observed that the range of variation among the 20 accessions was conserved between the two nitrogen levels. Globally HI is similar at high and low nitrogen, while NHI, which defines the resource allocation, is twice as high at low nitrogen compared to high nitrogen, indicating that grain NUE is higher when nitrogen fertiliser is limited.

Several studies have been conducted by investigating natural variation in crop species, which were focused mainly on yield improvement (Shewry et al. 2013) and grain composition (Li et al. 2011), but only a few have been done evaluating the interaction of those traits with fluctuation of nutrient content in the environment. This kind of study requires highly controlled conditions in order to clearly elucidate the impact of the genotype and the environment on the variance of traits. This is why such studies have been up until now conducted mainly in *Arabidopsis*, as shown in the review of Chardon et al. 2012. Variation between *Arabidopsis* accessions can be explored to discover ideotypes that can match with different crop specifications: four groups, corresponding to different agronomic indicators such as grain yield, vegetative biomass and composition of the grains at low or high nitrogen supply, can be defined. On the basis of their physiological performance for nitrogen uptake and remobilisation, some *Arabidopsis* accessions are presented as good models for the investigation of the agronomic performances required to fit with crops specifications. Some other examples come from the investigations of *Arabidopsis* accessions grown in limiting nutrition (Reymond et al. 2006; North et al. 2009) or complete nitrogen starvation (Richard-Molard et al. 2008; Ikram et al. 2012). The aim of these studies was to investigate the extent of variation of growth responses to nitrate limitation and starvation in *Arabidopsis* to identify accessions showing contrasted responses and eventually different growth adaptive strategies.

Uncovering Genes Involved in NUE Quantitative Loci by QTL Mapping in Segregating Populations

Principles of QTL

The establishment of the genetic basis of quantitative traits is commonly referred to as quantitative trait locus (QTL) mapping and has been hampered by their multigenic inheritance and the often strong interaction with the environment. The principle of QTL mapping in segregating populations is based on the genotyping of progenies derived from a cross of distinct genotypes for the trait under study. Phenotypic values for the quantitative traits are then compared with the molecular marker genotypes of the progeny to search for particular genomic regions showing statistically significant associations between polymorphism and the trait variation, which are then called QTL. QTL analysis makes use of the natural variation present within species. Once genetic variation is found among accessions, the aim is to identify how many loci account for it and where they are located in the genome (Koornneef et al. 2004).

Identifying the number and genome position of the segregating QTL in an experimental population requires the following steps: (a) the generation of an experimental mapping population; (b) its genotyping with markers throughout the genome and the phenotyping for the trait of interest; (c) the association analysis between phenotypic values of the trait and genotypic classes of the polymorphic markers. Thus, the number and genetic position of loci that control the trait variation in that population, their relative additive effect, the contribution of genetic interactions between loci (epistasis) and the mode of action of each QTL (dominance effects) are calculated depending on the population type (Koornneef et al. 2004). The number of loci identified per analysis varies from 1 to >10, depending on the complexity of the genetic variation under study, including parameters such as the true number of loci segregating, the relative additive effect of each QTL, and the effect of genetic interactions. In addition, this number depends on the heritability of the trait in the assay performed, *i.e.* the control of the environmental uniformity, the quality and density of genotypic data, the statistical method used to map QTL, and the size of the mapping population.

Mapping Populations

In plants, the use of "immortal" mapping populations consisting of homozygous individuals is preferred because it allows performance of replications and multiple analyses of the same population. Such populations known as recombinant inbred lines (RILs) or introgression lines (ILs), also referred to as near isogenic lines (NILs), are practically homozygous and therefore phenotypic values can be based on multiple replicates, reducing the environmental effects and increasing the power to detect QTL. They can be analysed in multiple environments without the need for further genotyping, and thus, the effects of each QTL in different environments can be precisely estimated and tested for QTL $\times$ environment interactions (Koornneef et al. 2004). Homozygous populations can be obtained by repeated selfing, as for RILs, but also by induced chromosomal doubling of haploids. In contrast, NILs consist of lines containing a single fragment or a small number of genomic introgression fragments from a donor parent into an otherwise homogeneous genetic background, which increases the power to detect a small-effect QTL. In plants, RILs and NILs are the most common types of experimental populations used for the analysis of quantitative traits. In both cases the accuracy of QTL localisation, referred to as mapping resolution, depends on population size. The mapping of QTL in segregating populations has limited resolution since loci associated with the expression of a quantitative trait can be mapped with a precision of about 5–20 cM depending on its relative effect and the quality of the QTL mapping assay (Keurentjes et al. 2007). The choice of one mapping population over another depends on the plant species and the specific parents of interest. In cases where different cultivars or wild accessions are studied, preference is often given to RILs. However, when different species or wild and cultivated germplasm are combined, NILs are preferred. In *Arabidopsis* for example, the ease with which fertile RIL populations with complete genome coverage can be generated, due to its fast generation time, has led to their extensive use in mapping quantitative traits. An overview of the steps undertaken to generate a set of RILs is represented in Fig. 2.1. The population derived from a European accession and a genetically distant one from central Asia, Bay-0 and Shahdara, is an example of a novel RIL population suitable for the investigation of traits such as the response to nitrogen availability, root architecture, seed germination, drought tolerance and virus resistance (Loudet et al. 2002). The phenotypic variation resulting from such a cross is expected to reflect the adaptation to the specific habitat and the genetic distance between the parental accessions. The first extensive study of N metabolism in Arabidopsis using QTL mapping was conducted by Loudet on 415 RILs derived from Bay-0 × Shahdara population (Loudet et al. 2003) to describe whole plant N physiology and growth at a vegetative stage. The study, conducted in controlled growth conditions, aimed at comparing two different N environments (10 mM and 3 mM nitrate) and identified several loci explaining the variability of growth and total N, nitrate, and free amino acid contents.

Other approaches involve the use of multiple parents, as in the multiple advanced generation intercross (MAGIC) and *Arabidopsis* multiparent RIL (AMPRIL) populations (Kover et al. 2009; Huang et al. 2011). The MAGIC design is more elaborate and generates more recombination events per line than the AMPRIL strategy, but the founder genomes are less evenly represented in the final lines. Mapping in either population is more complex than with RILs, but with a sufficiently high density of intermediate frequency markers, one can infer the most likely local founder genotype. Some of the advantages of using RIL-type populations will continue to apply in the future.

Maize is the crop species, which has traditionally been involved in QTL mapping, and numerous QTL studies for NUE are now available. Zhang et al. (2010) has published recently a study on QTL mapping for several enzyme activities. They detected 73 QTLs for the activity of 10 enzymes involved in carbon and nitrogen metabolism and eight QTLs for biomass in an intermating RIL population developed by randomly intermating plants for four generations following the F2, prior to the derivation of mapping progeny. A RIL population of rice has also been tested for tolerance to salinity, measuring the amount of Na^+ and K^+ ions in shoots and roots in three environmental conditions of 0, 100 and 120 mM NaCl

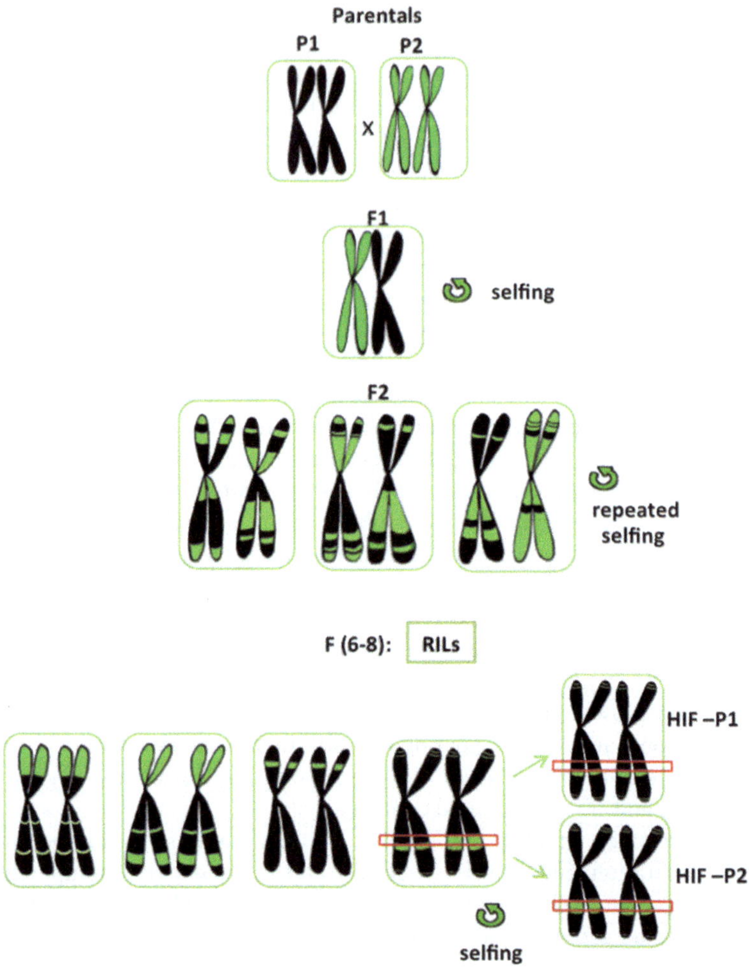

Fig. 2.1 Generation of RILs by successive selfings: two parental lines are crossed to produce an F1. The F1 is then selfed to obtain an F2. The selfing process continues until a certain level of homozygosity is reached. The end product consists of a set of RILs, each of wich is a fixed recombinant of the parental lines. HIFs individuals are derived from RILs in wich a small portion of the genome is still heterozygous (shown in red rectangle). Selfing such a RIL, it is possible to obtain individuals fixed for parental's 1 or parental's 2 allele. Only one chromosome pair is shown for each individual

(Wang et al. 2012). It is known that plants, which tolerate salinity are able to maintain a flux of Na^+ between shoots and roots in order to keep the ratio Na^+/K^+ as low as possible. The authors identified several major QTLs for salt tolerance and one of them, named qSNC11, will be suitable for use in marker-assisted selection when developing new salinity-tolerant cultivars.

QTL Fine Mapping

NILs have been used in various studies to confirm and fine map QTLs previously mapped in RIL populations (Koornneef and Smeekens 2005; Edwards et al. 2005) for which heterogeneous inbred families (HIFs) have also been used (Tuinstra et al. 1997). To construct an HIF-type NIL, a RIL is chosen that is still heterozygous around the QTL of interest but homozygous elsewhere. In fact, even after six generations of inbreeding, which is customary for RILs, a small percentage of the genome remains heterozygous. This RIL is then selfed and genotyped so that each homozygous genotype at the region of interest can be identified and studied in detail. HIFs should not to be compared with the reference parental genotype, but with one another within the descendants (family) of the chosen RIL. In contrast to "conventional" NILs, the genetic background of an HIF line is not homogeneous, but a mix of both parental genomes since these lines originate from one RIL of the population. The ease with which NILs can be extracted from a large population of HIFs may also allow a QTL mapped at low significance thresholds to be confirmed by subsequent examination of NILs. Second, NILs extracted from segregating HIFs are useful for the fine mapping of QTLs. Each segregating HIF is independent and contains unique recombination events in genomic regions flanking the QTL. An example of fine mapping using HIFs comes from the investigation of the four loci identified in the Arabidopsis population Bay-0 × Shahdara in low and high nitrogen environments by Loudet et al. (2003). The production of homogeneous plant material for a large number of lines was certainly the most challenging and limiting step of this work, but the authors also remarked that uncontrolled environmental effects can affect the evaluation of the quantitative traits and that environmental heterogeneity can occur between two different cultivation repetitions (even in the same growth chamber), as well as within one single growth chamber during a repetition. For these reasons, it is recommended (a) to always study all of the lines in the same cultivation repetition and (b) to compare different N environments in the same cultivation repetition. One of the limitations of HIF analysis for the evaluation of QTL is that the genetic background of NILs derived from HIFs are unique and cannot be easily replicated. NILs are not easily developed for evaluating the effects of more than one QTL in a single genetic background or for comparing the effects of QTL identified in different populations.

Candidate Genes

Co-localisation with candidate genes sometimes allows rapid molecular identification of QTL. This was the case in a study by Loudet et al. (2007) where QTL mapping for sulfate content in Arabidopsis leaves resulted in the identification of several minor QTLs and one major QTL on chromosome 1 in two contrasted N levels (3 mM and 10 mM). The major QTL co-localised with the gene APR2 coding for the adenosine-5′-phosphosulfate reductase, a key enzyme for assimilatory sulfate reduction pathway. The authors showed that the difference in sulfate contents between the two parental lines was not due to a different expression of the gene in the two genotypes, but to a change of alanine into glutamic acid in the APR2 protein. However, it is possible for there to be no overlap between the QTL and the candidate genes, as happened in the study on common bean conducted by Blair et al. (2010). A QTL for iron reductase activity in roots was identified under iron sufficiency (15 μg Fe), but it was mapped on a different chromosome from the one found under iron-limited growth (1 μM). Therefore, it was postulated that iron reductase activity was influenced by more than one locus, with a first iron reductase-related locus on chromosome b02, which contributed to the trait in iron-limited plants, and a second iron reductase-related locus present on chromosome b11 contributing in iron-sufficient plants. The authors also mapped loci for the FRO genes (iron-reductase homologues) but since there is no co-localisation with the QTL they concluded that some other gene may control iron reductase activity. Another resource for finding candidate genes is analysing gene expression in the vicinity of the QTL. This can be done using standard assays for a limited number of candidate genes, or using high-throughput genome-wide techniques, such as microarrays (Borevitz and Nordborg 2003). When the functional allelic variation results in gene expression differences, this may clearly indicate the candidate gene. Identifying an artificially induced mutant showing phenotypic effect in the trait of interest provides a unique functional argument to select a candidate gene. The availability of T-DNA insertion mutants for almost any Arabidopsis gene and the efficiency of TILLING (Targeting Induced Local Lesions in Genomes) procedures to identify mutations in numerous candidate genes provide efficient strategies to analyse knock-out phenotypes of (nearly) all genes in a QTL region. Nevertheless, most collections of mutants are in the laboratory backgrounds L*er* (Landsberg *erecta*) and Col (Columbia), which do not necessarily carry functional alleles at the gene of interest and, consequently, will not always show a distinct phenotype when mutated. Therefore, loss-of-function mutants of particular lines, such as NILs, carrying alleles different from the common laboratory accessions, can also be induced by mutagenesis with standard chemical or physical agents. This approach is especially useful when identifying novel alleles that are dominant over laboratory backgrounds (Koornneef et al. 2004). Ultimately, the proof for the identification of a QTL gene should come from complementation experiments by plant transformation.

QTL Validation

The presence of a QTL is validated when allelic variation of that QTL area has an effect on the studied trait. The search for polymorphic region-specific markers is crucial to fine map a QTL. NILs and HIFs are screened with molecular markers that are contrasted at the target region in parents. Finally, combining the analysis of a large number of segregating recombinants and the use of new polymorphic markers in the QTL area makes it possible to define a candidate region of less than 50 kb which will contain about 10 ORFs (open reading frames). If the region size is not sufficiently small for further analysis, new screening of rHIFs (possible recombinants within a heterozygous region of a QTL) is performed. When the desired region size contains only few genes, the QTL is considered as fine mapped and genes within it are examined for potential clues (Ikram and Chardon 2010). In addition to fine mapping, several functional strategies are available for plants whose complete genome sequence is available in order to select relevant candidate genes for the QTL. The knowledge of the complete genome sequence allows the search of such candidates on the bases of the predicted gene functions. Nevertheless, the function of many ORFs remains unknown at the cellular and/or phenotypic level and, therefore, it is not always possible to find obvious candidates from the genome sequence (Koornneef et al. 2004).

QTL cloning is a very efficient way to verify a new gene without *a priori*. For instance, Calenge et al. (2006) were interested in two genotypes, which accumulated soluble sugars at different rates in 10 and 3 mM nitrate nutrition. QTL analysis resulted in a major QTL for fructose content in leaves. The fine mapping restricted the QTL area to a 3 kb interval enclosing the single gene *SWEET17* (Chardon et al. 2013). The authors showed by functional analysis that variation in fructose content is uncoupled from further metabolic pathways, which result from sequestration of fructose into the vacuole, the main compartment for soluble sugars. They demonstrated that SWEET17 is a new vacuolar transporter of fructose in plant.

In contrast, simply resequencing a region with dozens or more genes is, on its own, not generally informative because of the high number of polymorphisms that distinguish an arbitrary pair of accessions, about 1 in every 200 bp (Weigel 2012). Fortunately, compared to other multicellular organisms in which natural variation is studied, *Arabidopsis* has the enormous advantage that almost all accessions are quite easily transformed by dipping flowering plants into a suspension of *Agrobacterium tumefaciens* containing a T-DNA vector with the transgene of interest.

QTL Meta-analysis

With data on multiple populations, it useful to know whether QTL identified for a given trait in one population correspond to those detected in other populations, or whether QTL locations identified in one species correspond to QTL or other types of loci detected in corresponding regions in other plant species. With this aim, a method has been developed by Goffinet and Gerber (2000) to estimate the minimum number of loci giving the observed QTL in individual studies and to combine the available information to precisely give the position of each individual QTL. Such an approach is called 'meta-analysis' and its usefulness is to pool information when raw data are not available. Comparative analysis of QTL between species reveals the existence of homologous QTL for traits involved in domestication, such as plant height and maturity, as well as tolerance to abiotic stress, within the cereals (Chardon et al. 2005; Hanocq et al. 2007; Li et al. 2013; Swamy et al. 2011). A recent example of meta-QTL analysis of a nutrition use efficiency-related trait comes from the dissection of an ortho-metaQTL in bread wheat by Quraishi et al. (2011). The authors identified a major NUE ortho-metaQTL conserved at orthologous positions in wheat, rice, sorghum and maize. Starting from three independent studies reporting QTL detection for traits related to NUE components in wheat, the authors proposed that a glutamate synthase (GoGAT) gene is conserved structurally and functionally at orthologous positions in rice, sorghum and maize genomes, and it that may contribute to NUE in wheat and other cereals.

Exploiting Genetic Variation in Wild Populations to Reveal NUE Genes by Association Mapping

First Results Obtained on Arabidopsis to Reveal NUE Genes

Over the past 10 years, traditional QTL mapping has led to the identification of sequence variants that modulate a range of physiological and developmental traits. Prior knowledge of the biological function of the affected genes was often helpful in identifying them, but increasingly the responsible locus is found to encode a protein without known biochemical function (Lempe et al. 2005). Apart from alleles that alter expression levels or protein function, a surprising number of drastic mutations such as deletions and stop codons underlie phenotypic variation. Some of these changes are found in many accessions (Weigel and Mott 2009) suggesting that they are adaptive. Nevertheless, despite some success stories, the number of known alleles responsible for phenotypic variation among accessions remains limited, mostly because fine mapping and dissection of QTLs are time consuming. *Arabidopsis thaliana* was the first plant species for which a genome sequence became available (Arabidopsis Initiative 2000). This initial sequence was from a single, high quality, inbred strain (accession) with each chromosome represented by only two contigs, one for each arm. In addition to functional analyses, the 120 Mb reference sequence of the Columbia (Col-0) accession proved to be a boon for evolutionary and ecological researches. A particular advantage in this respect is that the species is mostly self-fertilising, and most strains collected from the wild are homozygous throughout the genome (Weigel and Mott 2009). This distinguishes *Arabidopsis* from other model organisms such as the mouse or the fruit fly. In these systems, inbred strains have been derived, but they do not represent any individual actually found in nature. Numerous plants genomes have been completely sequenced and released recently.

Natural *Arabidopsis* accessions show tremendous genetic and phenotypic diversity. Thus far, significant natural variation has been reported for every phenotypic trait investigated (Koornneef et al. 2004). Moreover, assays of metabolite profiles by large-scale unbiased metabolomics methods have uncovered natural variation at the level of small molecules, suggesting that they reflect physiological phenotypes that could be under selection in nature (Keurentjes et al. 2006). Efforts to accelerate the discovery of functionally important variants began with a large-scale study in which some 1,000 fragments across the genomes of 96 accessions of *Arabidopsis thaliana* gathered from all over the world were compared by dideoxy sequencing (Rosenberg et al. 2005). A major conclusion from this work was that there has been considerable global gene flow, so that most sequence variants are found worldwide, although genotypes are not entirely random. There is isolation by distance, and even though population structure (which is a division of the population into distinct subgroups related by kinship) is relatively moderate, it can easily be a confounding factor in association studies. From this first set of 96 strains, 20 maximally diverse strains were chosen for much denser polymorphism discovery using array-based resequencing (Clark et al. 2007). This led to the identification of approximately one single nucleotide polymorphism (SNP) for every 200 base pairs of the genome, constituting one quarter or so of all SNPs estimated to be present. In addition, regions that are missing or highly divergent in at least one accession encompass about a quarter of the reference genome. For this reason it is becoming increasingly clear that it is inappropriate to think about 'the' genome of a species, even though this is what the initial sequencing papers stated in their titles just a few years ago (Weigel and Mott 2009). The previous emphasis on relatively minor changes between individuals, such as SNPs, was largely due to the fact that sequence variation had overwhelmingly been studied by PCR-based methods or hybridisation to known sequences. It is now known that Arabidopsis accessions can vary in hundreds of genes. Of particular importance is the observation that some genes with fundamental effects on life history traits such as flowering are not even functional in their reference accession, and thus could not have been discovered on the basis of the first genome sequence alone. Whilst knowledge about the origin and phenotypic effects of sequence polymorphisms is central to understanding how species adapt to their natural environment, most studies of genetic variation in Arabidopsis have probably been motivated by the desire to identify regulatory and other genes that are not present in the common laboratory accessions (Weigel 2012). A project begun in 2009 aimed to sequence the genome of 1001 accessions of *A. thaliana* (Weigel and Mott 2009), and the task is almost complete now. The main motivation for the 1001 Genomes project is, however, to enable genome-wide association studies (GWA) in this species. The seeds from the 1001 accessions are freely available from the *Arabidopsis* stock centres and each accession can be grown and phenotyped by scientists from all over the world (Weigel and Mott 2009). Importantly, because an unlimited supply of genetically identical individuals will be available for each accession, even subtle phenotypes and ones that are highly sensitive to the microenvironment, which is often difficult to control, can be measured with a high degree of confidence. The phenotypes can include morphological analyses, such as plant stature, growth and flowering; investigations of plant content, such as metabolites and ions; responses to the abiotic environment, such as resistance to drought or salt stress and to N deficiency; or resistance to disease caused by a host of prokaryotic and eukaryotic pathogens, from microbes to insects and nematodes.

Candidate Gene Association and Genome-Wide Association Studies

An explanation of how association mapping refers to the analysis of statistical association between genotypes (usually individual SNPs or SNP haplotypes, determined in a collection of individuals), and the phenotypes (traits) of the same individuals is undertaken here. Until recently, genetic mapping was usually done in purpose-created populations, such as progeny of parents chosen on the basis of the difference between them for the trait(s) of interest, or in defined pedigrees (families) (Rafalski 2010). By contrast, genetic association mapping involves using a collection of individuals, such as those derived from wild populations, germplasm collections or subsets of breeding germplasm. Consequently, at each locus several alleles may be simultaneously evaluated for association in a diverse population, while only two alleles segregate in any biparental population. Two association mapping methodologies are in use: Candidate Gene Association and Whole Genome Scan, also called Genome-Wide Association Study. In the candidate gene approach, the hypothesis that there is a correlation between DNA polymorphisms in gene A and the trait of interest is tested. For example, it is possible to test if in a diverse germplasm collection there is a correlation between DNA sequence alleles of phytoene synthase (or any other gene involved in carotenoid biosynthesis) and carotenoid content of seeds (Palaisa et al. 2003; Pozniak et al. 2007). This approach assumes good understanding of the biochemistry and genetics of the trait, but many genes may escape attention. Therefore, in the absence of detailed knowledge of the biochemical pathway of interest, including regulatory genes, whole genome scan (described below) is a better choice (Rafalski 2010). Genome scan involves testing most of the segments of the genome for association by genotyping densely distributed genetic marker loci over all chromosomes. The simple hypothesis that one of the genetic loci being considered is either causal for the trait or in linkage disequilibrium (LD, defined as association between genetic loci) with the causal locus is under consideration. The choice of population for association mapping, and of the appropriate marker density, are crucial decisions. One of the sources of false positives in association mapping is population structure. Complex population structure could be expected in crop species that were subject to a severe domestication bottlenecks followed by breeders' selection. Pronounced differences in the germplasm used in different regions of the world and maturity-related sets of allele frequencies for many genes may also be expected. Examples include the division of maize germplasm into heterotic groups (Reif et al. 2005) and a severe post-domestication bottleneck associated with adoption of soybean in North

America (Hyten et al. 2006). Before choosing the appropriate number of genetic markers (usually SNPs) for a genome scan, it is necessary to have some understanding of the LD in the population selected for the study. In general, LD decreases with distance between marker loci, more slowly in inbreds (soybean), faster in outbred species (maize), although breeding practices have a large impact (Flint-Garcia et al. 2003). LD is, however, very non-uniform across the genome, with both general trends (more LD in centromeric regions) and pronounced local fluctuation (Rafalski 2010). For instance, LD in Arabidopsis extends for roughly 10 kb, which is a nearly ideal distance for mapping since it extends up to the gene level (Bergelson and Roux 2010). Genetic resolution of any mapping methodology ultimately depends on the amount of recombination available in the experimental population, as measured by the rate of decay of LD (Rafalski 2010). In collections of distantly related individuals many generations have passed and much recombination occurred since the last common ancestor, therefore resolution of association mapping will, in general, be considerably higher than in simple biparental populations. The power of association mapping is strongly dependent upon the quality of phenotypic data. It is important to stress that in most cases it is necessary to use well-controlled environmental conditions, including, when possible, use of growth chambers, especially for the collection of samples for metabolomic or biochemical phenotypes. Relevance of such phenotypes for field performance will have to be separately established (Rafalski 2010). High throughput methods to precision phenotyping, frequently referred to as phenomics, are developing rapidly and automated facilities for high precision phenotyping are being established (E. Finkel 2009). The use of such a facility has been recently presented as a powerful tool to investigate plant response to drought stress by Tisné et al. (2013), allowing the precise control of watering condition of more than 700 plants. In this way, the environmental variance was strongly reduced allowing the identification of QTLs for complex traits related to drought response.

Validation and Applications

Validation of the hypotheses generated by association mapping constitutes an integral part of the experiment (Rafalski 2010). In one approach, NILs differing in the alleles at the candidate locus are constructed by repeated backcrossing into a reference genetic background (Vlad et al. 2010). The resulting NILs are then phenotyped side by side, and the amount of phenotypic variation ascribed to the presence of introgressed segment is estimated. Biparental populations segregating for the relevant alleles at the associated locus may also be used (Beló et al. 2008). Alternatively, the association experiment could be expanded by the inclusion of additional individuals in the expectation that the strength of the association should improve if the association hypothesis is correct. Association mapping is usually performed with the objective of applying the results for genotype-based selection of superior individuals in plant breeding, or as a step toward positional cloning

(Rafalski 2010). In marker assisted recurrent selection, breeders identify desirable alleles at one or more loci, basing on the outcome of a mapping experiment, and then use closely linked genetic markers for selecting individuals in breeding populations (Collard and Mackill 2008; Ribaut et al. 2010). This approach results in fixing the desirable allele(s) in the population(s) of interest.

Limitations

The detection power of association mapping greatly depends not only on the magnitude of the effect that can be ascribed to a locus, relative to other loci present in the population, but also on the allele frequency distribution. Rare alleles cannot be detected with good confidence, unless their effect is very large. Therefore, segregating biparental populations are more appropriate for the mapping of alleles rare in the germplasm pool of interest (Rafalski 2010). Genetic association mapping enriches the repertoire of tools available for the dissection of trait architecture in crop plants and model species. As high-density genotyping becomes increasingly accessible, this approach will gain power to identify with high-resolution genetic loci and in some cases causal polymorphism affecting agronomic and end-use traits in crop plants, as long as relevant alleles are present at high frequency. Mapping in defined biparental populations will remain the method of choice for rare alleles, especially those with moderate effects, and for the study of epistatic interactions. Independent validation of the associations found by both approach and evaluation of their effects in different genetic backgrounds remains an essential, even though sometimes neglected, aspect of a genetic experiment. Improvement in phenotyping remains a major challenge for mapping many agronomical important traits such as NUE or drought tolerance.

Future Perspectives

In addition to visually obvious phenotypes, natural variation has also been observed in genetic mechanisms such as cytosine methylation (Riddle and Richards 2002). A recent interest concerns the exploitation of natural variation in gene expression, leading to the first studies using expression QTL mapping (eQTL) proposed as a valuable approach to dissect the genetic basis of transcript variation, one of the prime causes of natural phenotypic variation. A recent study using eQTL conducted on 191-individual pseudo-F1 progeny of grape to dissect the genetic basis of berry colour formation (Huang et al. 2013), led to the identification of two major QTL explaining 20 % of genotypic variance and co-locating with a key enzyme for anthocyanin synthesis. With available genomic tools such as the whole genome

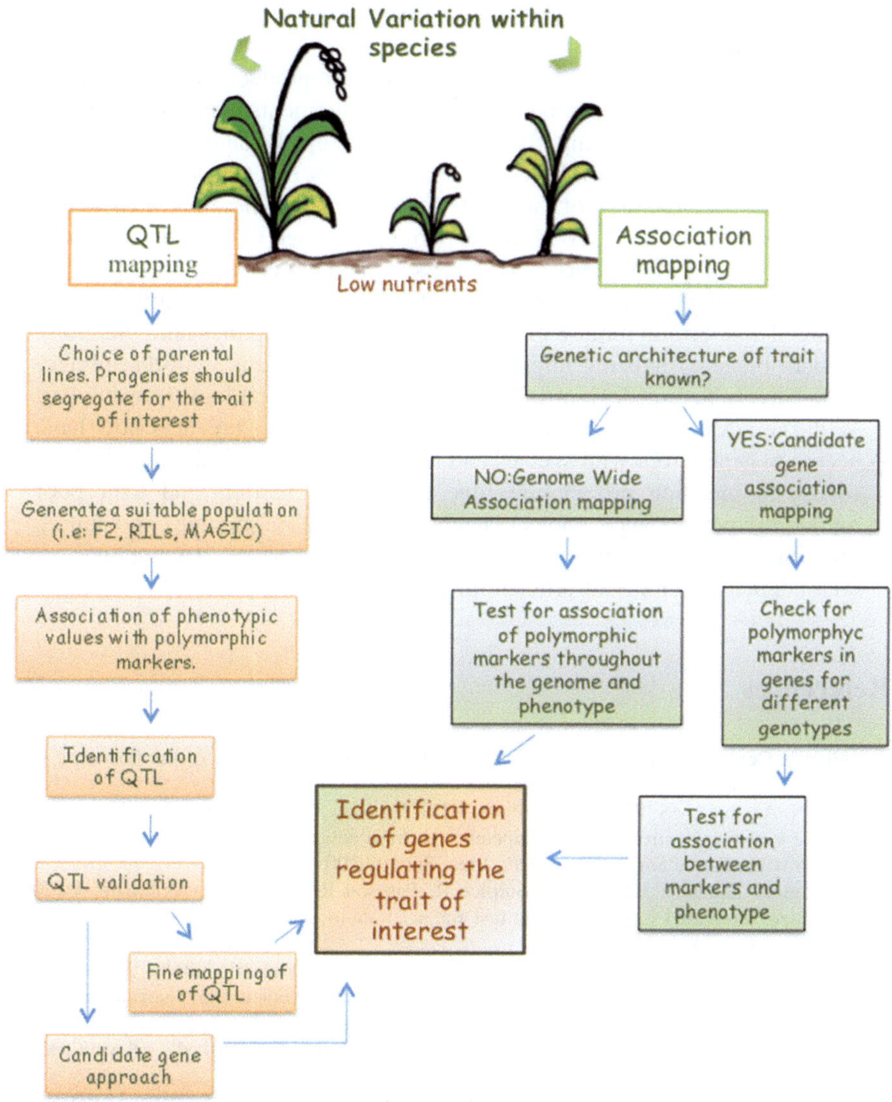

Fig. 2.2 A scheme of two possible strategies to be followed in the investigation of nutrient use efficiency related traits using natural variation in plants

sequence of many species, further investigation through genome-wide eQTL studies should bring a valuable contribution to understanding the molecular basis of traits of interest and should offer the opportunity for knowledge transfer to horticultural plants. In Fig. 2.2, a graphical overview of the possible strategies to investigate plant mineral use efficiency is displayed.

Conclusions

In light of scarce resources, increasing fertiliser production costs and the demand for greater crop production, the development of nutrient-efficient varieties is increasingly important. Both nutrient uptake and metabolic pathways are under the control of a complex regulatory network involving many genes. The identification of large-effect QTL/genes is therefore a challenge (Vinod and Heuer 2012). With the experiences gained in QTL mapping and the rapid development of genome-sequencing and molecular-marker technologies, more high-impact, large-effect QTL will surely be identified in the future. These efforts require expertise in different disciplines and, therefore, modern breeding is being implemented more and more in multidisciplinary teams involving breeders, physiologists and molecular biologists/geneticists. With the advances in molecular breeding technologies, breeders now have access to genes from wild species and unadapted genotypes that are difficult to use in breeding programmes due to crossing barriers and their poor agronomic performance. Molecular breeding therefore provides an exciting opportunity to use these gene pools effectively for the development of well-adapted and nutrient-efficient plants.

References

Arabidopsis Genome Initiative (2000) Analysis of the genome sequence of the flowering plant *Arabidopsis thaliana*. Nature 408:796–815

Arikita FN, Azevedo MS, Scotton DC, Pinto MDS, Figueira A, Peres LEP (2013) Novel natural genetic variation controlling the competence to form adventitious roots and shoots from the tomato wild relative *Solanum pennellii*. Plant Sci 199/200:121–130

Armengaud P, Zambaux K, Hills A, Sulpice R, Pattison RJ, Blatt MR, Amtmann A (2009) EZ-Rhizo: integrated software for the fast and accurate measurement of root system architecture. Plant J 57:945–956

Beló A, Zheng P, Luck S, Shen B, Meyer DJ, Li B, Tingey S, Rafalski A (2008) Whole genome scan detects an allelic variant of fad2 associated with increased oleic acid levels in maize. Mol Genet Genomics 279:1–10

Bergelson J, Roux F (2010) Towards identifying genes underlying ecologically relevant traits in *Arabidopsis thaliana*. Nat Rev Genet 11:867–879

Blair MW, Knewtson SJ, Astudillo C, Li CM, Fernandez AC, Grusak MA (2010) Variation and inheritance of iron reductase activity in the roots of common bean (*Phaseolus vulgaris* L.) and association with seed iron accumulation QTL. BMC Plant Biol 10:215

Borevitz JO, Nordborg M (2003) Update on genomics and natural variation in *Arabidopsis*. The impact of genomics on the study of natural variation in *Arabidopsis*. Plant Physiol 132:718–725

Calenge F, Saliba-Colombani V, Mahieu S, Loudet O, Daniel-Vedele F, Krapp A (2006) Natural variation for carbohydrate content in Arabidopsis. Interaction with complex traits dissected by quantitative genetics. Plant Physiol 141(4):1630–1643

Chardon F (2012) Exploring NUE in crops and in *Arabidopsis* ideotypes to improve yield and seed quality. J Exp Bot 63:3401–3412

Chardon F, Hourcade D, Combes V (2005) Mapping of a spontaneous mutation for early flowering time in maize highlights contrasting allelic series at two-linked QTL on chromosome 8. Theor Appl Genet 112:1–11

Chardon F, Noël V, Masclaux-Daubresse C (2012) Exploring NUE in crops and in Arabidopsis ideotypes to improve yield and seed quality. J Exp Bot 63(9):3401–3412. doi:10.1093/jxb/err353

Chardon F, Bedu M, Calenge F, Klemens PAW, Spinner L, Clement G, Chietera G, Léran S, Ferrand M, Lacombe B, Loudet O, Dinant S, Bellini C, Neuhaus HE, Daniel-Vedele F, Krapp A (2013) Leaf fructose content is controlled by the vacuolar transporter SWEET17 in *Arabidopsis*. Curr Biol 23:697–702

Clark RM, Schweikert G, Toomajian C, Ossowski S, Zeller G, Shinn P, Warthmann N, Hu TT, Fu G, Hinds DA, Chen H, Frazer KA, Huson DH, Schölkopf B, Nordborg M, Rätsch G, Ecker JR, Weigel D (2007) Common sequence polymorphisms shaping genetic diversity in *Arabidopsis thaliana*. Science 317:338–342

Collard BCY, Mackill DJ (2008) Marker-assisted selection: an approach for precision plant breeding in the twenty-first century. Philos Trans R Soc Lond B Biol Sci 363:557–572

Coque M, Martin A, Veyrieras JB, Hirel B, Gallais A (2008) Genetic variation for N-remobilization and postsilking N-uptake in a set of maize recombinant inbred lines. 3. QTL detection and coincidences. Theor Appl Genet 117:729–747

De Pessemier J, Chardon F, Juraniec M, Delaplace P, Hermans C (2013) Natural variation of the root morphological response to nitrate supply in *Arabidopsis thaliana*. Mech Dev 130:45–53

Dunbabin V, Rengel Z, Diggle AJ (2004) Simulating form and function of root systems: efficiency of nitrate uptake is dependent on root system architecture and the spatial and temporal variability of nitrate supply. Funct Ecol 18:204–211

Eduardo I, Chietera G, Bassi D, Rossini L, Vecchietti A (2010) Identification of key odor volatile compounds in the essential oil of nine peach accessions. J Sci Food Agric 90:1146–1154

Edwards KD, Lynn JR, Gyula P, Nagy F, Millar AJ (2005) Natural allelic variation in the temperature-compensation mechanisms of the *Arabidopsis thaliana* circadian clock. Genetics 170:387–400

El-Din El-Assal S, Alonso-Blanco C, Peeters AJ, Raz V, Koornneef M (2001) A QTL for flowering time in *Arabidopsis* reveals a novel allele of CRY2. Nat Genet 29:435–440

Finkel E (2009) With 'phenomics' plant scientists hope to shift breeding into overdrive. Science 325:380–381

Flint-Garcia SA, Thornsberry JM, Buckler ES (2003) Structure of linkage disequilibrium in plants. Annu Rev Plant Biol 54:357–374

Galkovskyi T, Mileyko Y, Bucksch A, Moore B, Symonova O, Price C, Topp CN, Iyer-Pascuzzi AS, Zurek PR, Fang S, Harer J, Benfey PN, Weitz JS (2012) GiA roots: software for the high throughput analysis of plant root system architecture. BMC Plant Biol 12:116

Goffinet B, Gerber S (2000) Quantitative trait loci: a meta-analysis. Genetics 155(1):463–473

Good AG, Shrawat AK, Muench DG (2004) Can less yield more? Is reducing nutrient input into the environment compatible with maintaining crop production? Trends Plant Sci 9:597–605

Hanocq E, Laperche A, Jaminon O, Lainé AL, Le Gouis J (2007) Most significant genome regions involved in the control of earliness traits in bread wheat, as revealed by QTL meta-analysis. Theor Appl Genetic 114:569–584

Hirai MY, Yano M, Goodenowe DB, Kanaya S, Kimura T, Awazuhara M, Arita M, Fujiwara T, Saito K (2004) Integration of transcriptomics and metabolomics for understanding of global responses to nutritional stresses in *Arabidopsis thaliana*. Proc Natl Acad Sci U S A 101:10205–10210

Huang X, Paulo MJ, Boer M, Effgen S, Keizer P, Koornneef M, van Eeuwijk FA (2011) Analysis of natural allelic variation in *Arabidopsis* using a multiparent recombinant inbred line population. Proc Natl Acad Sci U S A 108:4488–4493

Huang YF, Bertrand Y, Guiraud JL, Vialet S, Launay A, Cheynier V, Terrier N, This P (2013) Expression QTL mapping in grapevine-revisiting the genetic determinism of grape skin colour. Plant Sci 207:18–24

Hyten DL, Song Q, Zhu Y, Choi IY, Nelson RL, Costa JM, Specht JE, Shoemaker RC, Cregan PB (2006) Impacts of genetic bottlenecks on soybean genome diversity. Proc Natl Acad Sci U S A 103:16666–16671

Ikram S, Chardon F (2010) Plant quantitative traits. In: Encyclopedia of life sciences (ELS). John Wiley, Chichester

Ikram S, Bedu M, Daniel-Vedele F, Chaillou S, Chardon F (2012) Natural variation of *Arabidopsis* response to nitrogen availability. J Exp Bot 63:91–105

Jia Y, Yang X, Feng Y, Jilani G (2008) Differential response of root morphology to potassium deficient stress among rice genotypes varying in potassium efficiency. J Zhejiang Univ Sci B 9:427–434

Kellermeier F, Chardon F, Amtmann A (2013) Natural variation of *Arabidopsis* root architecture reveals complementing adaptive strategies to potassium starvation. Plant Physiol 161:1421–1432

Keurentjes JJB, Fu J, de Vos CHR, Lommen A, Hal LRD, Bino RJ, van der Plas LHW, Jansen RC, Vreugdenhil D, Koornneef M (2006) The genetics of plant metabolism. Nat Genet 38:842–849

Keurentjes JJB, Bentsink L, Alonso-Blanco C, Hanhart CJ, Blankestijn-De Vries H, Effgen S, Vreugdenhil D, Koornneef M (2007) Development of a near-isogenic line population of *Arabidopsis thaliana* and comparison of mapping power with a recombinant inbred line population. Genetics 175:891–905

Khan R, Srivastava R, Khan MA, Alam P, Abdin MZ, Mahmooduzzafar (2012) Variation in oil content and fatty acid composition of the seed oil of *Acacia* species collected from the northwest zone of India. J Sci Food Agric 92:2310–2315

Koornneef M, Smeekens S (2005) Sucrose-specific induction of anthocyanin biosynthesis in *Arabidopsis* requires the MYB75/PAP1 gene. Plant Physiol 139:1840–1852

Koornneef M, Alonso-Blanco C, Vreugdenhil D (2004) Naturally occurring genetic variation in *Arabidopsis thaliana*. Annu Rev Plant Biol 55:141–172

Kover PX, Valdar W, Trakalo J, Scarcelli N, Ehrenreich IM, Purugganan MD, Durrant C, Mott R (2009) A multiparent advanced generation inter-cross to fine-map quantitative traits in *Arabidopsis thaliana*. PLoS Genet 5:e1000551

Laegreid M, Bockman O, Kaarstad O (1999) Agriculture, fertilizers and the environment. CABI Publishing in association with Norsk Hydro ASA, New York

Łata B, Przeradzka M, Bińkowska M (2005) Great differences in antioxidant properties exist between 56 apple cultivars and vegetation seasons. J Agric Food Chem 53:8970–8978

Lempe J, Balasubramanian S, Sureshkumar S, Singh A, Schmid M, Weigel D (2005) Diversity of flowering responses in wild *Arabidopsis thaliana* strains. PLoS Genet 1:109–118

Li Y, Fan C, Xing Y, Jiang Y, Luo L, Sun L, Shao D, Xu C, Li X, Xiao J, He Y, Zhang Q (2011) Natural variation in GS5 plays an important role in regulating grain size and yield in rice. Nat Genet 43:1266–1269

Li WT, Liu CJ, Liu YX, Pu ZE, Dai SF, Wang JR, Lan XJ, Zheng YL, Wei YM (2013) Meta-analysis of QTL associated with tolerance to abiotic stresses in barley. Euphytica 189:31–49

Loudet O, Chaillou S, Camilleri C, Bouchez D, Daniel-Vedele F (2002) Bay-0 × Shahdara recombinant inbred line population: a powerful tool for the genetic dissection of complex traits in *Arabidopsis*. Theor Appl Genet 104:1173–1184

Loudet O, Chaillou S, Merigout P (2003) Quantitative trait loci analysis of nitrogen use efficiency in Arabidopsis. Plant Physiol 131:345–358

Loudet O, Saliba-Colombani V, Camilleri C, Calenge F, Gaudon V, Koprivova A, North KA, Kopriva S, Daniel-Vedele F (2007) Natural variation for sulfate content in *Arabidopsis thaliana* is highly controlled by APR2. Nat Genet 39:896–900

Masclaux-Daubresse C, Chardon F (2011) Exploring nitrogen remobilization for seed filling using natural variation in *Arabidopsis thaliana*. J Exp Bot 62:2131–2142

Masclaux-Daubresse C, Daniel-Vedele F, Dechorgnat J, Chardon F, Gaufichon L, Suzuki A (2010) Nitrogen uptake, assimilation and remobilization in plants: challenges for sustainable and productive agriculture. Ann Bot 105:1141–1157

North KA, Ehlting B, Koprivova A, Rennenberg H, Kopriva S (2009) Natural variation in *Arabidopsis* adaptation to growth at low nitrogen conditions. Plant Physiol Biochem 47:912–918

Palaisa KA, Morgante M, Williams M, Rafalski A (2003) Contrasting effects of selection on sequence diversity and linkage disequilibrium at two phytoene synthase loci. Plant Cell 15:1795–1806

Pozniak CJ, Knox RE, Clarke FR, Clarke JM (2007) Identification of QTL and association of a phytoene synthase gene with endosperm colour in durum wheat. Theor Appl Genet 114:525–537

Quraishi UM, Abrouk M, Murat F, Pont C, Foucrier S, Desmaizieres G, Confolent C, Rivière N, Charmet G, Paux E, Murigneux A, Guerreiro L, Lafarge S, Le Gouis J, Feuillet C, Salse J (2011) Cross-genome map based dissection of a nitrogen use efficiency ortho-metaQTL in bread wheat unravels concerted cereal genome evolution. Plant J 65:745–756

Rafalski JA (2010) Association genetics in crop improvement. Curr Opin Plant Biol 13:174–180

Reif JC, Hamrit S, Heckenberger M, Schipprack W, Maurer HP, Bohn M, Melchinger AE (2005) Trends in genetic diversity among European maize cultivars and their parental components during the past 50 years. Theor Appl Genet 111:838–845

Rengel Z, Damon PM (2008) Crops and genotypes differ in efficiency of potassium uptake and use. Physiol Plant 133:624–636

Reymond M, Svistoonoff S, Loudet O, Nussaume L, Desnos T (2006) Identification of QTL controlling root growth response to phosphate starvation in *Arabidopsis thaliana*. Plant Cell Environ 29:115–125

Ribaut JM, de Vicente MC, Delannay X (2010) Molecular breeding in developing countries: challenges and perspectives. Curr Opin Plant Biol 13:213–218

Richard-Molard C, Krapp A, Brun F, Ney B, Daniel-Vedele F, Chaillou S (2008) Plant response to nitrate starvation is determined by N storage capacity matched by nitrate uptake capacity in two *Arabidopsis* genotypes. J Exp Bot 59:779–791

Riddle NC, Richards EJ (2002) The control of natural variation in cytosine methylation in *Arabidopsis*. Genetics 162:355–363

Ristova D, Rosas U, Krouk G, Ruffel S, Birnbaum KD, Coruzzi GM (2013) RootScape: a landmark-based system for rapid screening of root architecture in Arabidopsis. Plant Physiol 161(3):1086–1096. doi:10.1104/pp. 112.210872

Robinson KM, Ingvarsson PK, Jansson S, Albrectsen BR (2012) Genetic variation in functional traits influences arthropod community composition in aspen (*Populus tremula* L.). PLoS One 7:e37679

Rosenberg NA, Shah C, Wall JD, Wang J, Zhao K, Kalbfleisch T, Schulz V, Kreitman M, Bergelson J (2005) The pattern of polymorphism in *Arabidopsis thaliana*. PLoS Biol 3:e196

Saisho D, Ishii M, Hori K, Sato K (2011) Natural variation of barley vernalization requirements: implication of quantitative variation of winter growth habit as an adaptive trait in east Asia. Plant Cell Physiol 52:775–784

Shewry PR, Hawkesford MJ, Piironen V, Lampi AM, Gebruers K, Boros D, Andersson AAM, Aman P, Rakszegi M, Bedo Z, Ward JL (2013) Natural variation in grain composition of wheat and related cereals. J Agric Food Chem. doi:dx.doi.org/10.1021/jf3054092

Shindo C, Bernasconi G, Hardtke CS (2007) Natural genetic variation in *Arabidopsis*: tools, traits and prospects for evolutionary ecology. Ann Bot 99:1043–1054

Sulpice R, Nikoloski Z, Tschoep H, Antonio C, Kleessen S, Larhlimi A, Selbig J, Ishihara H, Gibon Y, Fernie AR, Stitt M (2013) Impact of the carbon and nitrogen supply on relationships and connectivity between metabolism and biomass in a broad panel of Arabidopsis accessions. Plant Physiol 162(1):347–363. doi:10.1104/pp. 112.210104

Swamy BP, Vikram P, Dixit S, Ahmed HU, Kamar A (2011) Meta-analysis of grain yield QTL identified during agricultural drought in grasses showed consensus. BMC Genomics 12:319

Tisné S, Serrand Y, Bach L, Gilbault E, Ben Ameur R, Balasse H, Voisin R, Bouchez D, Durand-Tardif M, Guerche P, Chareyron G, Da Rugna J, Camilleri C, Loudet O (2013) Phenoscope: an automated large-scale phenotyping platform offering high spatial homogeneity. Plant J 74:534–544

Trontin C, Tisné S, Bach L, Loudet O (2011) What does *Arabidopsis* natural variation teach us (and does not teach us) about adaptation in plants? Curr Opin Plant Biol 14:225–231

Tuinstra MR, Ejeta G, Goldsbrough PB (1997) Heterogeneous inbred family (HIF) analysis: a method for developing near-isogenic lines that differ at quantitative trait loci. Theor Appl Genet 95:1005–1011

Ungerer MC, Johnson LC, Herman MA (2008) Ecological genomics: understanding gene and genome function in the natural environment. Heredity 100:178–183

Vinod KK, Heuer S (2012) Approaches towards nitrogen- and phosphorus-efficient rice. AoB Plants 2012: pls028

Vlad D, Rappaport F, Simon M, Loudet O (2010) Gene transposition causing natural variation for growth in *Arabidopsis thaliana*. PLoS Genet 6(5):e1000945

Wang X, Yan X, Liao H (2010) Genetic improvement for phosphorus efficiency in soybean : a radical approach. Ann Bot 106:215–222

Wang Z, Chen Z, Cheng J, Lai Y, Wang J, Bao Y, Huang J, Zhang H (2012) QTL analysis of Na + and K + concentrations in roots and shoots under different levels of NaCl stress in rice (*Oryza sativa* L.). PLoS One 7:e51202

Weigel D (2012) Natural variation in *Arabidopsis*: from molecular genetics to ecological genomics. Plant Physiol 158:2–22

Weigel D, Mott R (2009) The 1001 genomes project for *Arabidopsis thaliana*. Genome Biol 10:107

Xing JP, Jiang RF, Ueno D, Ma JF, Schat H, McGrath SP, Zhao FJ (2008) Variation in root-to-shoot translocation of cadmium and zinc among different accessions of the hyperaccumulators *Thlaspi caerulescens* and *Thlaspi praecox*. New Phytol 178:315–325

Zhang N, Gibon Y, Gur A, Chen C, Lepak N, Höhne M, Zhang Z, Kroon D, Tschoep H, Stitt M, Buckler E (2010) Fine quantitative trait loci mapping of carbon and nitrogen metabolism enzyme activities and seedling biomass in the maize IBM mapping population. Plant Physiol 154:1753–1765

Chapter 3
Macronutrient Use Efficiency – Sulfur in *Arabidopsis thaliana*

Patrycja Baraniecka and Stanislav Kopriva

Abstract Sulfur is an essential macronutrient required for proper growth of not only plants but also fungi and prokaryotes. It is present in a wide variety of metabolites such as amino acids: cysteine and methionine, coenzymes, vitamins and many others having distinctive biological functions. Plants take up sulfur from the soil in the form of sulfate via sulfate transporters. It is then reduced and assimilated in bioorganic compounds where cysteine is the first stable product. This process is very well described on both biochemical and molecular levels. Both reduction and assimilation are tightly regulated in demand-driven manner. The pathway has been extensively studied over last years because of important functions of sulfur in plant metabolism and stress defence. Here we summarise the up-to-date knowledge about the pathway and its regulation based mainly on the study on model plant *Arabidopsis thaliana*. We also emphasize areas in which little is known including the interconnection of sulfate metabolism with other nutrients.

Keywords Sulfur • Regulation • Arabidopsis • Sulfate transporters • Assimilation • Metabolism • APS reductase • Cysteine biosynthesis • OAS-TL • Glutathione • Methionine • Glucosinolates

Introduction

Apart from oxygen, carbon dioxide and water, plants require at least 14 mineral elements for sufficient nutrition. Six of them: nitrogen, phosphorus, potassium, calcium, magnesium and sulfur are required in large amounts and these are called macronutrients. The requirements for the rest, chlorine, boron, iron, manganese, copper, zinc, nickel, and molybdenum, are much lower and these are called micronutrients. Essential nutrients can be supplied to plants as fertilisers in case of deficiency to increase plant yield and quality. This applies mainly to crop production

P. Baraniecka
Department of Metabolic Biology, John Innes Centre, Norwich, UK
e-mail: patrycja.baraniecka@jic.ac.uk

S. Kopriva (✉)
Botanical Institute, University of Cologne, Cologne, Germany
e-mail: skopriva@uni-koeln.de

M.J. Hawkesford et al. (eds.), *Nutrient Use Efficiency in Plants*,
Plant Ecophysiology 10, DOI 10.1007/978-3-319-10635-9_3

(White and Brown 2010). However, this practice needs to be used optimally in agriculture as too high concentration of some of these elements in the soil solution may inhibit plant growth and reduce crop yield (White and Brown 2010).

Adequate sulfur supply is required for proper growth and fitness of all living organisms. Sulfur is present in a wide variety of metabolites with specific biological functions. Sulfur is cycled in the global ecosystem and can be converted to its organic compounds by photosynthetic organisms and microorganisms. The most common form of sulfur in nature is sulfate (SO_4^{2-}), which is the most oxidised form, in which sulfur is in the + VI redox state. Sulfate is taken up from the soil and reduced to sulfide in an energy-dependent reduction pathway. Sulfide can be incorporated into *O*-acetylserine to form cysteine, which is the first stable form of bound reduced sulfur in plants. Apart from cysteine, sulfur is present in methionine, sulfolipids, and cell walls. It also is contained in thiols, which are involved in redox control in plant cells and in vitamins and cofactors such as coenzyme A, thiamine and biotin. Noteworthy are the sulfur-containing secondary metabolites such as alliins and glucosinolates. Their breakdown products are responsible for characteristic smell and taste of many vegetables but also deter plant pathogens. Both of these classes of natural products are also very beneficial for human health. Alliins, found in large amounts in garlic, have antimicrobial properties, and glucosinolate degradation products induce enzymes that prevent tumour formation.

Sulfur availability in agricultural soils has been decreasing in many areas of Europe during the last two decades (McGrath et al. 1996; Zhao et al. 1996). The main reasons are: (i) reduction of sulfur dioxide emissions and subsequent reduction in sulfur depositions and (ii) changes in fertiliser practice, i.e. higher definition fertilisers without sulfate contamination (Blake-Kalff et al. 2001). Additionally, it was suggested that the demand for sulfur in many crops has increased due to intensive agriculture and optimisation during plant breeding programmes (Abdallah et al. 2010). The crop's requirement for sulfur varies in different species. Generally, wheat requires approximately 2–3 kg of sulfur for each tonne of grain produced (Zhao et al. 1999) whereas the production of 1 t of oilseed rape seeds requires about 16 kg of sulfur (McGrath and Zhao 1996). This high demand for sulfur of oilseed rape is probably due to an ineffective xylem-to-phloem transport mechanism in this species. A large amount of sulfate is accumulated in the vacuoles of mature oilseed rape leaves in sulfur sufficient conditions. This sulfate is not easily available for redistribution if sulfur is limited (Blake-Kalff et al. 1998). Therefore oilseed rape is particularly sensitive to sulfur deficiency or limitation which reduce both seed quality and yield (Malhi et al. 2007). Sulfur deficiency in crop plants has been recognised as a limiting factor not only for crop growth and seed yield but also for poor quality of products which is particularly important in the case of wheat and the maintenance of baking quality (Shahsavani and Gholami 2008). The protein fraction is known to play an essential role in the bread-making quality of wheat. The gluten proteins, gliadins and glutenins, represent about 80–85 % of total flour protein. These are responsible for elasticity and extensibility that are essential for functionality of wheat flours (Hussain et al. 2012; Kuktaite et al. 2004). However, in sulfur-limited conditions the synthesis of sulfur-poor storage proteins such as ω-gliadin and the high molecular weight subunits of glutenin is favoured at the expense of sulfur-rich proteins which may cause unpredictable and unwanted variations in wheat quality (Flæte et al. 2005; Moss et al. 1981). Another problem, discovered relatively recently, is the formation of acrylamide during high-temperature processing of potato and wheat products (Tareke et al. 2002). Acrylamide is potentially carcinogenic to human. It also has negative neurological and reproductive system effects (Friedman 2003). The major determinant of acrylamide forming potential is the concentration of free asparagine (Curtis et al. 2010). The accumulation of free asparagine in wheat grain in severe sulfur deprivation may increase to 30-fold higher levels compared to sulfur-sufficient conditions, which makes asparagine up to 50 % of the total free amino acid pool. For that reason, even very small amounts of such grain entering the food chain could have a significant effect on acrylamide formation, which makes the application of sulfur fertilisers over entire fields very important (Halford et al. 2012).

As sulfur deficiency in Europe appeared relatively recently, research on sulfur use efficiency still lags behind that on the other major nutrients. Therefore, exploring sulfur assimilation and the regulation of sulfur metabolism is of great interest for agriculture and plant science, because it is important to understand the process in order to optimise it for commercial use. However, there are still many gaps in our knowledge in this area. Studies on model plants provide a great tool for further exploration of the sulfur metabolic pathway mainly because of their rapid life cycle. The knowledge obtained from research on model plants can subsequently be transferred to crops and used for improving crop-breeding strategies. For these reasons this review is focused on the sulfur metabolism of the model plant *Arabidopsis thaliana*.

Sulfate Transport

Sulfate is the major form of sulfur in plants. Sulfate uptake from the soil is the first stage of plant sulfate metabolism. After entry into the plants, sulfate needs to be delivered to the plastids for assimilation or to the vacuoles for storage. Cell-to-cell transport as well as long-distance transport between organs required to fulfil the source/sink demands during plant growth, involve specific sulfate transporter proteins (Buchner et al. 2004b). Genes encoding these proteins belong to the sulfate transporter gene family and are divided into five groups. Members of different groups vary in kinetics of transport and in patterns of expression indicating different functions in the process of sulfate uptake and distribution. In general, high-affinity transporters are responsible for the initial uptake of sulfate by the root epidermis and cortex cells whereas low-affinity transporters are involved in the vascular transport. Sulfate transporters may be expressed constitutively or depending on sulfur availability. Decreased intracellular content of sulfate, cysteine and glutathione (GSH) results in increased transporter activity (Smith et al. 1997). The Group 3 sulfate transporters are not influenced by the sulfur status of the plant (Kataoka et al. 2004a; Takahashi et al. 2000). Sulfate transporters from Groups 1, 2, and 4 undergo a dual pattern of regulation: semi-constitutive and inducible. Under sufficient sulfur conditions and saturable uptake kinetics of the high- and low-affinity transporters, a semi-constitutive expression of one high-affinity transporter from Group 1, one low-affinity Group 2 transporter, undefined Group 3 transporters, and vacuolar sulfate efflux transporter from Group 4 are necessary for proper sulfate uptake and distribution (Smith et al. 1997; Takahashi et al. 2000). Under excess sulfur nutrition the expression of inducible sulfate transporters is repressed in order to prevent toxic internal sulfate accumulation. Feedback loops involving key metabolites from the sulfate reduction pathway are a common regulatory mechanism for expression of sulfate transporters (see further parts of this chapter).

The influx of sulfate through the plasma membrane is well characterised at the physiological and functional level. Plants essentially use a proton/sulfate co-transport system to mediate sulfate flux (Lass and Ullrich-Eberius 1984). A proton gradient is generated by plasma membrane proton ATPase (Saito 2004) and the transport process is pH dependent with $3H^+$/sulfate stoichiometry (Hawkesford et al. 1993; Smith et al. 1995a). Products encoded by sulfate transporter genes possess 12 membrane-spanning domains and belong to a large family of cation/solute co-transporters (Saito 2000). Additionally, the analysis of the C-terminal region of these transporters revealed the presence of STAS (sulfate transporters and antisigma factor antagonists) domains (Aravind and Koonin 2000). It was shown that deletion of the STAS domain affects the localisation of transporters to the membrane (Shibagaki and Grossman 2004) and results in a loss of sulfate transport activity (Rouached et al. 2005). These results indicate that STAS domains have a regulatory function in sulfate uptake and flux.

Functions of Sulfate Transporters in **Arabidopsis thaliana**

The first sulfate transporters were identified by functional complementation of a sulfate transporter-deficient yeast mutant (Smith et al. 1995a, b). Subsequently, yeast deletion mutants became the main tool for functional characterisation of sulfate transporters. To date, a number of sulfate transporters from a variety of plant species have been described (Buchner et al. 2004a; Howarth et al. 2003; Vidmar et al. 2000; Yoshimoto et al. 2002, 2003). The availability of fully sequenced genomes for *Arabidopsis thaliana* and rice has enabled analyses, which led to identification of 14 putative sulfate transporter genes in each genome. Based on phylogenetic analysis they were subdivided into four closely related groups, and a fifth group containing two smaller proteins lacking the STAS domain (Hawkesford 2003).

The analysis of the first isolated sulfate transporters identified high- and low-affinity transport characteristics (Smith et al. 1995a) and more detailed analysis revealed that the high-affinity components facilitate uptake of sulfate into the plant root (Barberon et al. 2008; Yoshimoto et al. 2007). The high-affinity sulfate transporters are very well studied. In *Arabidopsis thaliana* they comprise three genes (SULTR1;1–3) and belong to the clade which forms Group 1 of the sulfate transporters (Fig. 3.2). The primary sulfate acquisition in roots is mediated by two transporters: SULTR1;1 and SULTR1;2 (Fig. 3.1). Various studies confirmed the expression of these two genes in root hairs, root epidermal and cortical cells (Rae and Smith 2002; Takahashi et al. 2000; Yoshimoto et al. 2002) suggesting the capacity of these tissues for a high-affinity sulfate influx into the symplast. SULTR1;3 appears to be an exception in this group with specificity of localisation to the phloem in both roots and cotyledons (Yoshimoto et al. 2003). Sulfur

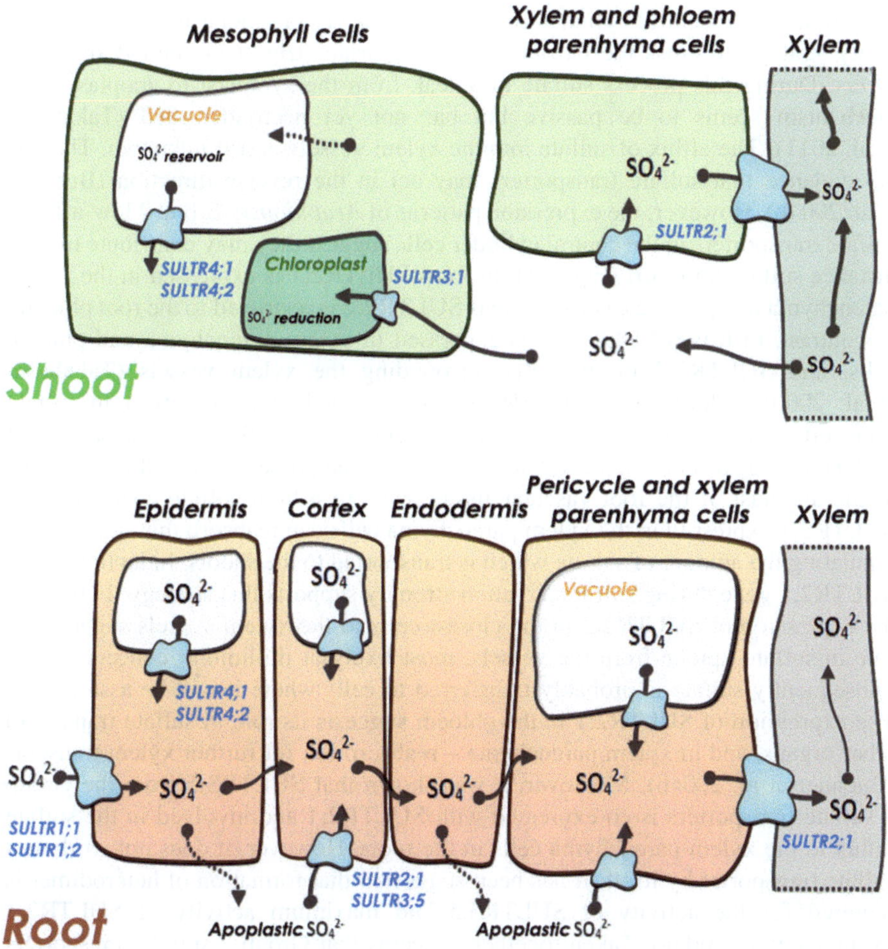

Fig. 3.1 Sulfate transport system in *Arabidopsis thaliana*. *Blue* shapes and labels indicate sulfate transporters (SULTRs) mediating transport across plasma membranes. *Dashed arrows* indicate yet unknown transport pathways

deficiency massively increased sulfate uptake (Clarkson et al. 1983). SULTR1;2 mediates sulfate uptake in normal conditions and in sulfur deficiency, and its expression is relatively independent of sulfate supply. In contrast, SULTR1;1 is strongly inducible by sulfate limitation but almost absent in normal sulfur conditions (Howarth et al. 2003; Yoshimoto et al. 2002). Moreover, another study showed that in mutants deficient in SULTR1;2, the expression of SULTR1;1 is slightly up-regulated. However, reduced growth suggests that SULTR1;1 is not able to compensate for the missing SULTR1;2. This might indicate that SULTR1;2 is the major component for sulfate acquisition (Maruyama-Nakashita et al. 2003).

Sulfate absorbed into the epidermis needs to be transferred across the root cells to the xylem. This step is necessary for delivery to the target cells in shoot organs for reduction or storage into the vacuole. The horizontal sulfate transfer from the epidermis to the central cylinder cells may occur via plasmodesmata. This is the likely strategy to cross the barrier of the Casparian strip at the endodermal cell layers. During this process sulfate may leak from the symplast to apoplast. This mechanism seems to be passive but has not yet been identified (Takahashi et al. 2011). The efflux of sulfate into the xylem vessels is still unknown. There is no evidence that sulfate transporters may act in the reverse direction (Buchner et al. 2004b). However, the expression patterns of *Arabidopsis* Group 2 low-affinity sulfate transporters in the central cylinder cells suggest they may contribute to long distance sulfate transport (Fig. 3.1). In roots SULTR2;1 is expressed in the xylem parenchyma and pericycle cells whereas SULTR2;2 is restricted to the root phloem. In contrast, in leaves SULTR2;1 is expressed in xylem parenchyma and phloem cells, and SULTR2;2 in the cells surrounding the xylem vessels (Takahashi et al. 2000). Additionally, SULTR2;1 was assumed to be involved in sulfate transport into developing seeds (Awazuhara et al. 2005). Localisation of SULTR2;2 suggests a role in sulfate transport via the phloem. The efflux of sulfate to the apoplast of the root vascular tissue leads to a high sulfate concentration. SULTR2;1 expressed in the xylem parenchyma cells can reabsorb this sulfate, thus regulating the amount of sulfate which is transported to the shoots. Induction of the SULTR2;1 gene during sulfur starvation strongly supports this strategy. In the leaf the expression of SULTR2;2 in the closest cells to the xylem vessels suggests its role in sulfate uptake from the vessels, most likely at millimolar concentrations. Subsequently sulfate is probably transferred to cells where it will be assimilated. The expression of SULTR2;1 in the phloem suggests its role in sulfate transfer to other organs, and in xylem parenchyma – reabsorption for further xylem transport (Buchner et al. 2004b). Moreover, it was shown that SULTR3;5 from the Group 3 sulfate transporters is co-expressed with SULTR2;1 and involved in the sulfate influx to the xylem parenchyma cells in the roots. However, it does not work as a sulfate transporter by itself. It has been suggested that formation of heterodimer is required for the activity of SULTR3;5 and maximum activity of SULTR2;1 (Kataoka et al. 2004a). Taken together, it seems that Group 2 sulfate transporters are involved in the balancing of the sulfate flux through the plant in various sulfate supply (Takahashi et al. 2000). It was also suggested that not only low- but also high-affinity sulfate transporters from Group 1 SULTR1;3 are involved in long distance sulfate transport. SULTR1;3 is expressed in phloem in all the *Arabidopsis* organs tested (Yoshimoto et al. 2003). Additionally, the analysis of a tomato homologue of this transporter showed that it is expressed under sulfur stress conditions (Howarth et al. 2003). These results indicate that plants are able to maintain balanced distribution of sulfate during limitation by inducing additional genes (Yoshimoto et al. 2003).

Plastids are the final destination for sulfate where it is assimilated. Alternatively, it can also be transported to the vacuoles, which play the role of sulfate reservoir in cells. It was shown that vacuoles isolated from the *Arabidopsis sultr4;1/sultr4;2* double mutant contain more sulfate than the wild type, suggesting that the efflux of sulfate from the vacuole is mediated by Group 4 sulfate transporters (Fig. 3.1). Moreover, enhanced expression of these two genes under sulfur stress conditions indicates that sulfur is released from the vacuole in response to sulfur demands in the cells (Kataoka et al. 2004b). The vacuole influx transporters have not yet been identified. A chloroplast sulfate transporter SULTR3;1 has been characterised only recently (Cao et al. 2013); chloroplast subcellular localisation was confirmed with the analysis of *SULTR3;1*-GFP constructs and the sulfate transport functionality with a validated *in organello* assay. However, the affinity of chloroplast for sulfate did not change despite the disruption of *SULTR3;1* suggesting the existence of other chloroplast sulfate transporters. The analysis of other members of the SULTR3 subfamily revealed that *SULTR3;2, SULTR3;3, SULTR3;4* but not *SULTR3;5* might be also chloroplast sulfate transporters (Cao et al. 2013). These results are consistent with those of Kataoka et al. (2004a) who demonstrated an essential role for SULTR3;5 in the vascular root-to-shoot transport.

As mentioned above, the members of Group 5 differ significantly from the other groups (Fig. 3.2). This group contains two isoforms, which are also dissimilar to each other. SULTR5;2 was described as involved in molybdenum transport (Baxter

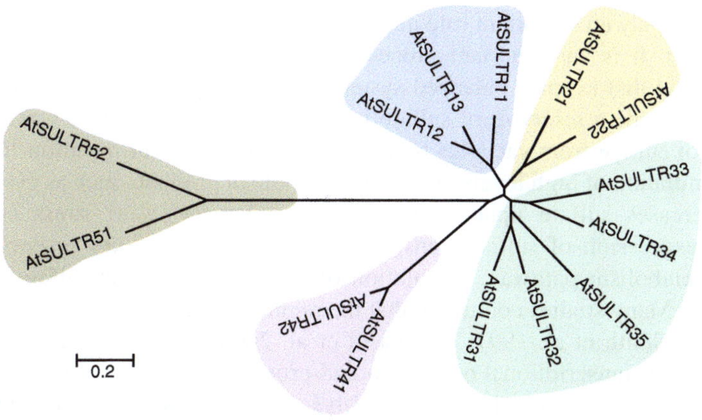

Fig. 3.2 Unrooted phylogenetic tree of the *Arabidopsis thaliana* members of the sulfate transporter family. *Different colours* represent different subfamilies. The tree was drawn using MEGA5.1 software

et al. 2008; Tomatsu et al. 2007) and renamed the MOT1 transporter. It is clear that MOT1 is essential for molybdenum accumulation. However, the exact subcellular localisation of this protein remains unclear. Tomatsu et al. (2007) localised this transporter to endomembrane system and plasma membrane whereas Baxter et al. (2008) suggested its localisation in the mitochondrial membrane. The explanation for these two predicted locations may be found in the different sites of GFP fusion, either to the N-terminal end of *MOT1* (Tomatsu et al. 2007) or to its C-terminus (Baxter et al. 2008). Nevertheless, the exact location of MOT1 requires further investigation. *SULTR5;1* was shown to be expressed in most plant tissues but it was not affected by sulfate supply (Shinmachi et al. 2010). Only recently its function as a vacuolar molybdate export protein was shown and it was renamed for MOT2 (Gasber et al. 2011). As there are no reports indicating a sulfate transport function for these two transporters and because of the absence of the STAS domain, which is present in all other sulfate transporters, these two genes could reasonably be excluded from the sulfate transporter family.

Regulation of Sulfate Transport at the Whole Plant Level

Growing plants require sulfate to synthesise amino acids, sulfolipids, thiols and many other sulfur-containing compounds. The demand for sulfur differs depending on the tissues, organs and developmental stage. Sulfate uptake and distribution is regulated in response to plant demand and changing environment. Specific patterns of expression of sulfate transporters genes in particular cells indicate the importance of proper sulfate distribution. Additionally, sulfate may be redistributed during development from mature leaves to roots, younger leaves or seeds. Redistribution requires changes in expression patterns. For example the up-regulation of SULTR2;2 and SULTR1;3 in leaves during sulfate starvation indicates that these transporters play an important role in sulfate allocation to other tissues (Yoshimoto et al. 2003). It is an important process that allows interconnection of different nutrients and that keeps a balanced system even during large environmental fluctuations. The main reservoir of sulfate is in the vacuoles of mature leaves. Redistribution of sulfate from vacuoles is particularly important during sulfur limitation.

Accumulation of high levels of sulfate metabolism products such as cysteine and GHS decreases sulfate uptake and transport whereas sulfur stress conditions increase expression of sulfate transporters and activities of key enzymes in the sulfate metabolism pathway. Regulation of sulfate transport is sulfate nutrition-dependent. Many studies confirmed that the regulation occurs primarily at the level of mRNA (Smith et al. 1997; Takahashi et al. 2000; Yoshimoto et al. 2002). In addition, post-transcriptional regulation and protein-protein interactions of sulfate transporters with *O*-acetylserine (thiol)lyase were described (Shibagaki and Grossman 2010; Yoshimoto et al. 2007). Their contribution to the overall control of sulfate uptake is, however, not clarified yet. Changes in expression patterns of sulfate transporter genes follow complex regulation responses. Firstly, tissue/organ-specific expression of transporters from Group 3 is not regulated by sulfur nutrition.

Secondly, cell-specific expression of some transporters from Groups 1, 2, and 4 is modulated by sulfur limitation. Finally, cell/tissue-specific, sulfur deficiency-related transporters from Groups 1 and 2 are derepressed (Buchner et al. 2004b). The question of how plants sense sulfate deficiency and how the signal is transduced to the transporter gene promoters still needs to be answered. Lejay et al. (2003) studying the regulation of ion transporter in roots by photosynthesis, concluded that some sulfate transporters (but also nitrate, phosphate, potassium and other metal transporters) show diurnal changes in expression and might be induced by sucrose, linking thus sulfate uptake with general primary metabolism.

In conclusion, plants are able to adjust sulfate transport to environmental changes and to the availability of the other nutrients. It is worth noting that these adaptations vary between species similarly to demands in sulfate metabolism. Different molecular regulation mechanisms are described in the following sections of this chapter.

Sulfate Assimilation and Metabolism in *Arabidopsis thaliana*

In nature, sulfur occurs in many oxidation states in inorganic, organic and bioorganic compounds. Various organisms such as algae, bacteria, plants and fungi are able to reduce sulfate and incorporate it into amino acids in the assimilatory sulfate reduction pathway. This process is very well described on both biochemical and molecular levels. In photosynthetic organisms it occurs in plastids (Brunold and Suter 1989). The only exception is *Euglena gracilis* where sulfate reduction takes place in mitochondria (Brunold and Schiff 1976). Since sulfate is chemically very stable, it requires activation before reduction. This occurs by adenylation to adenosine 5′-phosphosulfate (APS), which is catalysed by ATP sulfurylase (ATPS; EC: 2.7.7.4). ATPS in plants appears to be a homotetramer composed from 52 to 54 kDa polypeptides (Murillo and Leustek 1995). Most of total ATPS is plastid localised where it is responsible for sulfate activation for further reduction. However, activity was also detected in the cytosol (Lunn et al. 1990; Renosto et al. 1993; Rotte and Leustek 2000). During plant growth activity in chloroplasts declines, whereas it increases in the cytosol. ATPS is encoded by a small multigene family. Most plant species possess two ATPS isoforms. However, four different isoforms were isolated from *Arabidopsis*, which may indicate some level of genetic redundancy (Kopriva et al. 2009). Surprisingly all contain a chloroplast transit peptide (Hatzfeld et al. 2000a; Murillo and Leustek 1995). The most likely explanation for the existence of cytosolic ATPS isoforms is the use of a different translational start codons (Hatzfeld et al. 2000a). However, this hypothesis remains to be investigated.

APS forms a branching point in the sulfate reduction pathway (Fig. 3.3). APS can be directly reduced to sulfite by APS reductase (APR; EC: 1.8.99.2) or phosphorylated by APS kinase (APK; EC: 2.7.1.25) to form 3′-phosphoadenosine

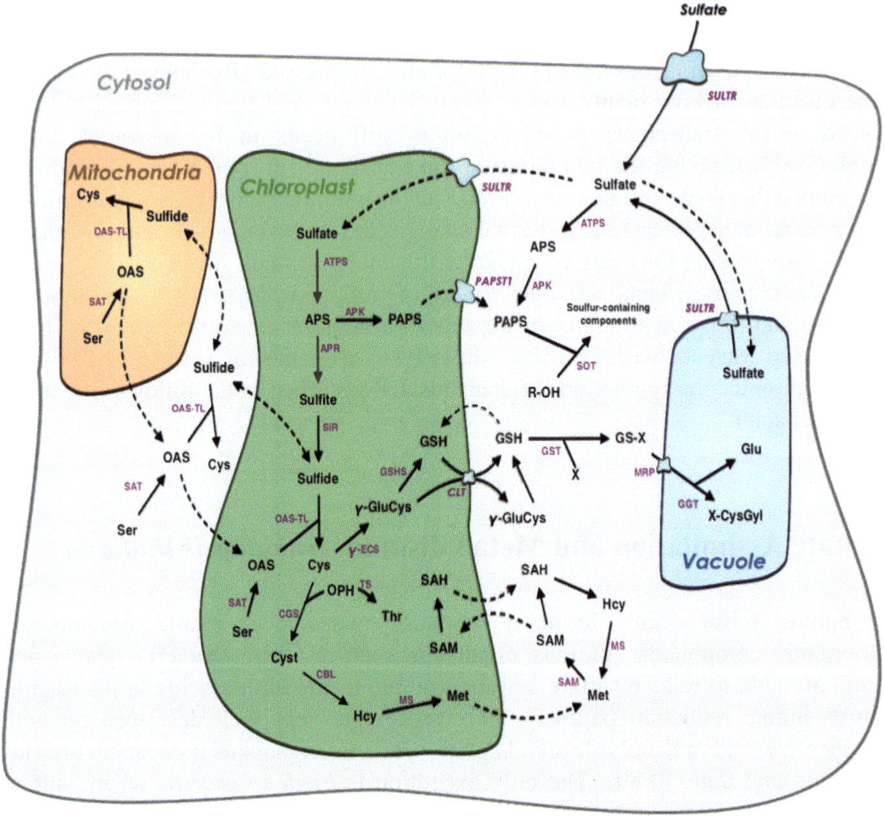

Fig. 3.3 Cellular organization of sulfate metabolism in *Arabidopsis thaliana*. Enzymes and transporters are indicated in *purple* characters. Abbreviations of enzymes and transporters: *ATPS* ATP sulfurylase, *APK* APS kinase, *APR* APS reductase, *SiR* sulfite reductase, *OAS-TL* OAS(thiol)lyase, *SAT* serine acetyltransferase, *CGS* cystathionine γ-synthase, *CBL* cystathionine β-lyase, *TS* threonine synthase, *MS* methionine synthase, *SAM* *S*-adenosylmethionine synthetase, *γ-ECS* γ-glutamylcysteine synthetase, *GSHS* glutathione synthetase, *CLT* thiol transporter, *GST* glutathione-S-transferase, *MRP* multidrug resistance-associated protein, *GGT* γ-glutamyltransferase, *SOT* sulfotransferase, *SULTR* sulfate transporter, *PAPST1* plastidic PAPS transporter. Abbreviations of metabolites: *APS* adenosine 5′-phosphosulfate, *Cys* cysteine, *Cyst* cystathionine, *Hcy* homocysteine, *OPH* *O*-phosphohomoserine, *Thr* threonine, *Met* methionine, *SAM* S-adenosylmethionine, *SAH* S-adenosylhomocysteine, *γ-GluCys* γ-glutamylcysteine, *GSH* glutathione, *GS-X* glutathione conjugate, *Glu* glutamate, *X-CysGyl* cysteinylglycine conjugate, *Ser* serine, *OAS* *O*-acetylserine, *PAPS* 3′-phosphoadenosine 5′-phosphosulfate, *R-OH* hydroxylated precursor

5′-phosphosulfate (PAPS) which serves as a donor of activated sulfate for various cellular reactions modifying proteins, saccharides or synthesis of secondary metabolites including glucosinolates. This process is called sulfation and it is important in regulating plant growth and development. However, sulfate reduction is a dominant route for assimilation (Leustek et al. 2000) and is fulfilled in two steps. In the first step APR transfers two electrons to APS to produce sulfite; the electrons are derived from GSH (Bick et al. 1998). Subsequently, sulfite is reduced to sulfide by ferredoxin-dependent sulfite reductase (SiR; EC: 1.8.7.1); this reaction requires a transfer of six electrons from ferredoxin to sulfite. Sulfide is then incorporated into the amino acid skeleton of *O*-acetylserine (OAS) to form cysteine. This reaction is catalysed by OAS thiol-lyase (OAS-TL; EC: 2.5.1.47) (Kopriva 2006; Leustek et al. 2000; Takahashi et al. 2011).

APR is a key enzyme of the sulfur assimilation pathway. It is regulated by various environmental factors and signalling molecules (Kopriva 2006; Koprivova et al. 2008). Similarly to ATPS, APR is encoded by a small multigene family and three isoforms are known in *Arabidopsis*. It was shown in various studies that APR1 and APR3 are co-regulated and share the highest sequence similarity. However, APR2 responds differently to hormone treatments (Koprivova et al. 2008) which indicates specific functions of particular APR isoforms. The amino acid sequence of APR suggests a multi-domain structure. Its precursor is synthesised with an N-terminal plastid transit peptide. In the mature protein the N-terminal domain is similar in predicted amino acid sequence to PAPS reductase (PAPR) from bacteria, and the C-terminal domain with thioredoxin (Trx). Due to the lack of the C-terminal domain, PAPR requires thioredoxin or glutaredoxin as a cofactor (Kopriva et al. 2007). This may suggest the role of C-terminal domain as redox cofactor (Leustek et al. 2000). *Arabidopsis* APR is a dimer of 45 kDa subunits (Kopriva and Koprivova 2004) binding a [Fe_4S_4] iron-sulfur cluster (Kopriva et al. 2001). PAPR consist of two 28 kDa subunits without any prostetic groups. A single conserved cysteine residue is responsible for its activity and dimerization.

APK catalyses the transfer of phosphate from ATP to APS to form PAPS which is a sulfate donor for sulfotransferases. Four genes coding APK are found in the *Arabidopsis* genome, all located on different chromosomes and with a high level of similarity. Three of the isoforms contain chloroplast transit peptides at the N-termini and these have been confirmed to be localised in plastids (Mugford et al. 2009). The APK3 isoform does not contain the N-terminal extension and it is likely to be responsible for cytosolic activity. Little is known about the biochemistry and the functions of the individual plant APKs. However, they have a significant effect on sulfur metabolism. It was shown that *apk1apk2* double mutant has a dramatically low glucosinolate level and also substantially higher levels of cysteine and GHS than wild type plants (Mugford et al. 2009). This suggests a compensation of low glucosinolate level by increases in cysteine and GSH. This subsequently may indicate that primary sulfate metabolism is up-regulated in this mutant, implying an important role of APK in controlling sulfur distribution in plants. Plants with APK1 as the only active APK isoform showed the wild type phenotype suggesting the major contribution of this isoform to total enzyme activity (Mugford et al. 2010). The analysis of mutants lacking various APR isoforms and differences in tissue-specific expression between the isoforms indicate specific roles of particular isoforms in plant sulfur metabolism (Kopriva et al. 2012; Mugford et al. 2009).

The six electron reduction of sulfite to sulfide is catalysed by SiR in plastids. Plant SiR is a 65 kDa monomer and requires the presence of a siroheme and an FeS cluster as cofactors, and ferredoxin as an electron donor (Nakayama et al. 2000). In contrast to other enzymes of the sulfur metabolism pathway, SiR is encoded by single gene in *Arabidopsis*. The amino acid sequence and protein structure are very similar to nitrite reductase (NiR) which catalyses a six electron reduction of nitrite to ammonia in nitrate assimilation pathway. The 19 % amino acid sequence identity suggests that these two enzymes have the same evolutionary origin.

Only recently a new enzyme in the sulfate assimilation pathway was identified (Eilers et al. 2001): *Arabidopsis* sulfite oxidase (SO; EC: 1.8.3.1) catalyses the oxidation of sulfite to sulfate in a two electron reaction. It is localised in peroxisomes (Nowak et al. 2004). Plant SO lacks the haem domain, which is known from animal SO's but it contains a molybdenum cofactor-binding domain. It was shown that the terminal electron acceptor for plant SO is molecular oxygen, the reaction converting O_2 into hydrogen peroxide. Hänsch et al. (2006) also provided evidence that hydrogen peroxide can oxidise sulfite non-enzymatically. It was shown that SO can protect plants from an excess of sulfite which is toxic in high concentrations, and SO_2 gas in the atmosphere (Brychkova et al. 2007). However, the exact role of SO in sulfur primary metabolism is still not clear.

Cysteine Biosynthesis

Cysteine is the key sulfur-containing compound in plants. It is synthesised by incorporation of sulfide into the β-position of the serine carbon skeleton in the terminal step of sulfur assimilation (Saito 2004). Before the incorporation of sulfide, serine needs to be activated to *O*-acetylserine (OAS). This process occurs by acetyl transfer from acetyl coenzyme A, which is catalysed by serine acetyltransferase (SAT; Serat; EC 2.1.3.30). Subsequently, OAS and sulfide are the substrates for O-acetylserine (thiol) lyase (OAS-TL; EC: 2.5.1.47) which catalyses the β-replacement reaction (Hell and Wirtz 2011; Takahashi et al. 2011). The enzymes involved in the cysteine biosynthesis process, SAT and OAS-TL, are localised in plastids, mitochondria and cytosol (Saito 2000; Fig. 3.3). Additionally, β-cyanoalanine synthase was identified in mitochondria, which has a similar catalytic activity to OAS-TL, and was previously identified as its mitochondrial form (Hatzfeld et al. 2000b). This enzyme is important for detoxification of cyanide produced e.g. during ethylene synthesis (Garcia et al. 2010). The analysis of whole plant protein extracts showed that entire SAT activity is always associated with OAS-TL, and that an excess of free active OAS-TL is present (Hell and Wirtz 2011). This and other results indicate that these two enzymes are associated and form a hetero-oligomeric cysteine synthase complex (Hell and Wirtz 2008, 2011; Saito 2004; Takahashi et al. 2011; Wirtz et al. 2001; Wirtz and Hell 2006). The binding of OAS-TL to SAT stabilises SAT. In the complex, SAT is the only active enzyme. The product OAS causes the release of OAS-TL from the complex which enables the conversion of OAS to cysteine by the free enzyme, and in a negative feedback, reduces the rate of OAS formation (Hesse et al. 2004b). Therefore, formation of the cysteine synthase complex appears to be a main regulatory step in cysteine synthesis (see the "Regulation of cysteine synthesis – protein-protein interactions" section in this chapter).

The crystallisation of SAT protein revealed that it is a hexamer composed from 29 kDa subunits which are folded in a left handed parallel β-helix, characteristic for this protein family (Olsen et al. 2004). The SAT gene family includes five members in *Arabidopsis*, two of which were recognised only recently (Hell and Wirtz 2008; Kawashima et al. 2005). Among the five SAT proteins three of them, SAT2, 4, and 5 (Serat 3.1, 3.2, and 1.1, respectively) are located in the cytosol whereas SAT1 (Serat 2.1) was found in plastids and SAT3 (Serat 2.2) in mitochondria. Additionally, because of the very low substrate affinity of SAT2 and 4 compared to SAT1, 3, and 5, it was suggested that these isoenzymes may actually process different substrates *in vivo* (Kawashima et al. 2005; Krueger et al. 2009). However, the analysis of multiple knock-out mutants for different SAT isoforms revealed that all of *Arabidopsis* SAT are able to complement, at least partially, for the loss of other isoforms (Watanabe et al. 2008).

OAS-TL belongs to the β-replacement enzyme family which requires pyridoxal-5′-phosphate as a cofactor. The protein is a homodimer composed from 35 kDa subunits. The *Arabidopsis* gene family contains nine members which encode eight functionally transcribed proteins (Hell and Wirtz 2008). Three of them, called OAS-TL A, B, and C, are thought to be the main OAS-TL proteins in plant cells. Similarly to SAT, they are localised in the cytosol, plastids and mitochondria, respectively (Wirtz et al. 2004). OAS-TL proteins seem to have wide range of functions. It is likely that apart from cysteine synthesis they are also responsible for other processes such as sulfide and cyanide detoxification in mitochondria (Alvarez et al. 2012) or determination of antioxidative capacity in cytosol (López-Martín et al. 2008). They also seem to be involved in the synthesis of secondary metabolites in various species.

Recent studies of SAT and OAS-TL mutants suggest the cytosol as the main cell compartment for cysteine production and mitochondria as the main place for OAS synthesis (Haas et al. 2008; Krueger et al. 2009; Watanabe et al. 2008). Plants with decreased mitochondrial SAT activity show strongly reduced OAS levels and reduced flux into cysteine and GHS (Haas et al. 2008). The analyses of OAS content and SAT activity in non-aqueous gradients showed the largest amount of OAS in mitochondria and the smallest in plastids, whereas the OAS-TL activity was localised in the cytosol and plastids (Krueger et al. 2009). However, OAS can be transferred between all three compartments. Additionally, the analysis of compartment-specific OAS-TL mutants revealed reduced cysteine content only in a mutant lacking the cytosolic OAS-TL isoform (Haas et al. 2008; Watanabe et al. 2008). Consequently, OAS has to be transported from mitochondria to the cytosol for efficient cysteine biosynthesis. Taking into account that plastids are the main compartment for sulfide production, the presence of sulfide in cytosol also requires sulfide transport across the chloroplast envelope membrane. Taken together, very low SAT activity in plastids, the presence of sulfide in the cytosol and the cytosolic localisation of cysteine strongly suggest the cytosol as the main cellular compartment for cysteine biosynthesis in *Arabidopsis* (Krueger et al. 2009).

Glutathione Biosynthesis and Functions

Formation of cysteine is the terminal step of sulfate assimilation pathway and the starting point for production of methionine, GHS and many other sulfur-containing compounds (Fig. 3.3). GSH is the main thiol-containing molecule in plant cells and is present in much higher concentrations than cysteine. It has a broad range of functions, which include removal of reactive oxygen species (ROS), detoxification of heavy metals and xenobiotics, sulfur donation, transport and storage (in catalytic reactions), redox signalling and many others. It is synthesised from glutamate, cysteine, and glycine by two enzymes: γ-glutamylcysteine synthetase (γ-ECS) and glutathione synthetase (GSHS). The reaction consumes two ATP molecules. γ-ECS is redox sensitive; only its oxidised form has high activity, whereas in reduced form the activity is much lower. The increase in γ-ECS transcript level in response to various environmental changes suggests its role as regulatory factor (Xiang and Oliver 1998). γ-ECS is also inhibited by higher concentrations of GSH. Synthesis of GSH is regulated by cysteine availability (Noctor et al. 2002). In *Arabidopsis thaliana* GSH is synthesised in cytosol and plastids, whereas γ-ECS is localised to plastids only and GSHS activity is distributed between plastids and cytosol. GSH from leaves is transported to roots, seeds and fruits via the phloem (Leustek et al. 2000), which supports an important role as a sulfur donor. In addition, GSH degradation is an important process in plants, however, the mechanism is not well understood. The main enzymes responsible for GSH degradation include glutathione reductase, γ-glutamyltransferase (GGT), glutathione S-transferase (GST), and glutaredoxin. It seems that GSH turnover in cells is maintained mainly by GGT activities (Takahashi et al. 2011). However, the intracellular degradation of GSH is independent of GGT and it is initiated by γ-glutamylcyclotransferase (Ohkama-Ohtsu et al. 2008).

Methionine Biosynthesis

Methionine is a sulfur-containing amino acid belonging to the aspartate family of amino acids together with lysine, threonine, leucine and isoleucine. It is essential amino acid for mammals and must be obtained entirely from the diet. Plants are able to synthesise it *de novo* from cysteine or homocysteine. Methionine plays important roles in plant metabolism. First of all it is a protein component. It is also involved in initiation of translation. *S*-adenosylmethionine (SAM), which is produced from methionine is a methyl-group donor and precursor of important secondary metabolites. It was shown that 80 % of methionine is used for SAM synthesis whereas 20 % is incorporated into proteins (Giovanelli et al. 1985). Synthesis of methionine requires products of three metabolic pathways: the carbon skeleton originates from aspartate, the sulfur atom from cysteine, and the methyl group from serine (Ravanel et al. 1998). In higher plants methionine synthesis starts from a γ-replacement reaction catalysed by cystathionine γ-synthase (CGS) leading to formation of cystathionine from cysteine and *O*-phosphohomoserine (OPH; Fig. 3.3). Cystathionine is converted to homocysteine by α,β-elimination catalysed by cystathionine β-lyase. The final step includes the transfer of the methyl group from N^5-methyl-tetrahydrofolate to homocysteine, which in plants is catalysed by cobalamin-independent methionine synthase (MS; Hesse et al. 2004a; Ravanel et al. 1998). In *Arabidopsis* three MS isoforms are known: two of them are in cytosol and a third one is localised to the plastids (Ravanel et al. 2004). Subsequently methionine is converted to SAM by SAM synthetase which requires ATP. It was shown that accumulation of SAM inhibits the enzyme activity (Ravanel et al. 1998). When SAM is used for synthesis of ethylene or polyamines, methylthioadenosine (MTA) is produced as intermediate. MTA can be used for synthesis of another methionine molecule increasing SAM availability as methyl-group donor (Burstenbinder et al. 2007).

Methionine synthesis undergoes a complex regulation. The pathway is regulated by feedback inhibition of aspartate kinase: the carbon flux inhibits the synthesis of methionine (Hesse et al. 2004a). It was also shown that CGS is involved in the control of the pathway on the level of sulfur and carbon-nitrogen skeleton delivery to the pathway (Hesse et al. 2004a). OPH is a common substrate for CGS and threonine synthase (TS). TS affinity for OPH is significantly higher than CGS affinity. TS is also positively regulated by SAM. Sufficient concentration of SAM enhances the activity of TS and affinity for OPH (Curien et al. 1996). Carbon skeletons are thus delivered to methionine biosynthetic pathway only when the SAM concentration decreases (see also "Control of methionine biosynthesis – posttranscriptional regulation" in the following sections of this chapter).

Sulfur-Containing Secondary Metabolites: Glucosinolates and Phytosulfokines

Sulfur is also present in plant metabolites as sulfo-group-modifying carbohydrates, proteins, and many natural products. Many sulfated metabolites play distinct roles in plant defence against biotic and abiotic stresses. Some of the best known groups of sulfated compounds are glucosinolates, which play an important role in protection against herbivores. They are also responsible for taste and flavour of many *Brassica* vegetables (e.g. cabbage, broccoli). Products of glucosinolate degradation, isothiocyanates, possess an anticarcinogenic activity in mammalian cells (Mithen

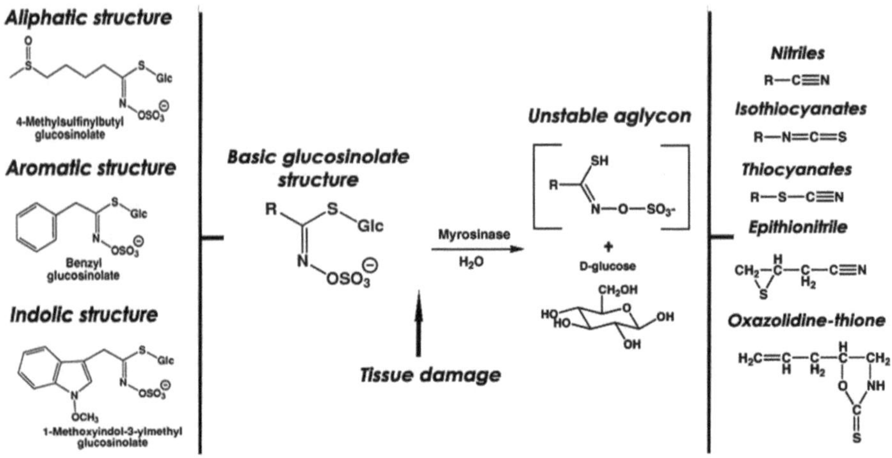

Fig. 3.4 Schematic representation of glucosinolate hydrolysis (*middle*) including examples of aliphatic, aromatic and indolic glucosinolates (*left*) and most common degradation products (*right*)

et al. 2003). In general, glucosinolates are synthesised in a three-block pathway. The elongation of certain amino acids, which are the precursors of aliphatic or aromatic glucosinolates, by sequential insertions of few (up to nine) methylene groups into the side chain is the first phase of the process. Subsequently, the amino acid moiety, elongated or not, is converted to form the core structure, which is common to all glucosinolates. The final block concerns various modifications, such as hydroxylation, *O*-methylation, desaturation, acylation and others, which leads to the formation of a broad range of structures (Halkier and Gershenzon 2006). To date, more than 140 different glucosinolates have been described (Fahey et al. 2001). In *Arabidopsis* nearly 30 different glucosinolates have been found in most organs, at various developmental stages (Brown et al. 2003). The chemical structure of most of them consists of a β-D-thioglucose group linked to a (Z)-N-hydroximinosulfate ester via a single sulfur atom (Halkier and Gershenzon 2006).

Although knowledge about the glucosinolate biosynthesis is important, it is the degradation products that are responsible for all their biological functions (Fig. 3.4). The process of hydrolysis begins with breakdown of the thioglucoside bond, which leads to the formation of glucose and unstable aglycone. The latter may then isomerise to different products depending on the structure of side chain and availability of various cofactors. The initiation process is catalysed by myrosinase (EC: 3.2.3.1) and has been investigated in a number of biochemical and molecular studies. Myrosinase is separated from glucosinolates in idioblast cells to avoid unnecessary glucosinolate hydrolysis. These two components, however, mix very quickly after the loss of cellular integrity as a result of wounding or insect or pathogen attack, to activate the binary glucosinolate – myrosinase system leading to the generation of glucosinolate hydrolysis products, which serve as plant defence molecules. The system, however, seems to work both ways: it deters usual pathogens but it may also attract some specialised herbivores (Renwick 2001). Many of them use glucosinolates as a food or use glucosinolates-containing plants for oviposit (Halkier and Gershenzon 2006).

Phytosulfokines (PSK) are another sulfur-containing metabolite of importance to plant growth and fitness, which promote various stages of plant growth such as somatic embryogenesis, adventitious bud and root formation and pollen germination (Chen et al. 2000; Kobayashi et al. 1999; Yamakawa et al. 1998). PSK-α is the first sulfated peptide found in plants. Its unique signal strongly promotes cell proliferation in plant cells at low concentrations. PSK-α is universally distributed in the plant kingdom (Yang et al. 2000) and is synthesised from an ~80 amino acid precursor peptide which has a secretion signal at its N-terminus (Yang et al. 1999). In *Arabidopsis* five genes encoding the precursor peptide have been identified (Yang et al. 2001). The expression of PSK genes in *Arabidopsis* is not limited to tissues characterised by active cell division and differentiation. It has been detected in most plant organs including mature leaves, stems, roots, and calluses, which indicate that it is not a simple mitogen or differentiation initiator. PSK precursor over-expression causes no apparent changes in plant growth or development under normal growth conditions (Matsubayashi and Sakagami 2006). It acts by binding to the specific binding sites which have been identified on the surface of suspension-cultured cells and in plasma membrane-enriched fractions of various plant species (Matsubayashi and Sakagami 1999). Based on the internal sequence of the PSK-binding protein, the component of the functional PSK receptor, LRR-RLK, was identified (Matsubayashi et al. 2002) and *LRR-RLK Arabidopsis* knock-out mutants are used to study the *in vivo* function of PSK (Matsubayashi and Sakagami 2006). The same receptor is used by another sulfated peptide called PSY1 which is also involved in the promotion of cell proliferation and expansion at nanomolar concentrations (Amano et al. 2007). Recently discovered RGF family peptides seem to be involved in root development (Matsuzaki et al. 2010; Meng et al. 2012), however, the exact role of these proteins and its sulfation remains unclear.

Regulation of Sulfate Assimilation

Understanding the tight regulation of the sulfate assimilation pathway is extremely important for two reasons: the essential role of sulfur for plant growth and quality, and the potential cytotoxicity of sulfite and sulfide, which are intermediates in sulfate assimilation. Various stages of the assimilatory pathway are regulated in both positive and negative feedback mechanisms in a demand-driven manner (Fig. 3.5).

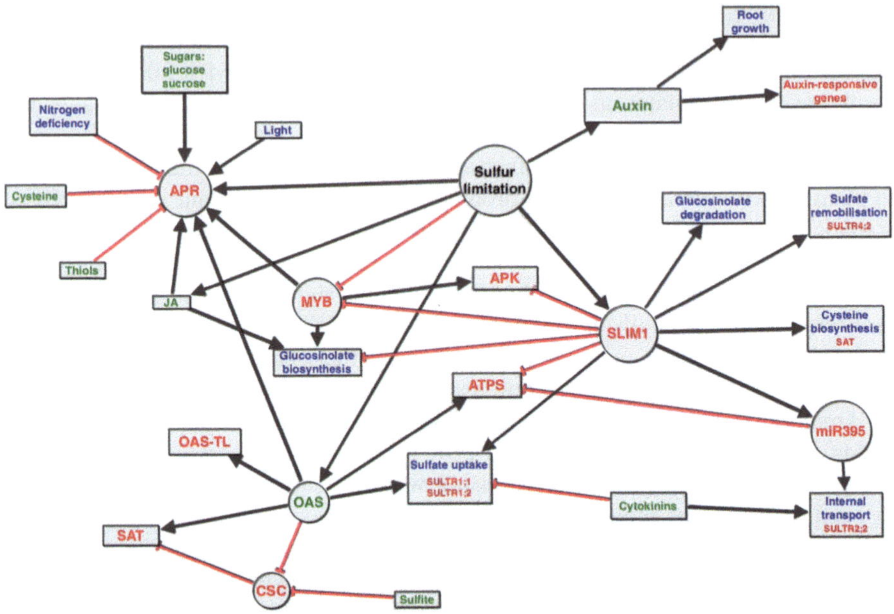

Fig. 3.5 Current understanding of the regulatory network of sulfate metabolism in *Arabidopsis thaliana*. *Black arrows* indicate positive regulation (induction) and *red arrows* indicate negative regulation (repression). The *grey circles* correspond to the key regulatory factors of the pathway, *grey boxes* correspond to all the other components involved in regulation. *Blue* colour indicates processes, *red* – genes and proteins, *green* – metabolites. Abbreviations: *APR* APS reductase, *MYB* MYB factors, *APK* APS kinase, *SLIM1* Sulfur Limiting factor 1, *SULTRs* sulfate transporters, *SAT* serine acetyltransferase, *ATPS* ATP sulfurylase, *OAS-TL* OAS(thiol)lyase, *OAS O*-acetylserine, *CSC* cysteine synthase complex

Regulation on the Level of Sulfate Uptake and Transport

The efficient acquisition of sulfate from the soil and its distribution in the plant is of great importance especially under sulfur limiting conditions. In a number of studies it was shown that the rate of sulfate transport during low sulfate supply is driven mainly by the regulation of the two high-affinity sulfate transporter genes, *SULTR1;1* and *SULTR1;2* (Shibagaki et al. 2002; Takahashi et al. 2000; Vidmar et al. 2000; Yoshimoto et al. 2002). The study of promoter-reporter constructs indicated that both are regulated in response to sulfate nutrition (Maruyama-Nakashita et al. 2004a). Further studies led to the identification of a transcription factor, SLIM1 (sulfur limitation 1), which is responsible for the regulation of sulfate uptake and metabolism during insufficient sulfur supply (Maruyama-Nakashita et al. 2006) (Fig. 3.5). S*lim1* mutants showed about 30 % reduction in root length and a 60 % decrease of sulfate uptake rates in sulfur-limiting conditions. The SLIM1 transcription factor belongs to the family of ethylene insensitive-like (EIL) transcription factors, from which EIL3 has a specific function in the regulation of sulfate uptake and metabolism (Maruyama-Nakashita et al. 2006). EIL3 was the only protein from the EIL-family able to restore the wild type phenotype of *slim1* mutants and was therefore renamed SLIM1. SLIM1 regulates the expression of the majority of sulfate-limitation responsive genes in the pathway, which suggests a hub-like function in the regulation system. Interestingly, APR, the key enzyme of sulfur metabolism, seems not to be subjected to the control of SLIM1, which suggests that it is not the only factor involved in the regulation process.

Most sulfur responsive genes are co-ordinately regulated for metabolism and stress reduction. A sulfur-responsive cis-acting element (SURE) was identified first in the sequence of the promoter of *SULTR1;1* sulfate transporter (Maruyama-Nakashita et al. 2005). SURE is a 7 nucleotide long specific sequence localised in the 5′-region of *SULTR1;1*. This sequence contains the core sequence of the auxin response factor (ARF) (Hagen and Guilfoyle 2002). However, SURE is sulfur-specific and is not relevant to the auxin response. The analysis of sulfur limitation inducible expression of *SULTR1;1* revealed that the SURE element was an essential target for the sulfur limitation response in *Arabidopsis* roots. Interestingly, SURE was identified in *SULTR1;1* but not in *SULTR1;2*, suggesting different mechanisms of regulation of these two transporters (Maruyama-Nakashita et al. 2005). Indeed, *SULTR1;1* is known to be controlled more specifically by sulfur limitation whereas the control of *SULTR1;2* is more dependent on metabolic demand and cellular status (Rouached et al. 2008). The *in silico* promoter analysis with GeneChip microarrays of 15 genes, which are known to be up-regulated by sulfur limitation, revealed the presence of a SURE core sequence (GAGAC or GTCTC) in the promoters of all of these genes. Similar sequences are present in the NIT3 nitrilase (Kutz et al. 2002) and β-conglycinin β-subunit (Awazuhara et al. 2002). Therefore it was concluded that the SURE core sequences are conserved in sulfur-limiting inducible promoters and may play a key role in sulfur limitation induction (Maruyama-Nakashita et al. 2005). However, a number of genes regulated by sulfur starvation do not have a SURE element. Despite the important role of SURE elements in the sulfur limitation-inducible response, there are still many gaps in our knowledge. Many important questions, such as specific SURE-binding candidates, still require further investigation.

In addition to transcriptional regulation, sulfur metabolism genes are also controlled post-transcriptionally. The main player in post-transcriptional regulation is microRNA395 (miR395; Fig. 3.5). MiRNAs are a group of small RNAs, which are formed from noncoding double-stranded RNA precursors. They are able to negatively regulate their target genes by cleavage or by binding to the complementary sequences in the target genes and repressing the translation (Bartel 2004). Analysis of the *Arabidopsis* genome revealed the low-affinity sulfate transporter SULTR2;1 and ATPS1 and ATPS4 isoforms of ATPS as target genes for miR395 (Jones-Rhoades and Bartel 2004). The accumulation of miR395 increases during sulfate starvation and this process is dependent on SLIM1 (Kawashima et al. 2009). Further studies revealed that miR395 regulates sulfate accumulation in shoots by cleavage of ATPS isoforms 1 and 4 during sulfate starvation. MiR395 was also shown to play a role in sulfate translocation from old to young leaves by targeting SULTR2;1 (Liang et al. 2010). Subsequently Kawashima et al. (2011) in experiments on SLIM1 and miR395-dependent regulation, reported an increase in *SULTR2;1* mRNA levels, a decrease in *ATPS4* levels and no changes in *ATPS1* mRNA levels in sulfate-limited conditions, suggesting three different mechanisms of miR395-mediated regulation. The increased expression of *SULTR2;1* in roots during sulfate starvation is limited to xylem parenchyma cells (Kawashima et al. 2009). The decrease in mRNA levels of *ATPS4* following induction of miR395 suggests a canonical regulation of *ATPS4* by miR395. The lack of response in *ATPS1* transcripts is in contrast to results obtained by Jones-Rhoades and Bartel (2004) who observed a decrease in *ATPS1* mRNA, however, these differences may be caused by different experimental setups. The increase in *SULTR2;1* expression in the xylem and reduction of flux through sulfate assimilation in the roots caused by miR395 indicate the importance of the SLIM-1-dependent induction of miR395 for the increased translocation of sulfate to the shoots when sulfate is limited. This results in an increase in the efficiency of sulfate assimilation in leaves (Kawashima et al. 2011). Very recently Matthewman and co-workers (2012) have shown that the complex regulation of miR395 is linked not only to SLIM1-dependent regulation during sulfate starvation but also to GHS and more generally thiol levels and/or cell redox state. Increased expression of miR395 in *fou8* and *sultr1;2* mutants affected in sulfate accumulation, suggests that miRNA395 is regulated by the internal sulfate level irrespective of external sulfate availability. All these results confirm that miR395 is an integral component of the sulfate assimilation regulatory network in a complex regulation mechanism.

Regulation of Sulfate Assimilation

Sulfate assimilation is tightly regulated in response to sulfate demand and environmental changes. The control mechanisms appear on different steps of the pathway and involve regulation of specific enzymes. Additionally, the whole pathway may be controlled as a process. Experiments focused on the first step of the pathway revealed the regulation of ATPS by sulfate availability (Logan et al. 1996). It was also shown that ATPS activity is inhibited by GSH. However, a number of studies revealed that the key regulatory step is sulfate reduction by APR (Fig. 3.5), as it is strongly affected by various treatments and environmental changes (Hesse et al. 2003; Koprivova et al. 2008; Vauclare et al. 2002). Vauclare et al. (2002) in the GSH feeding experiment showed that an increase of GSH concentration in the medium decreases APR expression and activity, indicating its regulation by GSH rather than cysteine. They also estimated the flux control coefficient for APR as 0.57, considering that the coefficient of all enzymes in the sulfate uptake and reduction pathway is up to 1, confirming the importance of APR in control of flux through sulfate assimilation pathway. The analysis of natural variation in sulfate content between two wild *Arabidopsis* accessions Bay-0 and Shahdara (Sha), performed to identify the genes controlling sulfate level, also revealed APR as a key control step in sulfate reduction pathway. The analysis of Bay-0 x Sha recombinant inbred lines (RILs) led to the identification of a single nucleotide polymorphism in an APR2 isoform of APR. The substitution of alanine with glutamate in a conserved domain of protein resulted in significant differences in enzyme activity leading to sulfate accumulation (Loudet et al. 2007). Additionally, reduction of APR activity and mRNA accumulation in low nitrogen availability shown in this study confirmed an interconnection of the two assimilatory pathways. Furthermore, APR over-expression leads to accumulation of sulfite and thiosulfate which are toxic for plants and strongly affect plant fitness (Martin et al. 2005). Such an effect is not observed when over-expressing other enzymes in the pathway suggesting that APR regulation affects the entire pathway. It was also shown that APR activity increases during the day and decreases during the night, which shows that it has a diurnal rhythm (Kopriva et al. 1999). A recently discovered transcription factor, Long Hypocotyl 5 (HY5), seems to be responsible for APR regulation by light. In dark adapted *Arabidopsis* seedlings, a rapid increase in the transcript levels of all three APR isoforms was observed to different extents, the highest being for the *APR2* isoform where it reached a 12-fold higher level after 90 min of illumination, compared to the control plants kept in the dark. However, in *hy5* mutant seedlings no light induction was observed for *APR1*, and *APR2* induction was lower. Further analysis revealed that HY5 is involved in APR regulation by OAS and nitrogen deficiency which alter the demand for reduced sulfur (Lee et al. 2011). Interestingly, the analysis of the *sir1-1* mutant revealed the downregulation of ATPS4, APR2, and SULTR2;1 in the mutant (Khan et al. 2010). The most likely reason for the downregulation of these genes, especially *ATPS4* and *APR2*, is to avoid the accumulation of sulfite which cannot be incorporated into cysteine as a result of reduced SiR activity in the mutant. These results suggest that SiR can contribute to the control of sulfate reduction pathway (Khan et al. 2010).

Regulation of Cysteine Synthesis – Protein-Protein Interactions

Cysteine synthesis plays an important role in regulation of sulfate metabolism. The regulation of SAT and OAS-TL, the main enzymes in cysteine biosynthesis, is mainly due to a protein-protein interaction in the cysteine synthase complex (CSC; Fig. 3.5). As mentioned above, SAT is strongly activated by OAS-TL which is inactive in the complex and has only a regulatory role (Droux et al. 1998). Formation of the complex is strongly dependent on the availability of OAS and sulfide. When SAT is bound to OAS-TL, the access of OAS to the complex is strongly inhibited (Francois et al. 2006). OAS, which cannot be bound to the complex, is released and metabolised by free OAS-TL dimers. During sulfur deficiency there is not enough sulfide for cysteine synthesis and therefore OAS accumulates in cells.

Accumulated OAS dissociates from the CSC complex, which rapidly decreases SAT activity (Hell and Wirtz 2008). High OAS concentration increases the expression of sulfate transporter genes, APR, SAT and OAS-TL. This leads to increased sulfate uptake and reduction and equilibrates the system (Hopkins et al. 2005; Koprivova et al. 2000; Smith et al. 1997). Hubberten et al. (2012) have shown recently that OAS may serve as a signalling molecule and change the transcription levels of specific genes irrespective of the sulfur status in the plant. The feedback inhibition of the cytosolic isoform of SAT by cysteine content serves as another form of regulation in *Arabidopsis*. It is important to note that in *A. thaliana*, plastidial and mitochondrial isoforms of SAT remain mostly insensitive to the changes of cysteine content (Noji et al. 1998). Cytosolic SAT activity may be considered as important for control of OAS concentration. The cysteine insensitive SAT isoforms in organelles may allow independent formation of cysteine (Noji et al. 1998). Although the regulation of cysteine biosynthesis is very important for sulfur homeostasis in plants and therefore it has been intensively studied over the past few years, many important aspects still require further investigation.

Control of Methionine Biosynthesis – Post-transcriptional Regulation

Methionine biosynthesis is subject to complex regulation. The whole pathway is controlled by feedback inhibition of aspartate kinase by lysine, threonine or lysine together with SAM. It seems that the competition between CGS and TS for their common substrate, *O*-phosphohomoserine, has an important regulatory role in the flux of carbon into methionine, which can be a limiting factor in the process (Hesse et al. 2004a). A recent study of transgenic *Arabidopsis* plants provided evidence that the regulation of methionine biosynthesis also occurs at the posttranscriptional level (Chiba et al. 1999, 2003). The analysis of this process is focused on the MTO1 region in exon 1 of CGS (Suzuki et al. 2001). The MTO1 mRNA region may act in *cis* and destabilise CGS mRNA in response to high concentrations of methionine or SAM (Chiba et al. 1999, 2003; Lambein et al. 2003). Computational analysis revealed the possible formation of a stable stem-loop structure in the MTO1 region which could support the posttranscriptional mechanism of regulation (Amir et al. 2002). More recently the presence of a truncated form of CGS transcript in *Arabidopsis* was demonstrated (Hacham et al. 2006). This transcript lacks about 90 nucleotides from the first exon. Over-expression of this CGS transcript causes even higher levels of methionine than the over-expression of full length CGS. This may suggest that this truncated transcript is not subject to feedback regulation by methionine (Hacham et al. 2006).

Control of Glucosinolate Biosynthesis – MYB Transcription Factors

Glucosinolate biosynthesis is highly regulated by a network of transcription factors in response to biotic and abiotic stress (Gigolashvili et al. 2007, 2008; Sønderby et al. 2007). They belong to two groups of the R2R3-MYB family of transcription factors. The first group consists of three members: MYB28, MYB76 and MYB29, and is involved in the control of aliphatic glucosinolate biosynthesis (Gigolashvili et al. 2008). The second group is responsible for control of biosynthesis of indolic glucosinolates and includes MYB51, MYB122, and MYB34 (Gigolashvili et al. 2007). Glucosinolates controlled by members of these two groups can be alternatively called high aliphatic glucosinolates (1–3) and high indolic glucosinolates (1–3), respectively (Fig. 3.5). T-DNA insertions, RNAi or over-expression of these factors affects the expression of glucosinolate biosynthesis genes and the level of glucosinolates. MYB34, MYB51 and MYB122 have different functions in the regulation of the pathway (Gigolashvili et al. 2007). Although all of them up-regulate the genes from the biosynthesis pathway, MYB34 and MYB122 also function as stimulators of auxin biosynthesis, whereas MYB51 additionally activates the glucosinolate biosynthesis pathway (Gigolashvili et al. 2007). Studies on MYB28, MYB29 and MYB 76 showed that they are able to transactivate each other in control of biosynthesis of aliphatic glucosinolates. They also downregulate the expression of enzymes involved in the synthesis of indolic glucosinolates (Gigolashvili et al. 2008). Analysis of MYB28 and MYB29 revealed that MYB28 is essential for the synthesis of aliphatic glucosinolates whereas MYB29 induces biosynthetic genes in response to plant hormone methyl jasmonate (Hirai et al. 2007). Maruyama-Nakashita and co-workers (2006) showed that mutation of SLIM1 additionally affects the expression of MYB34, suggesting that this factor may be negatively controlled by SLIM1 in response to sulfur deficiency. There is some evidence that SLIM1 also might affect the expression of MYB28 and MYB29, however the effect is not completely clear (Hirai et al. 2007). More recently, it was suggested that MYB factors are not only involved in the regulation of the glucosinolate biosynthesis genes, but also other genes in the sulfur reduction pathway. Yatusevich et al. (2010) concluded that *APR1* and *APR3* are under control of all MYB factors whereas *APR2* is regulated by only some of them. They suggested also that *APK1* and *APK2* are a part of the glucosinolate synthesis network controlled by MYB factors, whereas *APK3* and *APK4* have much lower input to the network. Glucosinolate biosynthesis may be also regulated by the DNA-binding-with-one-finger (DOF) transcription factor, OBP2, which is induced by herbivory and methyl jasmonate (Skirycz et al. 2006). Additionally, changes in expression of calmodulin binding IQD protein cause the variation in glucosinolate composition (Levy et al. 2005).

Regulation by Hormone Signals

Metabolic regulation is not the only means of controlling metabolic pathways. Plant hormones play very important roles in many developmental processes. Recent studies show involvement of plant hormones in regulation of different nutrient metabolic pathways (Fig. 3.5).

Cytokinins are adenine-derived plant hormones, which are responsible for regulation of cell division and differentiation in plants together with auxin. The best known example of regulation of the sulfate reduction pathway by cytokinin is the downregulation of high-affinity transporters SULTR1;1 and SULTR1;2 (Maruyama-Nakashita et al. 2004b). Addition of cytokinins to wild type plants decreases sulfate uptake and mRNA levels for the two transporters. It is interesting that SULTR1;2 is much more responsive to the cytokinins than SULTR1;1. It was suggested that in this regulation, the cytokinin response 1 (CRE1)/wooden leg (WOL)/*Arabidopsis* histidine kinase 4 (AHK4) cytokinin receptor is involved. The *Arabidopsis cre1-1* mutant was unable to regulate the high-affinity sulfate transporters in response to cytokinins (Maruyama-Nakashita et al. 2004b) suggesting the independent regulation of high-affinity sulfate transporters by cytokinin (repression) and by sulfate (induction). In contrast, Ohkama and co-workers (2002) have shown that sulfur responsive genes *APR1* and *SULTR2;2* were up-regulated by cytokinin. They have concluded that this induction is mediated by an increase in sucrose content.

Auxin-dependent signalling in the regulation of sulfate assimilation may be connected to the response to sulfate deficiency by indole glucosinolate hydrolysis (Kutz et al. 2002). During sulfate deficiency aglycon is released from indole glucosinolates and indole acetic acid (IAA) is generated from the remaining indole acetonitrile, catalysed by nitrilase NIT3. IAA may stimulate root growth, and indeed an increase in length and the numbers of lateral roots is a common phenotype of sulfur-deficient plants (Hell and Hillebrand 2001). Additionally, various studies indicate positive regulation of auxin-responsive genes during sulfate starvation (Maruyama-Nakashita et al. 2003; Nikiforova et al. 2003). Decrease in cysteine production leads to accumulation of OAS and its precursor serine and sulfate deficiency induces tryptophan synthase. Both of these events result in increase in tryptophan biosynthesis as in plants it is synthesised from indole and serine through the activity of the tryptophan synthase β-subunit. In consequence, increased biosynthesis of tryptophan increases production of auxin (Nikiforova et al. 2003).

Jasmonic acid (JA) is another plant hormone, which participates in regulation of sulfate metabolism. JA is involved in response to oxidative stress and synthesis of defence molecules. The cellular GSH content rapidly decreases during sulfur deficiency and this may lead to oxidative stress in cells. It was shown, however, that JA increases the expression of GSH synthesis pathway enzymes (Xiang and Oliver 1998). Additionally, microarray studies showed induced expression of JA genes during sulfur limitation and in *sultr1;2* mutants (Maruyama-Nakashita et al. 2003; Nikiforova et al. 2003). JA was also shown to induce the expression of the genes involved in glucosinolate biosynthesis (Brader et al. 2001; Doughty et al. 1995). JA-deficient mutants showed normal responses to sulfur limitation, similar to those when sulfate transporters and APR are induced and glucosinolate biosynthesis genes are repressed under sulfur starvation (Takahashi and Saito 2008). JA significantly induced expression and activity of APR, mainly isoforms 1 and 3 (Koprivova et al. 2008).

To summarise, JA acts positively for synthesis of both antioxidants and glucosinolates and has an important general role in plant pathogen defence and detoxification. It also co-ordinately induces multiple genes of sulfate assimilation suggesting its positive effect on sulfur homeostasis in plants (Jost et al. 2005).

Interconnection of Sulfate Assimilation with Other Nutrients

The sulfate assimilation pathway provides a range of sulfur-containing metabolites which play essential roles in a number of cellular processes. Many reports have shown that this pathway is very well co-ordinated with the assimilation of other nutrients such as nitrate, phosphate or carbon. This interconnection is crucial in dealing with the deficiency of one or more nutrients. It leads to activation or deactivation of important regulatory components. Such cross-talks provide flexibility in the adaptation to fluctuating environmental conditions, which helps plants to maintain nutrient homeostasis and thus complete the life cycle.

Regulation of Sulfate Assimilation by Nitrogen

The regulatory interactions between sulfate and nitrate assimilation in plants have been confirmed in a number of studies (Kim et al. 1999; Koprivova et al. 2000; Takahashi and Saito 1996). These two pathways are coordinated in the way that deficiency of one element represses the assimilation of the other. Studies on *Lemna minor* revealed that the activities of ATPS, APR and OAS-TL decreased during nitrogen deficiency (Brunold and Suter 1984). Addition of nitrate or ammonia to nitrate-deficient medium restored the activity of these enzymes very quickly. Additionally, when a normal nutrient solution was supplemented with ammonia or amino acids such as arginine, glutamate or aspartate, APR activity increased about 50–110 %. These nitrogen compounds also increased the flux through the sulfate assimilation pathway (Brunold and Suter 1984). Koprivova et al. (2000) have shown that after 72 h of nitrate deficiency in *Arabidopsis*, the activity of APR decreased about 30 % in leaves and 50 % in roots. This is correlated with the decrease of APR mRNA and protein accumulation, which suggests that nitrogen availability regulates sulfate assimilation on the transcriptional level. This finding is supported by the results of Yamaguchi and co-workers (1999) who have shown that nitrogen deficiency affects the mRNA accumulation of several other sulfate metabolism genes. Interestingly, the concentration of cysteine and GSH in both roots and leaves does not change significantly. Addition of ammonia or glutamate increased the sulfate flux through the *Arabidopsis* plants (Koprivova et al. 2000). Analysis of the *hy5* mutant revealed that the decrease in APR activity in nitrate deficiency was much more rapid in Col-0 than in the mutant suggesting HY5 involvement in regulation of APR in nitrate deficiency (Lee et al. 2011). Although this regulatory mechanism seems to play an important role in sulfate homeostasis, the molecular basis still requires elucidation. Koprivova and co-workers (2000) used feeding experiments to demonstrate a strong positive effect of nitrogen-containing compounds on the flux of sulfate through the pathway. Moreover, the closer the metabolic relationship between nitrogen source and OAS, the higher was the incorporation of ^{35}S into proteins. OAS also caused the accumulation of APR mRNA level and proteins (Koprivova et al. 2000). Therefore, it was proposed to play a signalling role in the co-ordination of nitrate and sulfate metabolic pathways. Other suggested signalling molecules are cytokinins: biosynthesis of cytokinins seems to be related to nitrate availability and they are known to affect nitrogen nutrition. As mentioned above, Ohkama et al. (2002) have shown that the *APR1* and *SULTR2;2* were up-regulated by cytokinins. Additionally, Wang et al. (2003) demonstrated that adding nitrate to plants grown on ammonium as the main source of nitrogen increased the accumulation of mRNA for high-affinity transporters and APR. Therefore, the nitrate effect on sulfate metabolism genes could be mediated by cytokinins (Kopriva and Rennenberg 2004). On the other hand, it was shown that sulfate deficiency decreased nitrate uptake and the activity of nitrate reductase and amino acid accumulation (Prosser et al. 2001). Abdallah et al. (2010) have shown that oilseed rape, which is known to have a high demand for sulfate, is able to maintain growth during sulfate deficiency by an optimisation of nitrogen uptake and remobilising sulfate from internal reservoirs. In field experiments with oilseed rape it was demonstrated that sulfur deficiency reduces nitrogen use efficiency and nitrogen deficiency reduces sulfate use efficiency (Fismes et al. 2000). This suggests the importance of further investigation of regulatory processes of nutrient metabolism in plants in order to improve crops yield and quality.

Interaction of Sulfate Assimilation with Carbon Metabolism

Carbohydrates provide the acceptor of sulfide for cysteine biosynthesis and serve as a source of reductants for sulfate reduction in non-photosynthetic tissues (Kopriva and Rennenberg 2004). However, despite this crucial role very little is known about regulation of sulfate assimilation by carbon. Kopriva et al. (1999) have shown that APR can be induced by light, suggesting a diurnal rhythm of regulation. However, when plants were subjected to continuous light or darkness the diurnal rhythm disappeared. This indicates that the light is not the direct signal for the changes in APR expression and the enzyme undergoes regulation by carbohydrates synthesised during photosynthesis. Furthermore, light induction of APR expression could be mimicked by addition of sucrose during the dark period (Kopriva et al. 1999). Subsequently, Hesse and co-workers (2003) have confirmed similar effect of glucose on APR activity. In addition, they have shown that sorbitol, mannitol and 2-deoxyglucose do not affect APR activity. Sugars seem to act in the same way as OAS: feeding plant with glucose and OAS simultaneously resulted in greater induction of APR activity than the sum of levels induced by individual treatments. Additionally, APR was also induced by glucose in plants subjected to nitrogen deficiency. Hesse et al. (2003) concluded that signals from nitrogen and carbon metabolism act synergistically on sulfate assimilation. Moreover, positive signals from sugars can exceed negative signals from nitrate assimilation. The reduction in APR mRNA level and activity was also observed in a study on *Lemna minor* grown in an atmosphere without CO_2 (Kopriva et al. 2002). The addition of sucrose prevented the decrease in APR activity indicating that this regulation is not dependent on CO_2 fixation. Additionally, it was shown that incubation in an atmosphere without CO_2 strongly inhibited sulfate uptake and the flux through the pathway, which also could be restored by addition of sucrose. However, sucrose was not as effective in restoring enzyme activities as normal air, suggesting that although it is the final product of carbon assimilation, it probably is not the molecular signal connecting the sulfate and carbon metabolism (Kopriva et al. 2002).

Coordination of Sulfate and Phosphate Homeostasis

Plants have evolved coordinated mechanisms to maintain the homeostasis of sulfur and phosphorus in response to changing environmental conditions, which confirms the importance of these two macroelements. It was shown that during sulfate limitation, sulfolipids are rapidly replaced by phospholipids and vice versa (Essigmann et al. 1998). Additionally, phosphate deficiency leads to an increase of GSH levels (Kandlbinder et al. 2004). It is interesting to note the similarity between the topology and regulatory mechanisms in the two pathways (Rouached 2011). In addition to the similar expression patterns of the main uptake transporters and the regulation of both of the pathways by cytokinins, the involvement of microRNAs in the regulation of both pathways seems to be particularly noteworthy in terms of cross-talk. As mentioned above, sulfate assimilation is regulated by miR395, whereas miR399 is known to be involved in regulation of phosphate uptake and translocation (Chiou et al. 2006; Fujii et al. 2005; Lin et al. 2008). It has been shown that miRNAs and the Phosphate Response 1 (PHR1) transcription factor are involved in the interconnection of sulfate and phosphate assimilation pathways (Rouached 2011). Additionally, it was suggested that miR395 might be suppressed in phosphate deficient plants (Hsieh et al. 2009), however, this observation requires further investigation. Recently Rouached et al. (2011) have shown that SULTR1;3, SULTR2;1 and SULTR3;4 sulfate transporters are up-regulated by phosphate deficiency. Induction of SULTR1;3 and SULTR3;4 was repressed by phosphite which is known to be involved in local phosphate sensing and long-distance signalling (Varadarajan et al. 2002). In contrast, SULTR2;1 was shown to belong to the class of phosphate-inducible genes not responding to phosphite (Rouached et al. 2011). Additionally, they have shown that PHR1 plays an important role in sulfate homeostasis in plants grown under phosphate deficient conditions by stimulating the expression of SULTR1;3 whereas SULTR2;1 and SULTR3;4 seem to be repressed by PHR1 in shoots only. These results suggest an important effect of phosphate as a signaling factor in maintaining source-sink sulfate distribution (Rouached et al. 2011).

To maintain sulfur homeostasis, plants need to cope with rapid environmental changes, which cause disturbances in levels of pathway intermediates and products. Therefore, a tight regulation of sulfate uptake and assimilation is required according to the demand for reduced sulfur. To ensure the precise modulation of sulfur metabolism, sulfate uptake and assimilation have to be coordinated. Indeed, sulfur limitation increases sulfate uptake and the expression and activity of key enzymes of the pathway. Other enzymes such as SiR and SAT are also involved in the control of sulfur metabolism, as disturbances in their activity cause growth defects. Sulfur uptake and assimilation are regulated on transcription, post-transcriptional and post-translational levels. However, some aspects of the control are specific only for uptake, such as regulation of sulfate transporters by SLIM1, and some only for assimilation. This suggests different mechanisms of achieving efficient control and maintaining system homeostasis. However, there are still gaps in the understanding of regulation of uptake, assimilation and distribution of sulfate at the whole plant level, the examination of which is a great challenge for the future research. Better understanding of molecular basis of control mechanisms will allow the production of engineered plants with improved sulfur use efficiency, with modulated amount of desirable sulfur-containing metabolites or with better resistance to low sulfur availability, drought and many other stresses.

Dealing with Sulfur Deficiency

As it was mentioned at the beginning of this chapter the reduction of sulfur dioxide emissions and changes in fertiliser practises have resulted in a widespread increase in the occurrence of sulfur deficiency in crops (McGrath et al. 1996). The application of fertilisers may be helpful in dealing with deficiency in many instances. However, the costs of fertilisers together with the possible negative effects of improper use with respect to timing and type of sulfur application in terms of its availability to the plant have led to the growing interest in looking for new solutions. Very often a substantial seasonal variation in sulfur availability occurs. Therefore, modern methods should be focused on uptake maximisation when sulfur is abundant in order to increase the tolerance to low sulfur availability periods (Hawkesford 2000).

Sulfur deficiency may not be easily recognised in the field as the symptoms are not obvious, except in severely deficient plants. In general, sulfur deficiency results in uniform pale green chlorosis through the plant. A considerable reduction in growth may be suffered without the appearance of any other visible symptoms. Clear symptoms are associated with severe stunting, reduced leaf size and activity of axillary buds, which results in less branching. Physiologically in wheat plants, sulfur deficiency affects CO_2 assimilation rates and Rubisco enzyme activity as well as protein abundance which results in general inhibition of *de novo* synthesis of the photosynthetic apparatus (Gilbert et al. 1997). Additionally, depression of root hydraulic conductivity was observed in sulfur-deficient barley plants. It was suggested that this response can have a role in signalling nutrient starvation from shoots to roots (Karmoker et al. 1991). Sulfur deficiency is dependent on the soil type and predominant climatic conditions and does not occur uniformly. Therefore, a reliable field based test is required to determine where it is likely to occur (Blake-Kalff et al. 2000). Soil testing can provide information about the amount of sulfur available to plants. However, there is no direct correlation between the content of sulfate in the soil solution and plant yield in field conditions (Zhao and McGrath 1994). Thus, the analysis of sulfur content in plant tissue might be a better indicator of sulfur available for the plant at the time of sampling. However the determination of a reliable diagnostic indicator is problematic because of constantly changing environmental conditions, variation in sulfur metabolism between species and plant demands (Blake-Kalff et al. 2000).

Genetically engineered plants are one of the most promising approaches to achieve an enhancement of sulfate use efficiency. Possible targets for genetic engineering might be focused on improved resource capture or efficient utilisation of increased uptake (Hawkesford 2000). The first goal could be achieved by modulation of transport systems. It is known that the expression of sulfate transporters is controlled by sulfur availability. During sulfur limitation the expression of sulfate transporter genes increases significantly, but it decreases when sulfate is abundant. Overriding this control might be achieved by expressing transporter genes under the control of an appropriate constitutive promoter. However, the control mechanisms would have to be removed only for sulfate transport and not for other steps of the pathway in order to prevent for example accumulation of sulfide, which is toxic for plants in high concentration. Alternatively, root structure and proliferation could be targeted. The second goal might be achieved by improving the mobilisation of vacuolar reserves or introducing an increased demand for sulfate to stimulate further sulfate uptake (Hawkesford 2000). In order to find the most efficient targets, further exploration of sulfate metabolism pathway and the regulatory mechanisms is required.

Natural Variation in Model Plants as a Tool to Improve Crop Quality

Restrictive regulations for commercial breeding of genetically modified plants have persuaded breeders and scientists to look for alternative methods to improve crop yield and quality. The result is a growing interest in naturally occurring genotypic variation. The enhancement of crop productivity is fundamental for human society. Therefore, exploring natural variation is of rapidly increasing importance to improve quality aspects that are associated with the chemical composition of agricultural products. The potential of wild species in the analysis of the molecular genetic basis of plant nutrition for crop improvement was recognised a century ago (Bessey 1906), however this area of research started to expand only recently (Fernie et al. 2006; McCouch 2004; Zamir 2001).

To investigate natural variation in model plants and in crops two main approaches are used. The first, the more traditional and still most popular way, is quantitative trait loci (QTL) analysis (Koornneef et al. 2004). It is based on the mapping of segregating populations such as a RILs derived from the crosses between two or more parental accessions. This leads to the localisation in the genome of loci exerting a control of a complex phenotypic trait in a given environment. A second approach is to use *Arabidopsis* natural accessions to identify polymorphisms associated with adaptation by exploiting a genome-wide association mapping (GWAS) also called linkage disequilibrium (LD) (Atwell et al. 2010). In contrast to QTL, in GWAS there are no crosses or pedigrees required, which makes it easier to collect the data. It also has advantages over QTL analysis because there are usually more alleles in the mixed population than in the two parents of the cross (Myles et al. 2009).

Mineral use efficiency, root uptake, translocation of nutrients from the roots to the shoots, and their accumulation in the seeds are the traits that, with many others, undergo a substantial natural variation. To date the QTL studies in crops have been focused mainly on seed or leaf mineral concentration in rice, wheat, barley, soybean, and *Brassica* species (Alonso-Blanco et al. 2009). Natural variation provides a complementary resource to discover novel gene functions as well as the variants of alleles that interact specifically with the genetic background or environmental changes, which is of great interest for the study focused on plant adaptation (Benfey and Mitchell-Olds 2008). This approach also allows creation of correlation networks between different nutrients that provide an overview of plant demands in various environments (Buescher et al. 2010). For example, a recent QTL study has identified a group of six genes which are involved in complex control of the mineral homeostasis in *Arabidopsis thaliana* (Ghandilyan et al. 2009). These results illustrate that the study of natural variation has significantly expanded our understanding of plant mineral nutrition, its regulation, and plant adaptation to environmental changes. The wide genetic variation allows the modification of quantitative traits that often could not be achieved through mutagenesis or transgenic approaches. They should therefore be of major interest for commercial agriculture. More detailed descriptions of natural variation approaches can be found in Chap. 2 of this book.

References

Abdallah M, Dubousset L, Meuriot F, Etienne P, Avice JC, Ourry A (2010) Effect of mineral sulphur availability on nitrogen and sulphur uptake and remobilization during the vegetative growth of *Brassica napus* L. J Exp Bot 61:2635–2646

Alonso-Blanco C, Aarts MG, Bentsink L, Keurentjes JJ, Reymond M, Vreugdenhil D, Koornneef M (2009) What has natural variation taught us about plant development, physiology, and adaptation? Plant Cell 21:1877–1896

Alvarez C, Garcia I, Romero LC, Gotor C (2012) Mitochondrial sulfide detoxification requires a functional isoform *O*-acetylserine(thiol)lyase C in *Arabidopsis thaliana*. Mol Plant 5:1217–1226

Amano Y, Tsubouchi H, Shinohara H, Ogawa M, Matsubayashi Y (2007) Tyrosine-sulfated glycopeptide involved in cellular proliferation and expansion in *Arabidopsis*. Proc Natl Acad Sci U S A 104:18333–18338

Amir R, Hacham Y, Galili G (2002) Cystathionine gamma-synthase and threonine synthase operate in concert to regulate carbon flow towards methionine in plants. Trends Plant Sci 7:153–156

Aravind L, Koonin EV (2000) The STAS domain – a link between anion transporters and antisigma-factor antagonists. Curr Biol 10:R53–R55

Atwell S, Huang YS, Vilhjalmsson BJ, Willems G, Horton M, Li Y, Meng D, Platt A, Tarone AM, Hu TT, Jiang R, Muliyati NW, Zhang X, Amer MA, Baxter I, Brachi B, Chory J, Dean C, Debieu M, de Meaux J, Ecker JR, Faure N, Kniskern JM, Jones JD, Michael T, Nemri A, Roux F, Salt DE, Tang C, Todesco M, Traw MB, Weigel D, Marjoram P, Borevitz JO, Bergelson J, Nordborg M (2010) Genome-wide association study of 107 phenotypes in *Arabidopsis thaliana* inbred lines. Nature 465:627–631

Awazuhara M, Kim H, Goto DB, Matsui A, Hayashi H, Chino M, Kim S-G, Naito S, Fujiwara T (2002) A 235-bp region from a nutritionally regulated soybean seed-specific gene promoter can confer its sulfur and nitrogen response to a constitutive promoter in aerial tissues of *Arabidopsis thaliana*. Plant Sci 163:75–82

Awazuhara M, Fujiwara T, Hayashi H, Watanabe-Takahashi A, Takahashi H, Saito K (2005) The function of SULTR2;1 sulfate transporter during seed development in *Arabidopsis thaliana*. Physiol Plant 125:95–105

Barberon M, Berthomieu P, Clairotte M, Shibagaki N, Davidian JC, Gosti F (2008) Unequal functional redundancy between the two *Arabidopsis thaliana* high-affinity sulphate transporters SULTR1;1 and SULTR1;2. New Phytol 180:608–619

Bartel DP (2004) MicroRNAs: genomics, biogenesis, mechanism, and function. Cell 116:281–297

Baxter I, Muthukumar B, Park HC, Buchner P, Lahner B, Danku J, Zhao K, Lee J, Hawkesford MJ, Guerinot ML, Salt DE (2008) Variation in molybdenum content across broadly distributed populations of *Arabidopsis thaliana* is controlled by a mitochondrial molybdenum transporter (MOT1). PLoS Genet 4(2):e1000004. doi:10.1371/journal.pgen.1000004

Benfey PN, Mitchell-Olds T (2008) From genotype to phenotype: systems biology meets natural variation. Science 320:495–497

Bessey CE (1906) Crop improvement by utilizing wild species. J Hered 2:112–118

Bick JA, Aslund F, Chen Y, Leustek T (1998) Glutaredoxin function for the carboxyl-terminal domain of the plant-type 5′-adenylylsulfate reductase. Proc Natl Acad Sci U S A 95:8404–8409

Blake-Kalff MM, Harrison KR, Hawkesford MJ, Zhao FJ, McGrath SP (1998) Distribution of sulfur within oilseed rape leaves in response to sulfur deficiency during vegetative growth. Plant Physiol 118:1337–1344

Blake-Kalff MMA, Hawkesford MJ, Zhao FJ, McGrath SP (2000) Diagnosing sulfur deficiency in field-grown oilseed rape (*Brassica napus* L.) and wheat (*Triticum aestivum* L.). Plant Soil 225:95–107

Blake-Kalff MMA, Zhao FJ, Hawkesford MJ, McGrath SP (2001) Using plant analysis to predict yield losses caused by sulphur deficiency. Ann Appl Biol 138:123–127

Brader G, Tas E, Palva ET (2001) Jasmonate-dependent induction of indole glucosinolates in *Arabidopsis* by culture filtrates of the nonspecific pathogen *Erwinia carotovora*. Plant Physiol 126:849–860
Brown PD, Tokuhisa JG, Reichelt M, Gershenzon J (2003) Variation of glucosinolate accumulation among different organs and developmental stages of *Arabidopsis thaliana*. Phytochemistry 62:471–481
Brunold C, Schiff JA (1976) Studies of sulfate utilization of algae: 15. Enzymes of assimilatory sulfate reduction in euglena and their cellular localization. Plant Physiol 57:430–436
Brunold C, Suter M (1984) Regulation of sulfate assimilation by nitrogen nutrition in the duckweed *Lemna minor* L. Plant Physiol 76:579–583
Brunold C, Suter M (1989) Localization of enzymes of assimilatory sulfate reduction in pea roots. Planta 179:228–234
Brychkova G, Xia Z, Yang G, Yesbergenova Z, Zhang Z, Davydov O, Fluhr R, Sagi M (2007) Sulfite oxidase protects plants against sulfur dioxide toxicity. Plant J 50:696–709
Buchner P, Prosser IM, Hawkesford MJ (2004a) Phylogeny and expression of paralogous and orthologous sulphate transporter genes in diploid and hexaploid wheats. [Research Support, Non-U S Gov't]. Genome 47:526–534
Buchner P, Takahashi H, Hawkesford MJ (2004b) Plant sulphate transporters: co-ordination of uptake, intracellular and long-distance transport. J Exp Bot 55:1765–1773
Buescher E, Achberger T, Amusan I, Giannini A, Ochsenfeld C, Rus A, Lahner B, Hoekenga O, Yakubova E, Harper JF, Guerinot ML, Zhang M, Salt DE, Baxter IR (2010) Natural genetic variation in selected populations of *Arabidopsis thaliana* is associated with ionomic differences. PLoS One 5:e11081
Burstenbinder K, Rzewuski G, Wirtz M, Hell R, Sauter M (2007) The role of methionine recycling for ethylene synthesis in *Arabidopsis*. Plant J 49:238–249
Cao MJ, Wang Z, Wirtz M, Hell R, Oliver DJ, Xiang CB (2013) SULTR3;1 is a chloroplast-localized sulfate transporter in *Arabidopsis thaliana*. Plant J 73:607–616
Chen YF, Matsubayashi Y, Sakagami Y (2000) Peptide growth factor phytosulfokine-alpha contributes to the pollen population effect. Planta 211:752–755
Chiba Y, Ishikawa M, Kijima F, Tyson RH, Kim J, Yamamoto A, Nambara E, Leustek T, Wallsgrove RM, Naito S (1999) Evidence for autoregulation of cystathionine gamma-synthase mRNA stability in *Arabidopsis*. Science 286:1371–1374
Chiba Y, Sakurai R, Yoshino M, Ominato K, Ishikawa M, Onouchi H, Naito S (2003) S-adenosyl-L-methionine is an effector in the posttranscriptional autoregulation of the cystathionine gamma-synthase gene in *Arabidopsis*. Proc Natl Acad Sci U S A 100:10225–10230
Chiou TJ, Aung K, Lin SI, Wu CC, Chiang SF, Su CL (2006) Regulation of phosphate homeostasis by MicroRNA in *Arabidopsis*. Plant Cell 18:412–421
Clarkson DT, Smith FW, Berg PJV (1983) Regulation of sulphate transport in a tropical legume, *Macroptilium atropurpureum*, cv. Siratro. J Exp Bot 34:1463–1483
Curien G, Dumas R, Ravanel S, Douce R (1996) Characterization of an *Arabidopsis thaliana* cDNA encoding an *S*-adenosylmethionine-sensitive threonine synthase. Threonine synthase from higher plants. FEBS Lett 390:85–90
Curtis TY, Powers SJ, Balagiannis D, Elmore JS, Mottram DS, Parry MAJ, Rakszegi M, Bedö Z, Shewry PR, Halford NG (2010) Free amino acids and sugars in rye grain: implications for acrylamide formation. J Agric Food Chem 58:1959–1969
Doughty KJ, Kiddle GA, Pye BJ, Wallsgrove RM, Pickett JA (1995) Selective induction of glucosinolates in oilseed rape leaves by methyl jasmonate. Phytochemistry 38:347–350
Droux M, Ruffet ML, Douce R, Job D (1998) Interactions between serine acetyltransferase and O-acetylserine (thiol) lyase in higher plants–structural and kinetic properties of the free and bound enzymes. Eur J Biochem 255:235–245
Eilers T, Schwarz G, Brinkmann H, Witt C, Richter T, Nieder J, Koch B, Hille R, Hansch R, Mendel RR (2001) Identification and biochemical characterization of *Arabidopsis thaliana* sulfite oxidase. A new player in plant sulfur metabolism. J Biol Chem 276:46989–46994

Essigmann B, Guler S, Narang RA, Linke D, Benning C (1998) Phosphate availability affects the thylakoid lipid composition and the expression of SQD1, a gene required for sulfolipid biosynthesis in *Arabidopsis thaliana*. Proc Natl Acad Sci U S A 95:1950–1955

Fahey JW, Zalcmann AT, Talalay P (2001) The chemical diversity and distribution of glucosinolates and isothiocyanates among plants. Phytochemistry 56:5–51

Fernie AR, Tadmor Y, Zamir D (2006) Natural genetic variation for improving crop quality. Curr Opin Plant Biol 9:196–202

Fismes J, Vong PC, Guckert A, Frossard E (2000) Influence of sulfur on apparent N-use efficiency, yield and quality of oilseed rape (*Brassica napus* L.) grown on a calcareous soil. Eur J Agron 12:127–141

Flæte NES, Hollung K, Ruud L, Sogn T, Færgestad EM, Skarpeid HJ, Magnus EM, Uhlen AK (2005) Combined nitrogen and sulphur fertilisation and its effect on wheat quality and protein composition measured by SE-FPLC and proteomics. J Cereal Sci 41:357–369

Francois JA, Kumaran S, Jez JM (2006) Structural basis for interaction of *O*-acetylserine sulfhydrylase and serine acetyltransferase in the *Arabidopsis* cysteine synthase complex. Plant Cell 18:3647–3655

Friedman M (2003) Chemistry, biochemistry, and safety of acrylamide, a review. J Agric Food Chem 51:4504–4526

Fujii H, Chiou T-J, Lin S-I, Aung K, Zhu J-K (2005) A miRNA involved in phosphate-starvation response in *Arabidopsis*. Curr Biol 15:2038–2043

Garcia I, Castellano JM, Vioque B, Solano R, Gotor C, Romero LC (2010) Mitochondrial beta-cyanoalanine synthase is essential for root hair formation in *Arabidopsis thaliana*. Plant Cell 22:3268–3279

Gasber A, Klaumann S, Trentmann O, Trampczynska A, Clemens S, Schneider S, Sauer N, Feifer I, Bittner F, Mendel RR, Neuhaus HE (2011) Identification of an *Arabidopsis* solute carrier critical for intracellular transport and inter-organ allocation of molybdate. Plant Biol 13:710–718

Ghandilyan A, Ilk N, Hanhart C, Mbengue M, Barboza L, Schat H, Koornneef M, El-Lithy M, Vreugdenhil D, Reymond M, Aarts MG (2009) A strong effect of growth medium and organ type on the identification of QTLs for phytate and mineral concentrations in three *Arabidopsis thaliana* RIL populations. J Exp Bot 60:1409–1425

Gigolashvili T, Berger B, Mock HP, Muller C, Weisshaar B, Flugge UI (2007) The transcription factor HIG1/MYB51 regulates indolic glucosinolate biosynthesis in *Arabidopsis thaliana*. Plant J 50:886–901

Gigolashvili T, Engqvist M, Yatusevich R, Muller C, Flugge UI (2008) HAG2/MYB76 and HAG3/MYB29 exert a specific and coordinated control on the regulation of aliphatic glucosinolate biosynthesis in *Arabidopsis thaliana*. New Phytol 177:627–642

Gilbert SM, Clarkson DT, Cambridge M, Lambers H, Hawkesford MJ (1997) SO_4^{2-} deprivation has an early effect on the content of ribulose-1,5-bisphosphate carboxylase/oxygenase and photosynthesis in young leaves of wheat. Plant Physiol 115:1231–1239

Giovanelli J, Mudd SH, Datko AH (1985) Quantitative analysis of pathways of methionine metabolism and their regulation in lemna. Plant Physiol 78:555–560

Haas FH, Heeg C, Queiroz R, Bauer A, Wirtz M, Hell R (2008) Mitochondrial serine acetyltransferase functions as a pacemaker of cysteine synthesis in plant cells. Plant Physiol 148:1055–1067

Hacham Y, Schuster G, Amir R (2006) An in vivo internal deletion in the N-terminus region of *Arabidopsis* cystathionine gamma-synthase results in CGS expression that is insensitive to methionine. Plant J 45:955–967

Hagen G, Guilfoyle T (2002) Auxin-responsive gene expression: genes, promoters and regulatory factors. Plant Mol Biol 49:373–385

Halford NG, Curtis TY, Muttucumaru N, Postles J, Elmore JS, Mottram DS (2012) The acrylamide problem: a plant and agronomic science issue. J Exp Bot 63:2841–2851

Halkier BA, Gershenzon J (2006) Biology and biochemistry of glucosinolates. Annu Rev Plant Biol 57:303–333

Hänsch R, Lang C, Riebeseel E, Lindigkeit R, Gessler A, Rennenberg H, Mendel RR (2006) Plant sulfite oxidase as novel producer of H_2O_2: combination of enzyme catalysis with a subsequent non-enzymatic reaction step. J Biol Chem 281:6884–6888

Hatzfeld Y, Lee S, Lee M, Leustek T, Saito K (2000a) Functional characterization of a gene encoding a fourth ATP sulfurylase isoform from *Arabidopsis thaliana*. Gene 248:51–58

Hatzfeld Y, Maruyama A, Schmidt A, Noji M, Ishizawa K, Saito K (2000b) beta-Cyanoalanine synthase is a mitochondrial cysteine synthase-like protein in spinach and *Arabidopsis*. Plant Physiol 123:1163–1171

Hawkesford MJ (2000) Plant responses to sulphur deficiency and the genetic manipulation of sulphate transporters to improve S-utilization efficiency. J Exp Bot 51:131–138

Hawkesford MJ (2003) Transporter gene families in plants: the sulphate transporter gene family – redundancy or specialization? Physiol Plant 117:155–163

Hawkesford M, Davidian J-C, Grignon C (1993) Sulphate/proton cotransport in plasma-membrane vesicles isolated from roots of *Brassica napus* L.: increased transport in membranes isolated from sulphur-starved plants. Planta 190:297–304

Hell R, Hillebrand H (2001) Plant concepts for mineral acquisition and allocation. Curr Opin Biotechnol 12:161–168

Hell R, Wirtz M (2008) Metabolism of cysteine in plants and phototrophic bacteria. In: Hell R, Dahl C, Knaff D, Leustek T (eds) Sulfur metabolism in phototrophic organisms. Springer, Netherlands, pp 59–91

Hell R, Wirtz M (2011) Molecular biology, biochemistry and cellular physiology of cysteine metabolism in *Arabidopsis thaliana*. The *Arabidopsis* book. Am Soc Plant Biol 9:e0154

Hesse H, Trachsel N, Suter M, Kopriva S, von Ballmoos P, Rennenberg H, Brunold C (2003) Effect of glucose on assimilatory sulphate reduction in *Arabidopsis thaliana* roots. J Exp Bot 54:1701–1709

Hesse H, Kreft O, Maimann S, Zeh M, Hoefgen R (2004a) Current understanding of the regulation of methionine biosynthesis in plants. J Exp Bot 55:1799–1808

Hesse H, Nikiforova V, Gakiere B, Hoefgen R (2004b) Molecular analysis and control of cysteine biosynthesis: integration of nitrogen and sulphur metabolism. J Exp Bot 55:1283–1292

Hirai MY, Sugiyama K, Sawada Y, Tohge T, Obayashi T, Suzuki A, Araki R, Sakurai N, Suzuki H, Aoki K, Goda H, Nishizawa OI, Shibata D, Saito K (2007) Omics-based identification of *Arabidopsis* Myb transcription factors regulating aliphatic glucosinolate biosynthesis. Proc Natl Acad Sci U S A 104:6478–6483

Hopkins L, Parmar S, Blaszczyk A, Hesse H, Hoefgen R, Hawkesford MJ (2005) *O*-acetylserine and the regulation of expression of genes encoding components for sulfate uptake and assimilation in potato. Plant Physiol 138:433–440

Howarth JR, Fourcroy P, Davidian JC, Smith FW, Hawkesford MJ (2003) Cloning of two contrasting high-affinity sulfate transporters from tomato induced by low sulfate and infection by the vascular pathogen *Verticillium dahliae*. Planta 218:58–64

Hsieh LC, Lin SI, Shih AC, Chen JW, Lin WY, Tseng CY, Li WH, Chiou TJ (2009) Uncovering small RNA-mediated responses to phosphate deficiency in *Arabidopsis* by deep sequencing. Plant Physiol 151:2120–2132

Hubberten HM, Klie S, Caldana C, Degenkolbe T, Willmitzer L, Hoefgen R (2012) Additional role of *O*-acetylserine as a sulfur status-independent regulator during plant growth. Plant J 70:666–677

Hussain A, Larsson H, Kuktaite R, Prieto-Linde ML, Johansson E (2012) Towards the understanding of bread-making quality in organically grown wheat: dough mixing behaviour, protein polymerisation and structural properties. J Cereal Sci 56:659–666

Jones-Rhoades MW, Bartel DP (2004) Computational identification of plant microRNAs and their targets, including a stress-induced miRNA. Mol Cell 14:787–799

Jost R, Altschmied L, Bloem E, Bogs J, Gershenzon J, Hahnel U, Hansch R, Hartmann T, Kopriva S, Kruse C, Mendel RR, Papenbrock J, Reichelt M, Rennenberg H, Schnug E, Schmidt A, Textor S, Tokuhisa J, Wachter A, Wirtz M, Rausch T, Hell R (2005) Expression profiling of metabolic genes in response to methyl jasmonate reveals regulation of genes of primary and secondary sulfur-related pathways in *Arabidopsis thaliana*. Photosynth Res 86:491–508

Kandlbinder A, Finkemeier I, Wormuth D, Hanitzsch M, Dietz KJ (2004) The antioxidant status of photosynthesizing leaves under nutrient deficiency: redox regulation, gene expression and antioxidant activity in *Arabidopsis thaliana*. Physiol Plant 120:63–73

Karmoker J, Clarkson D, Saker L, Rooney J, Purves J (1991) Sulphate deprivation depresses the transport of nitrogen to the xylem and the hydraulic conductivity of barley (*Hordeum vulgare* L.) roots. Planta 185:269–278

Kataoka T, Hayashi N, Yamaya T, Takahashi H (2004a) Root-to-shoot transport of sulfate in *Arabidopsis*. Evidence for the role of SULTR3;5 as a component of low-affinity sulfate transport system in the root vasculature. Plant Physiol 136:4198–4204

Kataoka T, Watanabe-Takahashi A, Hayashi N, Ohnishi M, Mimura T, Buchner P, Hawkesford MJ, Yamaya T, Takahashi H (2004b) Vacuolar sulfate transporters are essential determinants controlling internal distribution of sulfate in *Arabidopsis*. Plant Cell 16:2693–2704

Kawashima CG, Berkowitz O, Hell R, Noji M, Saito K (2005) Characterization and expression analysis of a serine acetyltransferase gene family involved in a key step of the sulfur assimilation pathway in *Arabidopsis*. Plant Physiol 137:220–230

Kawashima CG, Yoshimoto N, Maruyama-Nakashita A, Tsuchiya YN, Saito K, Takahashi H, Dalmay T (2009) Sulphur starvation induces the expression of microRNA-395 and one of its target genes but in different cell types. Plant J 57:313–321

Kawashima CG, Matthewman CA, Huang S, Lee B-R, Yoshimoto N, Koprivova A, Rubio-Somoza I, Todesco M, Rathjen T, Saito K, Takahashi H, Dalmay T, Kopriva S (2011) Interplay of SLIM1 and miR395 in the regulation of sulfate assimilation in *Arabidopsis*. Plant J 66:863–876

Khan MS, Haas FH, Allboje Samami A, Moghaddas Gholami A, Bauer A, Fellenberg K, Reichelt M, Hänsch R, Mendel RR, Meyer AJ, Wirtz M, Hell R (2010) Sulfite reductase defines a newly discovered bottleneck for assimilatory sulfate reduction and is essential for growth and development in *Arabidopsis thaliana*. Plant Cell Online 22:1216–1231

Kim H, Hirai MY, Hayashi H, Chino M, Naito S, Fujiwara T (1999) Role of *O*-acetyl-L-serine in the coordinated regulation of the expression of a soybean seed storage-protein gene by sulfur and nitrogen nutrition. Planta 209:282–289

Kobayashi T, Eun C-H, Hanai H, Matsubayashi Y, Sakagami Y, Kamada H (1999) Phytosulphokine-α, a peptidyl plant growth factor, stimulates somatic embryogenesis in carrot. J Exp Bot 50:1123–1128

Koornneef M, Alonso-Blanco C, Vreugdenhil D (2004) Naturally occurring genetic variation in *Arabidopsis thaliana*. Annu Rev Plant Physiol Plant Mol Biol 55:141–172

Kopriva S (2006) Regulation of sulfate assimilation in *Arabidopsis* and beyond. Ann Bot 97:479–495

Kopriva S, Koprivova A (2004) Plant adenosine 5′-phosphosulphate reductase: the past, the present, and the future. J Exp Bot 55:1775–1783

Kopriva S, Rennenberg H (2004) Control of sulphate assimilation and glutathione synthesis: interaction with N and C metabolism. J Exp Bot 55:1831–1842

Kopriva S, Muheim R, Koprivova A, Trachsel N, Catalano C, Suter M, Brunold C (1999) Light regulation of assimilatory sulphate reduction in *Arabidopsis thaliana*. Plant J 20:37–44

Kopriva S, Buchert T, Fritz G, Suter M, Weber M, Benda R, Schaller J, Feller U, Schurmann P, Schunemann V, Trautwein AX, Kroneck PM, Brunold C (2001) Plant adenosine 5-′-phosphosulfate reductase is a novel iron-sulfur protein. J Biol Chem 276:42881–42886

Kopriva S, Suter M, von Ballmoos P, Hesse H, Krähenbühl U, Rennenberg H, Brunold C (2002) Interaction of sulfate assimilation with carbon and nitrogen metabolism in *Lemna minor*. Plant Physiol 130:1406–1413

Kopriva S, Fritzemeier K, Wiedemann G, Reski R (2007) The putative moss 3-′-phosphoadenosine-5′-phosphosulfate reductase is a novel form of adenosine-5-′-phosphosulfate reductase without an iron-sulfur cluster. J Biol Chem 282:22930–22938

Kopriva S, Mugford S, Matthewman C, Koprivova A (2009) Plant sulfate assimilation genes: redundancy versus specialization. Plant Cell Rep 28:1769–1780

Kopriva S, Mugford SG, Baraniecka P, Lee B-R, Matthewman CA, Koprivova A (2012) Control of sulfur partitioning between primary and secondary metabolism in *Arabidopsis*. Front Plant Sci 3:163. doi:10.3389/fpls.2012.00163

Koprivova A, Suter M, Op den Camp R, Brunold C, Kopriva S (2000) Regulation of sulfate assimilation by nitrogen in *Arabidopsis*. Plant Physiol 122:737–746

Koprivova A, North KA, Kopriva S (2008) Complex signaling network in regulation of adenosine 5′-phosphosulfate reductase by salt stress in *Arabidopsis* roots. Plant Physiol 146:1408–1420

Krueger S, Niehl A, Lopez Martin MC, Steinhauser D, Donath A, Hildebrandt T, Romero LC, Hoefgen R, Gotor C, Hesse H (2009) Analysis of cytosolic and plastidic serine acetyltransferase mutants and subcellular metabolite distributions suggests interplay of the cellular compartments for cysteine biosynthesis in *Arabidopsis*. Plant Cell Environ 32:349–367

Kuktaite R, Larsson H, Johansson E (2004) Variation in protein composition of wheat flour and its relationship to dough mixing behaviour. J Cereal Sci 40:31–39

Kutz A, Muller A, Hennig P, Kaiser WM, Piotrowski M, Weiler EW (2002) A role for nitrilase 3 in the regulation of root morphology in sulphur-starving *Arabidopsis thaliana*. Plant J 30:95–106

Lambein I, Chiba Y, Onouchi H, Naito S (2003) Decay kinetics of autogenously regulated CGS1 mRNA that codes for cystathionine gamma-synthase in *Arabidopsis thaliana*. Plant Cell Physiol 44:893–900

Lass B, Ullrich-Eberius CI (1984) Evidence for proton/sulfate cotransport and its kinetics in Lemna gibba G1. Planta 161:53–60

Lee B-R, Koprivova A, Kopriva S (2011) The key enzyme of sulfate assimilation, adenosine 5′-phosphosulfate reductase, is regulated by HY5 in *Arabidopsis*. Plant J 67:1042–1054

Lejay L, Gansel X, Cerezo M, Tillard P, Muller C, Krapp A, von Wiren N, Daniel-Vedele F, Gojon A (2003) Regulation of root ion transporters by photosynthesis: functional importance and relation with hexokinase. Plant Cell 15:2218–2232

Leustek T, Martin MN, Bick JA, Davies JP (2000) Pathways and regulation of sulfur metabolism revealed through molecular and genetic studies. Annu Rev Plant Physiol Plant Mol Biol 51:141–165

Levy M, Wang Q, Kaspi R, Parrella MP, Abel S (2005) *Arabidopsis* IQD1, a novel calmodulin-binding nuclear protein, stimulates glucosinolate accumulation and plant defense. Plant J 43:79–96

Liang G, Yang F, Yu D (2010) MicroRNA395 mediates regulation of sulfate accumulation and allocation in *Arabidopsis thaliana*. Plant J 62:1046–1057

Lin S-I, Chiang S-F, Lin W-Y, Chen J-W, Tseng C-Y, Wu P-C, Chiou T-J (2008) Regulatory network of microRNA399 and PHO2 by systemic signaling. Plant Physiol 147:732–746

Logan HM, Cathala N, Grignon C, Davidian JC (1996) Cloning of a cDNA encoded by a member of the *Arabidopsis thaliana* ATP sulfurylase multigene family. Expression studies in yeast and in relation to plant sulfur nutrition. J Biol Chem 271:12227–12233

López-Martín MC, Becana M, Romero LC, Gotor C (2008) Knocking out cytosolic cysteine synthesis compromises the antioxidant capacity of the cytosol to maintain discrete concentrations of hydrogen peroxide in *Arabidopsis*. Plant Physiol 147:562–572

Loudet O, Saliba-Colombani V, Camilleri C, Calenge F, Gaudon V, Koprivova A, North KA, Kopriva S, Daniel-Vedele F (2007) Natural variation for sulfate content in *Arabidopsis thaliana* is highly controlled by APR2. Nat Genet 39:896–900

Lunn JE, Droux M, Martin J, Douce R (1990) Localization of ATP sulfurylase and O-Acetylserine (thiol)lyase in Spinach leaves. Plant Physiol 94:1345–1352

Malhi SS, Gan Y, Raney JP (2007) Yield, seed quality, and sulfur uptake of Brassica oilseed crops in response to sulfur fertilization. Agron J 99:570–577

Martin MN, Tarczynski MC, Shen B, Leustek T (2005) The role of 5′-adenylylsulfate reductase in controlling sulfate reduction in plants. Photosynth Res 86:309–323

Maruyama-Nakashita A, Inoue E, Watanabe-Takahashi A, Yamaya T, Takahashi H (2003) Transcriptome profiling of sulfur-responsive genes in *Arabidopsis* reveals global effects of sulfur nutrition on multiple metabolic pathways. Plant Physiol 132:597–605

Maruyama-Nakashita A, Nakamura Y, Watanabe-Takahashi A, Yamaya T, Takahashi H (2004a) Induction of SULTR1;1 sulfate transporter in *Arabidopsis* roots involves protein phosphorylation/dephosphorylation circuit for transcriptional regulation. Plant Cell Physiol 45:340–345

Maruyama-Nakashita A, Nakamura Y, Yamaya T, Takahashi H (2004b) A novel regulatory pathway of sulfate uptake in *Arabidopsis* roots: implication of CRE1/WOL/AHK4-mediated cytokinin-dependent regulation. Plant J 38:779–789

Maruyama-Nakashita A, Nakamura Y, Watanabe-Takahashi A, Inoue E, Yamaya T, Takahashi H (2005) Identification of a novel cis-acting element conferring sulfur deficiency response in *Arabidopsis* roots. Plant J 42:305–314

Maruyama-Nakashita A, Nakamura Y, Tohge T, Saito K, Takahashi H (2006) *Arabidopsis* SLIM1 is a central transcriptional regulator of plant sulfur response and metabolism. Plant Cell 18:3235–3251

Matsubayashi Y, Sakagami Y (1999) Characterization of specific binding sites for a mitogenic sulfated peptide, phytosulfokine-alpha, in the plasma-membrane fraction derived from Oryza sativa L. Eur J Biochem 262:666–671

Matsubayashi Y, Sakagami Y (2006) Peptide hormones in plants. Annu Rev Plant Biol 57:649–674

Matsubayashi Y, Ogawa M, Morita A, Sakagami Y (2002) An LRR receptor kinase involved in perception of a peptide plant hormone, phytosulfokine. Science 296:1470–1472

Matsuzaki Y, Ogawa-Ohnishi M, Mori A, Matsubayashi Y (2010) Secreted peptide signals required for maintenance of root stem cell niche in *Arabidopsis*. Science 329:1065–1067

Matthewman CA, Kawashima CG, Huska D, Csorba T, Dalmay T, Kopriva S (2012) miR395 is a general component of the sulfate assimilation regulatory network in *Arabidopsis*. FEBS Lett 586:3242–3248

McCouch S (2004) Diversifying selection in plant breeding. PLoS Biol 2:e347

McGrath SP, Zhao FJ (1996) Sulphur uptake, yield responses and the interactions between nitrogen and sulphur in winter oilseed rape (*Brassica napus*). J Agric Sci 126:53–62

McGrath SP, Zhao FJ, Withers PJA (1996) Development of sulphur deficiency in crops and its treatment. Proceedings of the Fertiliser Society

Meng L, Buchanan BB, Feldman LJ, Luan S (2012) CLE-like (CLEL) peptides control the pattern of root growth and lateral root development in *Arabidopsis*. Proc Natl Acad Sci U S A 109:1760–1765

Mithen R, Faulkner K, Magrath R, Rose P, Williamson G, Marquez J (2003) Development of isothiocyanate-enriched broccoli, and its enhanced ability to induce phase 2 detoxification enzymes in mammalian cells. Theor Appl Genet 106:727–734

Moss H, Wrigley C, MacRichie R, Randall P (1981) Sulfur and nitrogen fertilizer effects on wheat. II. Influence on grain quality. Aust J Agric Res 32:213–226

Mugford SG, Yoshimoto N, Reichelt M, Wirtz M, Hill L, Mugford ST, Nakazato Y, Noji M, Takahashi H, Kramell R, Gigolashvili T, Flugge UI, Wasternack C, Gershenzon J, Hell R, Saito K, Kopriva S (2009) Disruption of adenosine-5′-phosphosulfate kinase in *Arabidopsis* reduces levels of sulfated secondary metabolites. Plant Cell 21:910–927

Mugford SG, Matthewman CA, Hill L, Kopriva S (2010) Adenosine-5′-phosphosulfate kinase is essential for *Arabidopsis* viability. FEBS Lett 584:119–123

Murillo M, Leustek T (1995) Adenosine-5′-triphosphate-sulfurylase from *Arabidopsis thaliana* and Escherichia coli are functionally equivalent but structurally and kinetically divergent: nucleotide sequence of two adenosine-5′-triphosphate-sulfurylase cDNAs from *Arabidopsis thaliana* and analysis of a recombinant enzyme. Arch Biochem Biophys 323:195–204

Myles S, Peiffer J, Brown PJ, Ersoz ES, Zhang Z, Costich DE, Buckler ES (2009) Association mapping: critical considerations shift from genotyping to experimental design. Plant Cell 21:2194–2202

Nakayama M, Akashi T, Hase T (2000) Plant sulfite reductase: molecular structure, catalytic function and interaction with ferredoxin. J Inorg Biochem 82:27–32

Nikiforova V, Freitag J, Kempa S, Adamik M, Hesse H, Hoefgen R (2003) Transcriptome analysis of sulfur depletion in *Arabidopsis thaliana*: interlacing of biosynthetic pathways provides response specificity. Plant J 33:633–650

Noctor G, Gomez L, Vanacker H, Foyer CH (2002) Interactions between biosynthesis, compartmentation and transport in the control of glutathione homeostasis and signalling. J Exp Bot 53:1283–1304

Noji M, Inoue K, Kimura N, Gouda A, Saito K (1998) Isoform-dependent differences in feedback regulation and subcellular localization of serine acetyltransferase involved in cysteine biosynthesis from *Arabidopsis thaliana*. J Biol Chem 273:32739–32745

Nowak K, Luniak N, Witt C, Wustefeld Y, Wachter A, Mendel RR, Hansch R (2004) Peroxisomal localization of sulfite oxidase separates it from chloroplast-based sulfur assimilation. Plant Cell Physiol 45:1889–1894

Ohkama N, Takei K, Sakakibara H, Hayashi H, Yoneyama T, Fujiwara T (2002) Regulation of sulfur-responsive gene expression by exogenously applied cytokinins in *Arabidopsis thaliana*. Plant Cell Physiol 43:1493–1501

Ohkama-Ohtsu N, Oikawa A, Zhao P, Xiang C, Saito K, Oliver DJ (2008) A gamma-glutamyl transpeptidase-independent pathway of glutathione catabolism to glutamate via 5-oxoproline in *Arabidopsis*. Plant Physiol 148:1603–1613

Olsen LR, Huang B, Vetting MW, Roderick SL (2004) Structure of serine acetyltransferase in complexes with CoA and its cysteine feedback inhibitor. Biochemistry 43:6013–6019

Prosser IM, Purves JV, Saker LR, Clarkson DT (2001) Rapid disruption of nitrogen metabolism and nitrate transport in spinach plants deprived of sulphate. J Exp Bot 52:113–121

Rae A, Smith F (2002) Localisation of expression of a high-affinity sulfate transporter in barley roots. Planta 215:565–568

Ravanel S, Gakière B, Job D, Douce R (1998) The specific features of methionine biosynthesis and metabolism in plants. Proc Natl Acad Sci U S A 95:7805–7812

Ravanel S, Block MA, Rippert P, Jabrin S, Curien G, Rébeillé F, Douce R (2004) Methionine metabolism in plants: chloroplasts are autonomous for de novo methionine synthesis and can import s-adenosylmethionine from the cytosol. J Biol Chem 279:22548–22557

Renosto F, Patel HC, Martin RL, Thomassian C, Zimmerman G, Segel IH (1993) ATP sulfurylase from higher plants: kinetic and structural characterization of the chloroplast and cytosol enzymes from spinach leaf. Arch Biochem Biophys 307:272–285

Renwick JA (2001) Variable diets and changing taste in plant-insect relationships. J Chem Ecol 27:1063–1076

Rotte C, Leustek T (2000) Differential subcellular localization and expression of ATP sulfurylase and 5′-adenylylsulfate reductase during ontogenesis of *Arabidopsis* leaves indicates that cytosolic and plastid forms of ATP sulfurylase may have specialized functions. Plant Physiol 124:715–724

Rouached H (2011) Multilevel coordination of phosphate and sulfate homeostasis in plants. Plant Signal Behav 6:952–955

Rouached H, Berthomieu P, El Kassis E, Cathala N, Catherinot V, Labesse G, Davidian J-C, Fourcroy P (2005) Structural and functional analysis of the C-terminal STAS (sulfate transporter and anti-sigma antagonist) domain of the *Arabidopsis thaliana* sulfate transporter SULTR1.2. J Biol Chem 280:15976–15983

Rouached H, Wirtz M, Alary R, Hell R, Arpat AB, Davidian J-C, Fourcroy P, Berthomieu P (2008) Differential regulation of the expression of two high-affinity sulfate transporters, SULTR1.1 and SULTR1.2, in *Arabidopsis*. Plant Physiol 147:897–911

Rouached H, Secco D, Arpat B, Poirier Y (2011) The transcription factor PHR1 plays a key role in the regulation of sulfate shoot-to-root flux upon phosphate starvation in *Arabidopsis*. BMC Plant Biol 11:19

Saito K (2000) Regulation of sulfate transport and synthesis of sulfur-containing amino acids. Curr Opin Plant Biol 3:188–195

Saito K (2004) Sulfur assimilatory metabolism. The long and smelling road. Plant Physiol 136:2443–2450

Shahsavani S, Gholami A (2008) Effect of sulphur fertilization on breadmaking quality of three winter wheat varieties. Pak J Biol Sci 11:2134–2138

Shibagaki N, Grossman AR (2004) Probing the function of STAS domains of the *Arabidopsis* sulfate transporters. J Biol Chem 279:30791–30799

Shibagaki N, Grossman AR (2010) Binding of cysteine synthase to the STAS domain of sulfate transporter and its regulatory consequences. J Biol Chem 285:25094–25102

Shibagaki N, Rose A, McDermott JP, Fujiwara T, Hayashi H, Yoneyama T, Davies JP (2002) Selenate-resistant mutants of *Arabidopsis thaliana* identify Sultr1;2, a sulfate transporter required for efficient transport of sulfate into roots. Plant J 29:475–486

Shinmachi F, Buchner P, Stroud JL, Parmar S, Zhao FJ, McGrath SP, Hawkesford MJ (2010) Influence of sulfur deficiency on the expression of specific sulfate transporters and the distribution of sulfur, selenium, and molybdenum in wheat. Plant Physiol 153:327–336

Skirycz A, Reichelt M, Burow M, Birkemeyer C, Rolcik J, Kopka J, Zanor MI, Gershenzon J, Strnad M, Szopa J, Mueller-Roeber B, Witt I (2006) DOF transcription factor AtDof1.1 (OBP2) is part of a regulatory network controlling glucosinolate biosynthesis in *Arabidopsis*. Plant J 47:10–24

Smith FW, Ealing PM, Hawkesford MJ, Clarkson DT (1995a) Plant members of a family of sulfate transporters reveal functional subtypes. Proc Natl Acad Sci U S A 92:9373–9377

Smith FW, Hawkesford MJ, Prosser IM, Clarkson DT (1995b) Isolation of a cDNA from Saccharomyces cerevisiae that encodes a high affinity sulphate transporter at the plasma membrane. Mol Gen Genet 247:709–715

Smith FW, Hawkesford MJ, Ealing PM, Clarkson DT, Vanden Berg PJ, Belcher AR, Warrilow AG (1997) Regulation of expression of a cDNA from barley roots encoding a high affinity sulphate transporter. Plant J 12:875–884

Sønderby IE, Hansen BG, Bjarnholt N, Ticconi C, Halkier BA, Kliebenstein DJ (2007) A systems biology approach identifies a R2R3 MYB gene subfamily with distinct and overlapping functions in regulation of aliphatic glucosinolates. PLoS One 2:e1322

Suzuki A, Shirata Y, Ishida H, Chiba Y, Onouchi H, Naito S (2001) The first exon coding region of cystathionine gamma-synthase gene is necessary and sufficient for downregulation of its own mRNA accumulation in transgenic *Arabidopsis thaliana*. Plant Cell Physiol 42:1174–1180

Takahashi H, Saito K (1996) Subcellular localization of spinach cysteine synthase isoforms and regulation of their gene expression by nitrogen and sulfur. Plant Physiol 112:273–280

Takahashi H, Saito K (2008) Molecular biology and functional genomics for identification of regulatory networks of plant sulfate uptake and assimilatory metabolism. In: Hell R, Dahl C, Knaff D, Leustek T (eds) Sulfur metabolism in phototrophic organisms. Springer, Netherlands, pp 149–159

Takahashi H, Watanabe-Takahashi A, Smith FW, Blake-Kalff M, Hawkesford MJ, Saito K (2000) The roles of three functional sulphate transporters involved in uptake and translocation of sulphate in *Arabidopsis thaliana*. Plant J 23:171–182

Takahashi H, Kopriva S, Giordano M, Saito K, Hell R (2011) Sulfur assimilation in photosynthetic organisms: molecular functions and regulations of transporters and assimilatory enzymes. Annu Rev Plant Biol 62:157–184

Tareke E, Rydberg P, Karlsson P, Eriksson S, Tornqvist M (2002) Analysis of acrylamide, a carcinogen formed in heated foodstuffs. J Agric Food Chem 50:4998–5006
Tomatsu H, Takano J, Takahashi H, Watanabe-Takahashi A, Shibagaki N, Fujiwara T (2007) An *Arabidopsis thaliana* high-affinity molybdate transporter required for efficient uptake of molybdate from soil. Proc Natl Acad Sci U S A 104:18807–18812
Varadarajan DK, Karthikeyan AS, Matilda PD, Raghothama KG (2002) Phosphite, an analog of phosphate, suppresses the coordinated expression of genes under phosphate starvation. Plant Physiol 129:1232–1240
Vauclare P, Kopriva S, Fell D, Suter M, Sticher L, von Ballmoos P, Krahenbuhl U, den Camp RO, Brunold C (2002) Flux control of sulphate assimilation in *Arabidopsis thaliana*: adenosine 5′-phosphosulphate reductase is more susceptible than ATP sulphurylase to negative control by thiols. Plant J 31:729–740
Vidmar JJ, Tagmount A, Cathala N, Touraine B, Davidian JE (2000) Cloning and characterization of a root specific high-affinity sulfate transporter from *Arabidopsis thaliana*. [Research Support, Non-U S Gov't]. FEBS Lett 475:65–69
Wang R, Okamoto M, Xing X, Crawford NM (2003) Microarray analysis of the nitrate response in *Arabidopsis* roots and shoots reveals over 1,000 rapidly responding genes and new linkages to glucose, trehalose-6-phosphate, iron, and sulfate metabolism. Plant Physiol 132:556–567
Watanabe M, Mochida K, Kato T, Tabata S, Yoshimoto N, Noji M, Saito K (2008) Comparative genomics and reverse genetics analysis reveal indispensable functions of the serine acetyltransferase gene family in *Arabidopsis*. Plant Cell 20:2484–2496
White PJ, Brown PH (2010) Plant nutrition for sustainable development and global health. Ann Bot 105:1073–1080
Wirtz M, Hell R (2006) Functional analysis of the cysteine synthase protein complex from plants: structural, biochemical and regulatory properties. J Plant Physiol 163:273–286
Wirtz M, Berkowitz O, Droux M, Hell R (2001) The cysteine synthase complex from plants. Mitochondrial serine acetyltransferase from *Arabidopsis thaliana* carries a bifunctional domain for catalysis and protein-protein interaction. Eur J Biochem 268:686–693
Wirtz M, Droux M, Hell R (2004) O-acetylserine (thiol) lyase: an enigmatic enzyme of plant cysteine biosynthesis revisited in *Arabidopsis thaliana*. J Exp Bot 55:1785–1798
Xiang C, Oliver DJ (1998) Glutathione metabolic genes coordinately respond to heavy metals and jasmonic acid in *Arabidopsis*. Plant Cell Online 10:1539–1550
Yamaguchi Y, Nakamura T, Harada E, Koizumi N, Sano H (1999) Differential accumulation of transcripts encoding sulfur assimilation enzymes upon sulfur and/or nitrogen deprivation in *Arabidopsis thaliana*. Biosci Biotechnol Biochem 63:762–766
Yamakawa S, Sakuta C, Matsubayashi Y, Sakagami Y, Kamada H, Satoh S (1998) The promotive effects of a peptidyl plant growth factor, phytosulfokine-α, on the formation of adventitious roots and expression of a gene for a root-specific cystatin in cucumber hypocotyls. J Plant Res 111:453–458
Yang H, Matsubayashi Y, Nakamura K, Sakagami Y (1999) Oryza sativa PSK gene encodes a precursor of phytosulfokine-alpha, a sulfated peptide growth factor found in plants. Proc Natl Acad Sci U S A 96:13560–13565
Yang H, Matsubayashi Y, Hanai H, Sakagami Y (2000) Phytosulfokine-alpha, a peptide growth factor found in higher plants: its structure, functions, precursor and receptors. Plant Cell Physiol 41:825–830
Yang H, Matsubayashi Y, Nakamura K, Sakagami Y (2001) Diversity of *Arabidopsis* genes encoding precursors for phytosulfokine, a peptide growth factor. Plant Physiol 127:842–851
Yatusevich R, Mugford SG, Matthewman C, Gigolashvili T, Frerigmann H, Delaney S, Koprivova A, Flugge UI, Kopriva S (2010) Genes of primary sulfate assimilation are part of the glucosinolate biosynthetic network in *Arabidopsis thaliana*. Plant J 62:1–11
Yoshimoto N, Takahashi H, Smith FW, Yamaya T, Saito K (2002) Two distinct high-affinity sulfate transporters with different inducibilities mediate uptake of sulfate in *Arabidopsis* roots. Plant J 29:465–473

Yoshimoto N, Inoue E, Saito K, Yamaya T, Takahashi H (2003) Phloem-localizing sulfate transporter, Sultr1;3, mediates re-distribution of sulfur from source to sink organs in *Arabidopsis*. Plant Physiol 131:1511–1517

Yoshimoto N, Inoue E, Watanabe-Takahashi A, Saito K, Takahashi H (2007) Posttranscriptional regulation of high-affinity sulfate transporters in *Arabidopsis* by sulfur nutrition. Plant Physiol 145:378–388

Zamir D (2001) Improving plant breeding with exotic genetic libraries. Nat Rev Genet 2:983–989

Zhao F, McGrath SP (1994) Extractable sulphate and organic sulphur in soils and their availability to plants. Plant Soil 164:243–250

Zhao FJ, Hawkesford MJ, Warrilow AGS, McGrath SP, Clarkson DT (1996) Responses of two wheat varieties to sulphur addition and diagnosis of sulphur deficiency. Plant Soil 181:317–327

Zhao FJ, Hawkesford MJ, McGrath SP (1999) Sulphur assimilation and effects on yield and quality of wheat. J Cereal Sci 30:1–17

Chapter 4
Efficient Mineral Nutrition: Genetic Improvement of Phosphate Uptake and Use Efficiency in Crops

Astrid Gruen, Martin R. Broadley, Peter Buchner, and Malcolm J. Hawkesford

Abstract Phosphorus (P) is an essential macronutrient for plants, the lack of which can be a major constraint for agricultural productivity. Economic, political and environmental factors have prioritized the need for research on P acquisition efficiency (PAE), P utilization efficiency (PUE) and P fertiliser uptake efficiency in crops. P has critical functions in plants and complex interactions in soils. Appropriate screening approaches and implications of improvement in crop production are discussed. P acquisition is mediated by members of phosphate transporter families and the roles of these phosphate transporters as well as enzymes involved in P partitioning and re-translocation are complex. There is also a critical importance of regulatory genes including transcription factors, signalling pathways and apparently other P-responsive genes with unknown function. Furthermore, morphological and biochemical responses enhance P solubility in the soil and facilitate uptake and include root plasticity, secretion processes and symbioses. Exploitation of genetic variation, classical breeding and biotechnological gene modification of target genes are future routes for crop improvement. There is a need for selection not just for uptake but also focussing on P storage pools within cells and tissues, and additionally a consideration of crop P requirements during the different growth stages of crops. The review concludes with a summary giving an outlook to future questions related to crop PAE/PUE improvement.

Keywords Phosphorus • Acquisition efficiency • Utilization efficiency • Phosphate transporters • Mycorrhiza • Signaling • Diversity • Adaptation • QTL

A. Gruen • P. Buchner • M.J. Hawkesford (✉)
Rothamsted Research, Harpenden AL5 2JQ, Hertfordshire, UK
e-mail: malcolm.hawkesford@rothamsted.ac.uk

M.R. Broadley
Plant Science Division, School of Biosciences, University of Nottingham, Loughborough LE12 5RD, UK

M.J. Hawkesford et al. (eds.), *Nutrient Use Efficiency in Plants*,
Plant Ecophysiology 10, DOI 10.1007/978-3-319-10635-9_4

Efficient Mineral Nutrition: Phosphate as Nutritional Trait of Interest

Implications for Phosphorus Uptake

Phosphorus (P) is an essential macronutrient with multiple functions in plant macromolecular structures as a component of nucleic acids and phospholipids, with crucial roles in energy metabolism, participation in signal transduction pathways via phosphorylation/dephosphorylation and controlling key enzyme reactions (Theodorou and Plaxton 1993; Schachtman et al. 1998; Marschner 2012). In many agricultural systems, P is one of the most limiting nutrients for crop production and is a major constraint for yield (Vance et al. 2003; Raghothama 2005; Kirkby and Johnston 2008), with shoot growth, tiller number and tiller weight all influenced by P availability (Römer and Schilling 1986; Bollons and Barraclough 1997). Critical P concentrations in wheat range from 0.5 to 0.4 % in dry matter (DM) at tillering and 0.2–0.3 % in DM at booting (Finck 1991; Marschner 2012). The amount of readily available P may be raised through P fertilisation and agronomic strategies such as fertiliser placement or the application of the "critical value" concept (Bahl and Singh 1986; Strong et al. 1997; Kirkby and Johnston 2008; Syers et al. 2008).

Phosphorus fertiliser derived from rock phosphate is a finite and non-renewable resource (Cordell et al. 2009). Furthermore, P derived from P fertiliser may cause environmental problems associated with eutrophication (Gaxiola et al. 2001), especially as a result of overuse (Fig. 4.1). Hence, a major challenge for future crop production will be to produce higher yields with fewer inputs such as P fertiliser (Gregory and George 2011). One strategy would be to develop crop genotypes that require smaller amounts of fertiliser and therefore using nutrients more efficiently, bred based on trait-focused screens of germplasm collections (Gregory and George 2011). Nevertheless, little progress has been made in breeding cultivars with high P utilisation efficiency or P acquisition efficiency (Calderón-Vázquez et al. 2011; Rose et al. 2011), and the recovery (uptake) of applied P, which ranges between 25 and 60 % depending on the method used, is still modest (Syers et al. 2008).

Nutrient acquisition via the plant root system is a crucial factor for agricultural productivity and crop yield (Lynch 1995). Consequently, the amount of available P is not only determined by the ability of the soil to replenish P ions, but is also influenced by the extent and efficiency of uptake by the plant roots. Thus, cropping system-specific and plant-specific approaches must be taken into account in order to raise P acquisition efficiency (PAE) and P use efficiency (PUE) and a consideration of root traits in genetic selection might result in significant improvements for crops (Vance et al. 2003).

Soil phosphorus is the most immobile, inaccessible and unavailable of all macronutrient elements (Holford 1997) and is taken up by plants mainly in its inorganic form as phosphate (P_i) (George and Richardson 2008).

Edaphic and climate factors and cropping systems have a strong impact on the proliferation and rate of replenishment of the available P_i pool, which can be

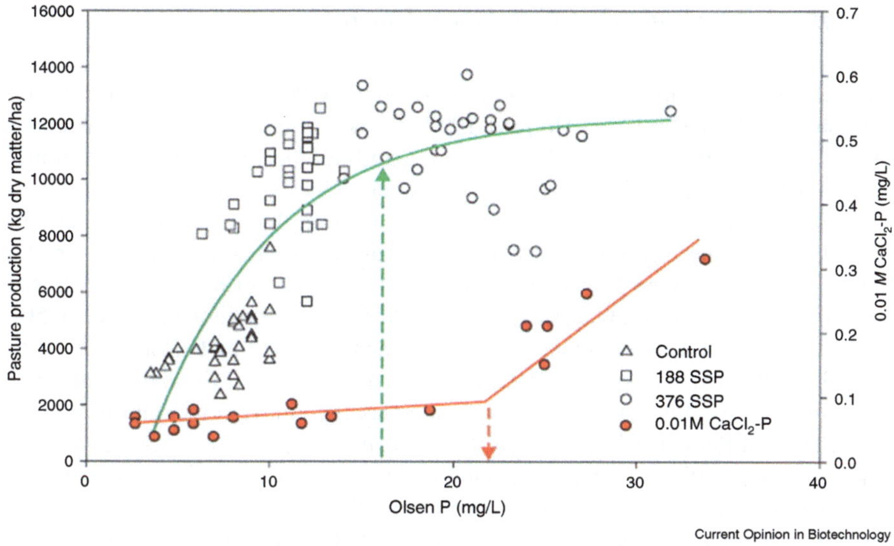

Fig. 4.1 The gap between the agronomic optimum in Olsen P for pasture production (*green up arrow*) and a potential threshold in Olsen P for P loss in subsurface drainage (as estimated by 0.01 M $CaCl_2$ -P; *red down arrow*) shows there is little justification in exceeding the agronomic optimum. Data are from plots receiving different rates of superphosphate (SSP; kg ha^{-1} $year^{-1}$) in a trial in Canterbury, New Zealand (Data from McDowell (2012) and used with permission)

recovered by the plant. P_i transfer from the soil to the root proceeds mainly by diffusion rather than mass flow, with slow diffusion rates at around 10^{-15} m s^{-1} and a concentration gradient as the driving force towards the roots (Hinsinger 2001; Rausch and Bucher 2002) resulting in a depletion zone of 1–2 mm around the root (Jungk 2001). Soil P_i concentrations are extremely low, being generally <10 μM and typically around 2 μM (Bieleski 1973; Barber 1984; Holford 1997), whereas in plants, concentrations of over 40 mM can be achieved (Bollons and Barraclough 1997). Hence, plants take up P_i faster than it is supplied by diffusion (Bieleski 1973).

Soil pH, buffer capacity, soil moisture and soil structure affect P_i solubility and sorptivity (Holford 1997; Syers et al. 2008); P_i can be absorbed on the surface of clay minerals, Fe- and Al-hydrous oxide surfaces and organic matter complexes or be fixed in acidic soils as Al-/Fe phosphates or Ca/Mg-phosphates in alkaline soils (Barber 1984; Bahl and Singh 1986; Holford 1997; Hinsinger 2001).

Significant amounts of soil phosphorus (20–80 %) is bound in organic forms such as nucleic acids, phospholipids and predominately monophosphate esters, as phytic acid and derivatives (Richardson 1994). These organic forms have to be mineralised and/or solubilised into inorganic forms in order to be available for plants; a process which is either microbiological or plant mediated.

Phosphate Transporters: Uptake and Translocation

Plants acquire P predominately as orthophosphate ($H_2PO_4^-/HPO_4^{2-}$; P_i) from the soil solution (Bieleski 1973; Holford 1997; Schachtman et al. 1998; Marschner 2012). This process is mediated by plasma membrane-localised phosphate transporters which have been suggested to operate as H^+ co-transporters (Daram et al. 1998; Smith et al. 1999; Mimura 2001; Rae et al. 2003; Raghothama 2005). Consistent with the physiological pH in many agricultural soils, maximal uptake rates occur in a pH range 5–6 (Ullrich-Eberius et al. 1981; Furihata et al. 1992; Rae et al. 2003). A constitutively expressed low-affinity uptake system with a K_m of 50–300 μM and a high-affinity uptake system, which is regulated by P_i availability with K_m of 3–7 μM have been proposed (Ullrich-Eberius et al. 1981; Furihata et al. 1992; Preuss et al. 2010). Phosphate transporters are classified into distinct families: Pht1, Pht2, Pht3 and Pht4 (Bucher et al. 2001; Liu et al. 2011).

Pht1 are high-affinity transporters homologous to the yeast PHO84 P_i transporter and other fungal high-affinity P_i transporters (Pao et al. 1998). However, the functional characterisation of the Pht1 transporter family in rice and barley (Rae et al. 2003; Ai et al. 2009) revealed kinetic properties, which are within both high- and low-affinity ranges. Pht1 transporters belong to the distinct phosphate:H^+ symporter (PHS) family which is a member of the major facilitator superfamily (MFS) of proteins (Pao et al. 1998). All these transporters exhibit high sequence similarity with each other, being similar in size and having 12 predicted transmembrane domains (TMs) with a large hydrophilic loop between TM6 and TM7 that results in a 6 + 6 configuration (Liu et al. 2011). The N- and C-termini are oriented towards the inside of the cell and they contain potential sites for phosphorylation and N-glycosylation (Smith et al. 1999).

Transporters of the Pht2 family have been cloned in *Arabidopsis*, and have been suggested to have roles as constitutively expressed low-affinity proton symporters (H^+/P_i cotransporter) for P_i loading in green shoot organs and predominantly leaf tissues (Daram et al. 1999; Versaw and Harrison 2002). The Pht2 protein is structurally similar to the Pht1 members, but is more closely related to the putative P_i transporters from bacteria and mammalian Na^+/P_i transporters. Pht2 amino acid sequences are distinct from Pht1 transporters and have a large hydrophilic loop between TM8 and TM9 (Daram et al. 1999). In wheat, expression analysis of TaPht2;1 revealed a predominant expression in a photoperiod-dependent manner in the leaves, which was significantly enhanced during P starvation (Guo et al. 2013). GFP fusion studies localised TaPht2;1 to the chloroplast envelope, suggesting a regulatory role mediating Pi translocation from the cytosol to the chloroplast as low-affinity transporter with a K_m of 225 μM P_i (Guo et al. 2013).

The Pht3 transporters belong to the mitochondrial transporter family (Rausch and Bucher 2002) and the Pht4 family has been suggested to play a role in P_i translocation between the cytosol, chloroplast, plastids and the Golgi apparatus (Guo et al. 2008; Liu et al. 2011).

Targeting the P_i transporters to the plasma membrane via the secretory trafficking pathway is mediated by PHF genes (phosphate transporter traffic facilitator 1), which are expressed strongly in root tissues and in leaf mesophyll cells, mimicking the expression pattern of the Pht1 gene family (González et al. 2005; Chen et al. 2011). OsPHF1 and OsPHF1L (Chen et al. 2011) and TaPHF1 (Wang et al. 2013) in rice and wheat, which are homologous genes to AtPHF1 in *Arabidopsis* (González et al. 2005; Bayle et al. 2011), are localised specifically in the endoplasmic reticulum (González et al. 2005; Bayle et al. 2011). AthPHF1 encodes for a plant-specific protein structurally related to SEC12 proteins of the early secretory pathway (González et al. 2005). The exit of Pht1 transporter from the endoplasmic reticulum and the targeting through the endosomal compartments are modulated by phosphorylation (Bayle et al. 2011). The *phf1* mutation in *Arabidopsis* impairs P_i uptake and P_i transport (González et al. 2005). In rice, it decreases excessive shoot P_i accumulation and P_i concentrations in leaves and roots driven by OsPHR2 and OsPHF1 over-expression (Chen et al. 2011). Even under P-replete experimental conditions, expression of phosphate starvation-induced genes (PSI) was induced in *Osphf1-1* mutants due to impaired plasma membrane location of the low-affinity P transporter OsPT2 and a high-affinity P transporter OsPht1;8 (Cheng et al. 2011).

Increasing expression of Pht1 transporters was observed as a response to P starvation or mycorrhizal infection, exhibiting a large diversity of expression patterns throughout the plant tissues. However, the lack of complete genome sequence information has hindered detailed investigation in wheat. Pht1 transporters, which are preferentially or exclusively expressed in roots, were found in barley (Smith et al. 1999; Rae et al. 2003), wheat (Davies et al. 2002; Teng et al. 2013; Wang et al. 2013), *Arabidopsis* (Mudge et al. 2002; Muchhal et al. 1996), rice (Paszkowski et al. 2002) maize (Nagy et al. 2006) and tomato (Liu et al. 1998; Muchhal and Raghothama 1999). Furthermore, Pht1 expression in shoot tissues has been reported e.g. in phloem of older leaves, flag leaves or sheaths of barley (Rae et al. 2003), in the panicle at the heading stage and flag leaves after heading in rice (Liu et al. 2011), in mature pollen of *Arabidopsis* (Mudge et al. 2002) or during mycorrhizal infection in roots of rice (Paszkowski et al. 2002; Yang et al. 2012), wheat (Glassop et al. 2005), *Brachypodium* (Hong et al. 2012) and *Medicago* (Gaude et al. 2012).

The specificity of Pht1 expression to arbuscular mycorrhiza (AM) colonisation has been investigated in cereal species and was considered symbiosis specific (Harrison et al. 2002; Gutjahr et al. 2008; Glassop et al. 2005; Paszkowski et al. 2002), as for instance for the expression of OsPT1;11 (Gutjahr et al. 2008; Paszkowski et al. 2002) and OsPT1;13 (Yang et al. 2012; Güimil et al. 2005) in rice. Homologues of both genes occur in *Brachypodium* (Hong et al. 2012) and maize (Nagy et al. 2006), where transcripts also accumulated in non-colonised roots and leaves, suggesting additional roles during P starvation (Yang et al. 2012).

The Pht1 promoters contain the target elements for transcription factors of the P signalling network, for instance the P1BS *cis*-element as a target for PHR1 (Schünmann et al. 2004; Ren et al. 2012a) or the WRKY-binding W-box as a target for WRKY75 (Devaiah et al. 2007a, b; Miao et al. 2009), suggesting their embedding in the cross-talk of P signalling during P starvation to maintain P homeostasis.

It may be concluded that the regulation of P_i uptake and P_i transport mediated by phosphate transporters, for which multiple roles in P_i acquisition and P_i remobilisation have been suggested, are very complex. Roles of transporters in the genetically diverse trait of tolerance to low P or in the PUE context will be discussed further but remain elusive.

Morphological and Biochemical Adaptations of Plants During P Limitation

When sensing nutrient depletion, plants have developed broad morphological and biochemical strategies to deal with the heterogeneous availability of soil resources. Root plasticity (morphology, topology and architecture) is a crucial but neglected factor which is linked with immobile nutrients such as P (Lynch 1995). Plant roots typically respond to P deficiency through allocation of more carbohydrates towards the roots, which enhances root growth to maximise the soil volume exploited and increases root to shoot ratio (Hermans et al. 2006; Hammond and White 2008). Root hair formation (number, length and surface area) is strongly related to P depletion (Bates and Lynch 1996; Gahoonia et al. 1997; Jungk 2001; Zhu et al. 2005a, b) emphasizing a strong role in P_i acquisition from the soil (Gahoonia and Nielsen 1998; Gahoonia et al. 2001). Topsoil foraging, which is characterised by enhanced lateral root branching over primary root growth, contributes to efficient P acquisition (Lynch and Brown 2001; Williamson et al. 2001; Pérez-Torres et al. 2008).

P acquisition is enhanced through symbioses with arbuscular mycorrhizal (AM) fungi (Barber 1984; Fitter 2006), by substantially increasing the P absorbing surface for P uptake (Jakobsen et al. 1992) and the ability to access mineralised organic P sources (Koide and Kabir 2000) and increased expression and secretion of plant acid phosphatases (Tarafdar and Marschner 1994).

Strigolactones have a role in facilitating symbiosis formation (Gomez-Roldan et al. 2008), and also stimulate tiller formation (Hong et al. 2012), which is an important parameter for yield. As pH has a strong influence on the bioavailability of soil P_i (Barber 1984; Hinsinger 2001) root excreted protons tend to acidify the rhizosphere and, along with organic acids like malic acid, citric acid or phenolic compounds that also act as chelators (Raghothama 1999; Vance et al. 2003), will all help to solubilise P_i in the rhizosphere. Organic acids displace bound P_i from Al^{3+}-, Fe^{3+}- and Ca^{2+}-phosphates (Dinkelacker et al. 1989; Gerke et al. 1994). In particular, cluster roots (brush-like root formations) are an adaptation strategy to low soil P_i availability of many plant species such as white lupin and other members of the Proteaceae family and induce such chemical changes in the rhizosphere (Neumann and Martinoia 2002). Plants also respond to P deprivation through the induction of various metabolic processes e.g. the induction of a bypass pathway of glycolysis and mitochondrial electron transport to replace ATP as an energy resource (Theodorou and Plaxton 1993; Duff et al. 1989). All these metabolic changes contribute to a better internal P utilisation when P_i is limiting.

The genetic background for these responses and their potential as targets for crop genetic improvement based on studies with model plants will be described below.

Efficient Phosphate Nutrition in Model Plants

Phosphate acquisition and use efficiency (PAE, PUE) represent capabilities to cope with either P limiting conditions or growth maintenance under P deficiency stress, utilising morphological, biochemical and molecular changes without sacrificing yields (Chiou and Lin 2011). The P_i-starvation adaptation response (PSR) consisting of the regulation or coordination of the P_i-starvation inducible genes (PSI) in order to maintain P homeostasis has been investigated predominantly in the model plant *Arabidopsis thaliana*, in white lupine, in rice and in maize using forward (mutants) and reverse genetic approaches (Alexova and Millar 2013; Calderón-Vázquez et al. 2008; Li et al. 2010; Nilsson et al. 2007, 2010; Bustos et al. 2010; Oono et al. 2011; Uhde-Stone et al. 2003; Rubio et al. 2001; Aung et al. 2006; Sánchez-Calderón et al. 2006; Franco-Zorrilla et al. 2007; Miura et al. 2011; Shin et al. 2006; Duan et al. 2008).

PSI genes and enzymes, which are induced or suppressed at the molecular level, and commonly used as markers for monitoring PSR, are described below. Biochemical and morphological changes have been elucidated to a much larger extent, whereas information on phosphate sensing and signal transmission is more limited (Chiou and Lin 2011). Nonetheless, the lack of studies investigating genotypic variation in a broad spectrum of cultivars makes it difficult to draw conclusion for enhancing PUE in crops (Alexova and Millar 2013; Calderón-Vázquez et al. 2011; Veneklaas et al. 2012).

Root Morphology

Among common responses to P starvation are changes in root morphology such as increasing root hair density, reduction of primary root growth and promoted lateral root initiation, which are well described in *Arabidopsis* (Williamson et al. 2001; López-Bucio et al. 2000; Sánchez-Calderón et al. 2006) and have been recently studied in cereals (Hochholdinger and Zimmermann 2008). For instance, a root-hairless mutant of *Arabidopsis* (Bates and Lynch 2001) and a barley root-hair-deficient mutant (Gahoonia et al. 2001) grew poorly under low P_i. The root cap is the site of sensing local P_i concentrations initiating spatial changes, comprising inhibition of cell division activity of primary meristematic cells and root cell elongation (Ticconi et al. 2004; Sánchez-Calderón et al. 2005; Franco-Zorrilla et al. 2007; Svistoonoff et al. 2007). Phytohormone-related genes have been reported to mediate root architectural changes under low P growing environments, particular auxin-responsive genes (Bates and Lynch 1996; Hammond et al. 2004; Pérez-Torres et al. 2008; Jain et al. 2007; Miura et al. 2011). Expansins are involved in cell wall extension (Zhao et al. 2012), including root-hair formation (Yu et al. 2011) and are stimulated by indole-3-acetic acid and abscisic acid under abiotic stress (Zhao et al. 2012).

For instance, Miura et al. (2011) assumed that genes coding for expansin 17, glycosyl hydrolase 19 and UDP-glycosyltransferase are involved in the regulation of cell wall-loosening and elongation in response to P starvation in *Arabidopsis*. Distinct responsiveness to P_i availability among *Arabidopsis* ecotypes (Chevalier and Rossignol 2011) and the identification of quantitative trait loci (QTLs) (Reymond et al. 2006) suggests a genetically determined control of the root growth response to low P_i. Genetic factors controlling root plasticity have been investigated using *low phosphorus insensitive* (*lip*) mutants (Sánchez-Calderón et al. 2006). The mutation disrupted not only the root developmental response but also altered the induction of P deprivation responsive genes, which are relevant for adaptation to low-P_i, including acid phosphatases (AtPAP1, AtACP5) and phosphate transporters (AtPT1, AtPT2) (Sánchez-Calderón et al. 2006). Furthermore, their findings suggest that the root architectural response is mediated by a specific nutrient (P) sensing signalling network (Sánchez-Calderón et al. 2006). For instance, WRKY75 (Devaiah et al. 2007a) and ZAT6 (Devaiah et al. 2007b) are among several transcription factors, which have a regulatory effect on root architecture of *Arabidopsis* and were suggested to have an impact on P starvation responses. However, genetic selection based on root parameters has been difficult due to their multigenic nature and the lack of appropriate evaluation methods (Vance et al. 2003).

P Starvation Signaling

Elicited responses to internal and external nutritional status involve local and systemic signalling (Chiou and Lin 2011); signalling molecules, their mode-of-action and interacting pathways are summarised below.

A strong increase of Induced by P starvation gene transcripts (IPS) under P_i starvation has been reported in *Arabidopsis* and rice (Rubio et al. 2001; Oono et al. 2011) and members of the IPS gene family have been widely used as molecular markers of plant P_i nutritional status (Zhou et al. 2008; Tian et al. 2012; Wang et al. 2013). IPS genes are involved in the miR399-PHO2 regulatory loop as ribo-regulators (Doerner 2008) and function as miRNA399 antagonists, which negatively alter PHO2 expression at the post-transcriptional level; a regulatory process which is described as "target mimicry" (Franco-Zorrilla et al. 2007). It seems likely that they stabilise the initial decrease of PHO2 transcript to prevent P_i toxicity via P_i accumulation in the shoots (Bari et al. 2006; Chitwood and Timmermans 2007). Five IPS genes have been found in the *Arabidopsis* genome (Franco-Zorrilla et al. 2007), and two in rice, maize and barley (Hou et al. 2005). Promoters of the pho4-regulon in yeast have two *cis*-regulatory elements, which were also found in *Arabidopsis* (At4/AtIPS4), tomato (TPSI1), *Medicago truncula* (Mt4) and rice (OsPI1) (Hammond et al. 2003). At4 and AtIPS4 in *Arabidopsis* are involved in P_i allocation between roots and shoot and enhance lateral root development (Shin et al. 2006; Franco-Zorrilla et al. 2007).

The *at4* mutant exhibited P accumulation in shoots (Shin et al. 2006), whereas over-expression decreased P accumulation (Franco-Zorrilla et al. 2007). AtIPS1 modulates PHR expression, a MYB-CC type transcription factor which is involved in P starvation responses (Rubio et al. 2001). PHR1 (phosphate starvation responsive 1) plays a pivotal role in sensing P availability (Chiou and Lin 2011) and has been examined in detail. PHR1 is a member of the MYB-transcription factor family (15 members) and seems to be a key regulator for downstream P responsive genes through binding to a P1BS (PHR1 specific binding sequence) *cis*-element, which is an imperfect palindromic sequence (GNATATNC) (Rubio et al. 2001; Nilsson et al. 2007, 2010; Bustos et al. 2010). An important downstream target of AtPHR1 and possible homologues is miRNA399, which is involved in the PHO2 regulation as previously mentioned (Miura et al. 2005; Schachtman and Shin 2007). Over-expression of AthPHR1 increased the transcript level of miRNA399 and decreased expression of PHO2, increased further the P_i content and enhanced root hair density in rice and *Arabidopsis* (Nilsson et al. 2007; Zhou et al. 2008; Bustos et al. 2010). Promoters of several P starvation-induced and repressed genes, including IPS and a high-affinity P-transporters, contain the P1BS *cis*-element (Oono et al. 2011; Hammond et al. 2003; Rubio et al. 2001; Schünmann et al. 2004; Guo et al. 2013; Bustos et al. 2010). In wheat, over-expression of TaPHR1 did not change transcript levels of TaPHF1, TaPHO2 or TaSPX3, whereas TaIPS and TaPht1;2 exhibited increased expression levels in the transgenic lines suggesting that transcriptional factors additional to TaPHR1 may be functional in the P starvation signalling (Wang et al. 2013). However, the *Athphr1* mutant impairs a broad range of P starvation responses and shows impaired root growth and root hair length (Rubio et al. 2001; Bustos et al. 2010; Nilsson et al. 2007).

In rice, two homologues of AtPHR1, OsPHR1 and OsPHR2, are involved in P starvation signalling (Zhou et al. 2008; Wang et al. 2009a, b). However only over-expression of OsPHR2 resulted in increased shoot P_i and altered root morphology (Zhou et al. 2008; Wu and Wang 2008; Bustos et al. 2010). OsPHR2 positively regulated the low-affinity phosphate transporter OsPT2 in roots resulting in excessive P_i accumulation in the shoot tissue (Liu et al. 2010). Further, a root-associated purple acid phosphatase (10a) in rice, OsPAP10a, is controlled and induced by OsPHR2 (Tian et al. 2012).

SPX proteins (which contain a SPX domain, SYG1, PHO81, XPR1 at the N-termini) are involved in the downstream responses of PHR1 in *Arabidopsis* (Duan et al. 2008) and OsPHR2 and PHO2 in rice (Wang et al. 2009a, b). Members of the SPX protein family in rice (OsSPX3 and SPX1/2/6) have been shown to be highly induced (preferentially) in rice roots and shoots where they are involved in the regulation of PSI and OsIPS1 (Wang et al. 2009a, b; Oono et al. 2011). OsSPX1 over-expression suppressed IPS gene induction, miRNA399 and phosphate transporter Pht1 expression (Wang et al. 2009a, b) and in yeast, an SPX domain limited the phosphate uptake velocity (Hürlimann et al. 2009). Similar results were obtained with *Arabidopsis* mutants for AtSPX1-AtSPX4 affecting the expression pattern of purple acid phosphatases genes (Duan et al. 2008). Furthermore, OsSPX1 is positively regulated by OsPHR2, involved in the feedback P_i signalling network in roots by suppressing OsPT2 and other PSI genes in the PHR2/Pho2 background (Liu et al. 2010), and negatively regulates shoot P accumulation (Wang et al. 2009a, b). OsSPX1 over-expression counteracted the effect of PHR2 over-expression in rice, which mimics P starvation and induces PSI gene expression but not the function of Ospho2 regulating OsPT2 expression (Liu et al. 2010). In summary, SPX proteins seem to be essential players for maintaining P homeostasis and P signalling in plants (Rouached et al. 2010; Nilsson et al. 2012; Secco et al. 2012).

The regulatory mechanism of P allocation among different organs during plant development under P stress remains relatively elusive and investigations have been mainly focused on screening *Arabidopsis* mutants with abnormal P distribution. The *Athpho1* mutant showed severe P deficiency in above-ground shoot tissues due to influencing transfer of P_i to the xylem vessels for subsequent transport to the shoot and leaves (Poirier et al. 1991; Liu et al. 2012). PHO1 is a membrane-spanning protein but there is no evidence that it is a transporter itself and it does not have homology to any other previously known transporter (Hamburger et al. 2002). Eleven members of the AthPHO1 transporter family are known which share the same topology (Wang et al. 2004); a SPX tripartite domain in the N-terminal (SYG1/PHO81/XPR1) and an EXS domain at the C-terminal (ERD1/XPR1/SYG1). AthPHO1 has been localised to the ER and the Golgi (Liu et al. 2012). The EXS/SPX domains, and particularly the N-terminal region of PHO1, have been identified in yeast as being involved in either phosphate transport or in sorting proteins to endomembranes (Liu et al. 2012; Wang et al. 2004). AtPHO1 seem to mediate P_i efflux out of root stellar cells along its electrochemical gradient (Hamburger et al. 2002) and AtPHO1;H1, seems to be regulated by PHR1 (Stefanovic et al. 2007). The roles of the other members, AtPHO1;H2 to AtPHO1;H9, are unknown (Secco et al. 2010) but show a distinct expression pattern from that of AtPHO1 and AtPHO1;H1 (Hamburger et al. 2002). The AthPHO1 family clusters into two clades, which are expressed in a broad range of tissues, including leaves and predominately in vascular tissues of roots, leaves, stems or flowers. Only one clade which contains AtPHO1 and AtPHO1;H1 clusters with the three OsPHO1 proteins found in rice (Secco et al. 2010). OsPHO1;2 was mostly expressed in roots and was relatively lowly and constantly expressed in other tissues. A mutation affected root-to-shoot Pi transfer (Secco et al. 2012). OsPHO1;1 was predominately expressed in flowers before and during pollination and OsPHO1;3 was the lowest expressed, and higher in leaves and flowers (Secco et al. 2010). AthPHO1 is a downstream component of the AthPHO2 regulatory pathway (Liu et al. 2012).

PHO1 degradation or down-regulation is not P_i and PHO2 dependent and was suggested to occur at the post-translational level (Liu et al. 2012; Aung et al. 2006).

The *Arabidopsis pho2* mutant exhibits excessively high P concentrations in the shoots, displaying symptoms of toxicity as a result of enhanced uptake and root-to-shoot translocation (Delhaize and Randall 1995; Liu et al. 2012). PHO2 is predominantly expressed in the root (Chiou et al. 2006) and its localisation within the cell has been determined as being with the ER and Golgi (Liu et al. 2012). PHO2 is a member of the E2 ubiquitin conjugase family (UBC24) (Bari et al. 2006) and is a target for miRNA399, which suppresses PHO2 expression (Aung et al. 2006; Chiou et al. 2006; Fujii et al. 2005). Over-expression of miRNA399 or loss of function of UBC24 resulted both in enhanced P_i accumulation and impairment of P_i remobilisation from old to young leaves in *Arabidopsis* (Chiou et al. 2006). Expression of miRNA399 is an early response to P deficiency and a systemic signal of P_i deficiency derived from P deplete root and shoot tissues (Lin et al. 2009; Bari et al. 2006; Aung et al. 2006; Chiou et al. 2006). Its expression is regulated by PHR1 (Bari et al. 2006; Aung et al. 2006) in a mechanism called "target mimicry" (Franco-Zorrilla et al. 2007) as previously described. In addition to miRNA399, various other miRNAs could be identified by deep sequencing in *Arabidopsis* (Hsieh et al. 2009; Lundmark et al. 2010) and other plant species (Chiou and Lin 2011), including white lupin (Zhu et al. 2010), soybean (Zeng et al. 2010), tomato (Gu et al. 2010) and wheat (Zhao et al. 2013). It has been assumed that high-affinity phosphate transporters, AthPht1;8 and AthPht1 cause the phenotype of the *pho2* mutant and miRNA399 over-expressors because they are up-regulated in these mutants (Aung et al. 2006; Bari et al 2006). In rice shoots, Os*pho2* mediated P accumulation was assumed being the result of induced expression of high- and low-affinity phosphate transporters OsPht1;2, OsPht1;9 and OsPht1;10 (Liu et al. 2010). It has been suggested that PHO2 encodes an additional regulator for the low-affinity P_i translocator protein, AthPT2;1, in green shoot tissues of *Arabidopsis* (Daram et al. 1999). Nonetheless, downstream responses of PHO2 are still not completely understood (Liu et al. 2012). PHO1 and PHO2 are both expressed in vascular root tissues (Hamburger et al. 2002) as is miRNA399 (Aung et al. 2006).

There is evidence that sucrose is a global regulator of plant P starvation responses, interacting with P starvation signals (Lloyd and Zakhleniuk 2004; Karthikeyan et al. 2007; Lei et al. 2011) and root architecture alterations (Niu et al. 2012). Sucrose phosphate synthase (SPS) has been reported to be more abundant in P deficiency tolerant cultivars of *Brassica napus* (Yao et al. 2011). Exogenously applied sugars or sucrose-enriched growing media could stimulate P starvation inducible genes such as acid phosphatases (APase), AtIPS1 (Müller et al. 2005) or an APGase subunit in tobacco (Nielsen et al. 1998).

Phosphate transporters, for instance TaPht1;2 (Miao et al. 2009), seem to belong to the group of sugar-modulated genes under P-starvation (Jain et al. 2007; Karthikeyan et al. 2007; Hammond and White 2008).

The *Arabidopsis pho3* mutant exhibits a restricted sucrose translocation from root to shoot caused by a defective sucrose transporter, SUC2, which is involved in phloem loading (Lloyd and Zakhleniuk 2004). *Pho3* shows altered APase induction/secretion on the root surface, reduced P_i accumulation in both leaves and roots (Zakhleniuk et al. 2001) and a strongly induced Glc-6-P/phosphate translocator. This phenomenon is again consistent with the observation that sucrose accumulates in Pi-starved leaves of various crops (Hammond and White 2008). Microarray analysis investigating consequences of SUC2 over-expression in the transcriptome of the *hypersensitive to phosphate starvation1* (*hps1*) *Arabidopsis* mutant revealed the induction of P_i starvation induced genes under P replete conditions (Lei et al. 2011). In conclusion, sugar sensing and signalling is involved in adaptation responses to P starvation even if the exact mechanism is not understood.

There are other transcription factors induced by P starvation and involved in P_i availability responses including OsPTF1 (Yi et al. 2005), WRKY75 and ZAT6 (Devaiah et al. 2007a, b), BHLH32 (Chen et al. 2007), WRKY6 (Chen et al. 2009a, b) and MYB62 (Misson et al. 2005). Two WRKY box (W-box) elements have been found in the promoter of genes involved in P_i retranslocation and scavenging, including AtPht1 transporters, AtIPS, acid phosphatase genes (AtPS2), purple acid phosphatases (PAP11) and PHR1 (Devaiah et al. 2007a, b).

BHLH32 acts a negative regulator for PPCK (phosphoenolpyruvate carboxylase kinase) expression in P sufficiency, root hair formation and anthocyanin production (Chen et al. 2007). Plant hormones have been implicated with P signalling mechanisms as changes in P_i availability alters the expression of genes involved in the biosynthesis of phytohormones which in turn may influence PSR (Morcuende et al. 2007; Chiou and Lin 2011). P_i itself may be important in P starvation signalling, as is the case for nitrate acting as a signal stimulating root growth (Zhang and Forde 2000). The P_i flow across the peri-arbuscular cortex membrane may be among the mechanisms which allow plants to recognise AM fungi from less beneficial microbes (Yang and Paszkowski 2011). Compelling evidence is provided by studies using phosphite, a phosphate analogue which is taken up by plants but cannot be oxidised once inside the cell, mimicking sufficient P_i supply in P starving plants, which result in an interference and attenuation with PSR at the transcriptional and post-transcriptional level (Chiou and Lin 2011).

Metabolic Changes

Several proteomics studies are contributing to the understanding of the metabolic responses to P starvation in model plants and crops (Yao et al. 2011; Li et al. 2008a, b; Chevalier and Rossignol 2011). However, proteins responsible for distinct PAE/PUE in various species and cultivars have not been identified (Alexova and Millar 2013).

P limitation alters the TCA cycle metabolism and has been shown in a broad range of studies. Enzymes involved in the TCA cycle and glycolysis produce organic acids required for P recycling from phosphorylated glycolytic intermediates as well as releasing P_i from organic P sources or inorganic bound P_i in the soil

(Oono et al. 2011). The over-production of citrate in transgenic tobacco (López-Bucio et al. 2000) as well as mitochondrial citrate synthase in *A. thaliana* (Koyama et al. 2000) enhanced P_i uptake. Key enzymes which have been studied in *A. thaliana* are citrate synthase, malic enzyme and aconitase, which exhibited variation in protein abundance between ecotypes during P deficiency (Chevalier and Rossignol 2011). In other species, the activity of aconitase correlated with organic acid secretion (Neumann and Römheld 1999) and in alfalfa, the overexpression of malate dehydrogenase (MDH) resulted in increased P accumulation (Tesfaye et al. 2001). Another approach was the expression of phytase genes of alfalfa or of a fungal origin in *Arabidopsis* and tobacco plants, resulting in improved acquisition of organic P sources (George et al. 2005). However, the length of P starvation influences the synthesis and degradations of proteins which are potentially involved in enhancing the plants adaptation to P deficiency. For instance, genes encoding for isocitrate dehydrogenase were suppressed in rice roots only after a certain time of exposure to P starvation, resulting in a suppression of citrate degradation (Oono et al. 2011).

The replacement of phospholipids by galactolipids or sulpholipids is a well-known adaptation process in plants during P deficiency (Andersson et al. 2003; Hammond et al. 2003; Byrne et al. 2011), even if phospholipid degradation is differently mediated in different species (Calderón-Vázquez et al. 2011). For instance, in potato, an array study identified novel roles for the main storage protein in potato tubers, the patatin-like proteins, which also have lipase activity and are potentially involved breakdown of phospholipids for P_i recycling (Hammond et al. 2011). Numerous studies in model plants or crops reported the induction of genes related to an altered lipid metabolism, for example UDP-sulfoquinovose synthase 1 (SQD1) or glycerophosphoryl diester phosphodiesterase (GDPD) or lipid transfer proteins (Hammond et al. 2011; Oono et al. 2011; Morcuende et al. 2007; Calderón-Vázquez et al. 2008; Wasaki et al. 2003). Glycerophosphodiester phosphodiesterases (GPX-PDE) catalyse the hydrolysis of phospholipids to glycerol-3-phosphate and the corresponding alcohol. Recently, GPX-PDE genes were identified which were highly expressed in cluster roots of white lupine under P_i-deficiency (Cheng et al. 2011; Uhde-Stone et al. 2003).

To date, the knowledge about functional consequences of replacing phospholipids in membranes is very limited (Veneklaas et al. 2012).

Post-translational Modifications

Post-translational modifications are important factors in P signalling and metabolic pathways involved in PUE (Alexova and Millar 2013; Plaxton and Tran 2011), highlighted when comparing proteome with transcriptome studies in P-starved maize (Calderón-Vázquez et al. 2008; Li et al. 2008a, b) and *Arabidopsis* (Morcuende et al. 2007). A striking example underpinning the importance of post-translational modifications within adaptation to low P exposure is the potential role of a protein kinase, OsPupK46-2, within the *Pup1* locus (*Phosphorus uptake 1*), which is a major QTL for low P tolerance in rice (Gamuyao et al. 2012).

Pht1 phosphate transporter genes are also regulated at the post-transcriptional level by the recently discovered PHF1 in rice and *Arabidopsis* (Chen et al. 2011; Bayle et al. 2011; Chiou and Lin 2011), which is connected to a kinase, RAPTOR1B. PHR1 is up-regulated at the transcript level and both are up-regulated at the protein-level in P-starved roots of *Arabidopsis* (Lan et al. 2012). The enhanced synthesis of organic acids allows P recycling from phosphorylated glycolytic intermediates, particularly phosphoenolpyruvate (PEP), which is mediated via the enzyme phosphoenolpyruvate carboxylase (PEPC). PEPC is in turn activated by reversible phosphorylation via PPCK (PEPC kinase) (Gregory et al. 2009). PPCK has been reported as being among up-regulated genes during P starvation (Morcuende et al. 2007; Müller et al. 2007; Chen et al. 2007). Therefore, PEP and PEPC are among the metabolites and enzymes, which are more abundant or active in P starved *A. thaliana, B. nigra, O. sativa* and *T. aestivum* (Morcuende et al. 2007; Johnson et al. 1996; Duff et al. 1989; Oono et al. 2011; Neumann and Römheld 1999).

In *Arabidopsis*, PHR1, which initiates P starvation signalling responses, was hypothesised to be a target for the conjugation of the SUMO superfamily of proteins via SUMO ligases (sumoylation; Miura et al. 2005; Wang et al. 2013). The sumoylation is mediated by the small ubiquitin like modifier SUMO E3 ligase, AtSIZ1, which is more root than shoot abundant and is localised in the nucleus of the cells (Miura et al. 2005). The homologue, OsSIZ1, have been also detected among transcripts of P starved rice (Oono et al. 2011). The observation that AtIPS gene induction was reduced in *siz1* mutants, indicating a positive regulation with AtSIZ1, (Miura et al. 2005), strengthens the hypothesis that post-translational modifications as essential for the initial stages of P starvation signalling cascades. Additionally, AtSIZI negatively regulates P starvation-dependent primary root growth inhibition (increased root hair number and length) through the control of auxin patterning (Miura et al. 2011), whereas the *phr1* mutant does not exhibit affected root architecture (Rubio et al. 2001).

In conclusion, post-translational modifications may be essential leaders for P economy in plants (Raven 2008) and potentially for enhancing PUE in crops.

PAE and PUE in Crops

Definitions and Implications

Grain crops like rice, maize, wheat and oilseed rape are essential major staple foods (FAO 2011) and major contributors to the global phosphorus cycle (Rose and Wissuwa 2012). Global P flows resulting from, for instance, the rock mineral fertiliser trade, the global grain trade or manure-delivering livestock production

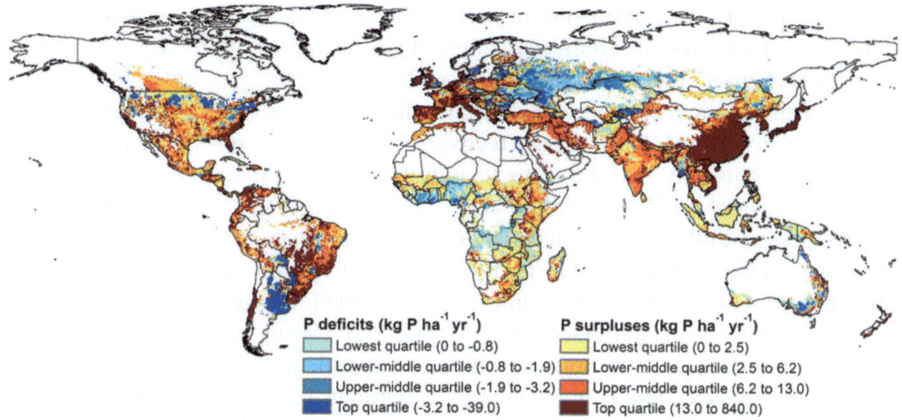

Fig. 4.2 Global map of agronomic P imbalances for the year 2000 per unit of cropland area in each 5 °grid cell. The surpluses and deficits are each classified according to quartiles globally (0–25th, 25–50th, 50–75th, and 75–100th percentiles) (Data taken from MacDonald et al. (2011) and used with permission)

are unevenly distributed, resulting in massive agronomic imbalances and spatial surplus or deficit patterns across regions and countries (Tiessen 2008; MacDonald et al. 2011; Fig. 4.2). Additionally, large amounts of applied P are removed in harvested products from the fields and across the globe (Lott et al. 2009), and nutrient recycling, especially organic P sources, from urban areas or returning biomass is rare (Cordell et al. 2009). Rock P fertilisers are extracted and exported from only a few countries worldwide, which are geopolitically controversial and potentially unstable (Cordell et al. 2009; Tiessen 2008; FAO 2011). Demand is still increasing and estimated to be 1.9 % yearly from 2011 to 2015 (FAO 2011). In order to meet the requirement for the future demands of an increasing world population, bearing in mind that P reserves are limited and deposits cannot be exploited infinitely (Kirkby and Johnston 2008), it is a necessity to enhance P fertiliser use efficiency (Gregory and George 2011).

There is an essential need for agronomic P efficiency criteria to be defined in order to be able to exploit and screen genetic variation in grain crops (Rose and Wissuwa 2012), one of the key strategies for breeding more P efficient crops (Gregory and George 2011). Phosphorus efficiency is a complex trait and the genetic determinants, which are involved in enhanced low P tolerance or P efficiency are still not clearly understood. As previously described, the focus of molecular research lies mainly on P stress responses of individual genotypes or model plants and it is questionable if these findings are generally applicable for crops in an agricultural context. Genotypic phosphate efficiency differences that exist might only rarely be conserved between and within species (Hammond et al. 2009; Alexova and Millar 2013; Calderón-Vázquez et al. 2008, 2011; Niu et al. 2012) and need to be identified before they can be exploited for breeding. Due to the previously described interactions of P accessibility in soils (Holford 1997), field experiments or experiments in soil might be more suitable for assessing or identifying traits for P efficiency rather than using solution culture (Hayes et al. 2004; Gunes et al. 2006; George et al. 2004, 2005).

For agronomic and economic reasons, P efficiency is based on the cropping area leading to improved P fertiliser recovery and use of soil P (Sattelmacher et al. 1994). This concept takes into account the low P availability in e.g. tropical soils (Wissuwa et al. 2009), limited access to P fertilisers in some regions of the world (Tiessen 2008), costs for fertilisation (FAO 2011) and environmental aspects including impaired water quality by run-off or drainage due to agricultural intensification (McDowell 2012). This unit is based on the P efficiency properties of the plant itself, which can be divided into PAE and PUE (Wang et al. 2010a, b). It is of strong economic interest to not only enhance P acquisition, which might result in over-mining the soil, but also enhance PUE without increasing P export from the field via the grain (Batten 1992; Rose and Wissuwa 2012). Both traits are usually linked, negatively associated with each other and are hard to distinguish (Su et al. 2006; Su et al. 2009; Rose and Wissuwa 2012), indicating a need to choose a selection technique achieving an appropriate distinction (Batten 1992) and a clear positive correlation of biomass ratios with increasing PUE (Rose and Wissuwa 2012). In contrast to PAE, PUE is much less well understood, lacks clearly defined and consistent terminologies or screening methodologies across the literature which is a bottleneck for P efficiency improvement in crops (Hammond et al. 2009; Wang et al. 2010a, b; Rose and Wissuwa 2012). To achieve an increase of PUE is especially important in regions of high cropping intensity facing plateauing yields during the last decade.

Screening Approaches

When P efficiency is to be evaluated across a range of distinct or heterogeneous genotypes e.g. high-yielding modern varieties and low-yielding land races, a tissue-specific approach is probably required (Rose and Wissuwa 2012). Using this approach presupposes knowledge about the different P pools as well as about the changing P requirements depending on the growth stage of the crop (Veneklaas et al. 2012).

Phosphate is compartmentalised within plant cells and exists in two main P pools (Veneklaas et al 2012). The first P pool consists of free inorganic orthophosphate (P_i), which is either metabolically active in the cytoplasm or stored in the vacuole to buffer P demands of the cytoplasm (Mimura et al. 1996; Lauer et al. 1989). The storage P_i can have a diagnostic value (Bollons and Barraclough 1997, 1999), although shoot growth seems to be reduced before severe P_i depletion of the vacuolar storage pool occurs (Rouached 2011; Mimura et al. 1996). The second P pool represents organic forms as P esters, comprising nucleic acids, phospholipids, phosphorylated proteins and low relative molecular mass metabolites (Veneklaas et al. 2012). In the nucleic acid pool, RNA is usually the largest with 40-60 % of this P pool (Bieleski 1968) with ribosomal RNA (rRNA) having the biggest share, adjusting with growing pattern (Suzuki et al. 2010; Hensel et al. 1993; Kanda et al. 1994). Nucleic acid and protein turnover and repair has a large P cost (Raven 2012). Hence, plants cannot dispense DNA or RNA without affecting growth (Raven 2008); a target for use efficiency would be the optimising of the ribosomal pool size, protein biosynthesis and especially protein turnover (Veneklaas et al. 2012).

Phospholipids in cell membranes fulfil structural roles, serve as substrates for biochemical signals and are required in abundance by photosynthetic tissues and cell-expanding/-dividing tissues. The replacement of phospholipids by glycolipids (galactolipids, sulfolipids) in plastids, as result of P deficiency are known (Wasaki et al. 2003; Andersson et al. 2003; Essigmann et al. 1998; Oono et al. 2011), in order to economise the use of P, as the total area of membranes can hardly be decreased (Veneklaas et al. 2012). In cyanobacteria and algae their replacement can be complete (Van Mooy et al. 2006, 2009) or partial in rhizobial symbionts (Gaude et al. 2004), but in plants consequences of such a replacement remain speculative (Veneklaas et al. 2012).

Phosphate uptake seems to be most critical during vegetative growth whereas a shift occurs in the late vegetative or early reproductive stage and remobilisation and optimal allocation becomes another resource of P (Rose et al. 2007; Römer and Schilling 1986; Veneklaas et al. 2012). There are rankings in the literature focusing on early growth (Liao et al. 2008) or contrastingly, on grain yield (Jones et al. 1992). P concentrations in the grain of crops are usually much higher than in vegetative tissues (Veneklaas et al. 2012), and seed P reserves occur predominately as phytate (Lott et al. 2009; Raboy 2009; White and Veneklass 2012), which are the salts of phytic acid with high affinity to Zn and Fe (Michael et al. 1980). Seed P content as well as phytate content differs between wheat genotypes (Batten 1992) and declines with decreasing P supply (Mengel and Kirkby 2001). High P grain content, especially phytate, is not particularly desirable, as it acts as an anti-nutrient in animal and human diet which aggravates the global problem of mineral malnutrition (White et al. 2012) and causes environmental problems in the form of phosphate-rich manure or sewage (Raboy 2009) once it is removed with the grain as the harvested product from the cropping area. However, whilst low seed P content or concentration could be appropriate selection criteria for improved PUE, it is controversial. Grain or seed P reserves support initial seedling growth until it is supplemented through P_i uptake by the developing root system (White and Veneklaas 2012) and correlated with the initial root biomass (Zhu et al. 2005a, b). It is questionable if seed coating or P fertiliser placement could compensate low grain P (Rebafka et al. 1993), even if lower root development due to lower seed P reserves can be overcome due to mycorrhizal infection (Zhu and Smith 2001).

Exploiting Genetic Differences

There are studies showing genotypic differences for P deficiency tolerance suggesting that P efficiency mechanisms may differ among wheat, rye and triticale genotypes (Ozturk et al. 2005; Manske et al. 2001; Osborne and Rengel 2002; Gunes et al. 2006). Hammond et al. (2009) observed a considerable diverse species-wide variation for shoot P concentrations and several PUE measures in *Brassica oleracea* landraces and commercial varieties. Manske et al. (2001) argued that under P deficient conditions P uptake efficiency plays the key determinant for yield, whereas P utilisation efficiency was more relevant under well P supplied conditions. There are also several investigations in other *Brassica* plants aiming to exploit genetic diversity (Akhtar et al. 2008; Solaiman et al. 2007; Yang et al. 2010, 2011).

Batten (1992) pointed out that selection of more P efficient wheat occurs unconsciously during breeding when selecting for higher yields at sub-optimal P levels. Chin et al. (2010) mentioned a similar observation of an unconscious selection for a major P deficiency tolerant rice QTL, *Pup1*, in drought tolerant varieties developed under unfavourable conditions. Hammond et al. (2009) similarly implicated inadvertent selection in *B. oleracea* breeding programmes.

The selection process resulting in modern (*Brassica*) crop varieties with increased grain yield per mg shoot P (Hammond et al. 2009) might be just a result of higher HI (Batten and Khan 1987), rather than a selection of physiological traits (Rose and Wissuwa 2012). Chin et al. (2010) support that hypothesis by observing a lower frequency of the low P-tolerant QTL *Pup1* in modern irrigated rice varieties compared to traditional varieties. There are further studies showing that enhanced low P tolerance in wheat was due to enhanced P uptake rather than enhanced P utilisation (Gahoonia et al. 1996; Gahoonia et al 1999).

A comparison of new CIMMYT wheat varieties with an older Mexican variety showed that PUE did not differ (Egle et al. 1999). However, P acquisition under low P conditions was improved mainly due to a better root length density, especially during the period during anthesis and grain filling (Egle et al. 1999). With P fertilisation, root length density was not significantly different between the older and the modern varieties, but higher P uptake rates, especially during grain filling, of the modern varieties seemed to have contributed to higher P acquisition ability at appropriate P supply. A higher shoot growth and a subsequently higher sink capacity of more kernels might have contributed to that effect (Egle et al. 1999). In maize, several studies have identified distinguishable P uptake amongst genotypes (Da Silva and Gabelman 1992; Zhu and Lynch 2004; Chen et al. 2009a, b; de Sousa et al. 2012), without providing large amounts of potential target genes for exploiting crop biodiversity (Calderón-Vázquez et al. 2008).

Potential Targets for Genetic Improvement

The design of phenotypic screens for dissecting P acquisition or P utilisation differences in crops requires an understanding of the underlying molecular mechanism of low-P tolerance. This section highlights mechanisms which are known in crops and that may be potential targets for a breeding approach, integrating physiological, molecular and genetic strategies. Genes will be listed that come from model organisms and were subsequently investigated in crop species. A potential candidate target would be characterised by being a key factor in the molecular mechanism of the P starvation response, adaptation and genetic diversity responsible for low P tolerance, keeping in mind that there is a need for different strategies in low-input and high-input systems, focusing more on PAE or PUE respectively. Teng et al. (2013) investigated the expression profiling of known P starvation-induced genes in wheat under different levels of P fertiliser and soil Olsen P, proving that the turning point for the genetic response was the critical P level. This observation leads to a more general model (Fig. 4.3) which raises the question: which strategy would be most suitable to shift the onset of gene expression for the P starvation signalling response into lower P levels and therefore decreasing the crop demand for equal yield performance? There were three main scientific approaches to this problem.

The first approach comprises the majority of studies which are based on comparisons of low P-tolerant genotypes with more susceptible genotypes or cultivars (Pariasca-Tanaka et al. 2009; Li et al. 2008a, b; Hammond et al. 2009; Zhang

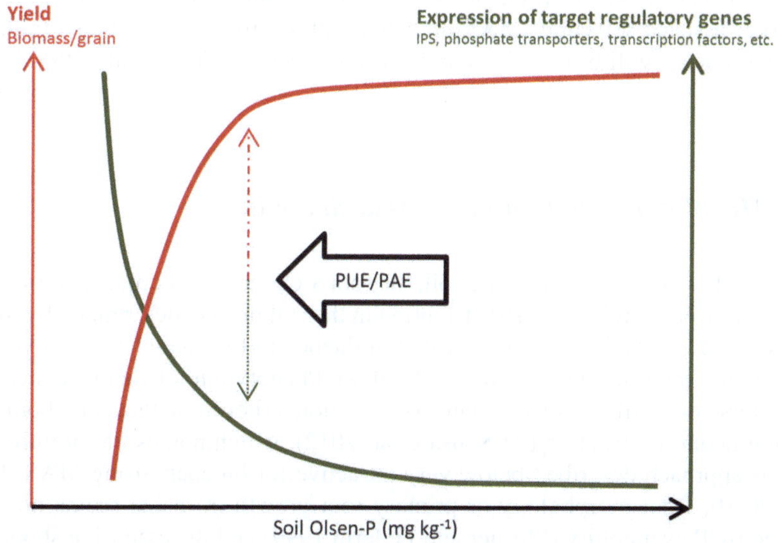

Fig. 4.3 Model for of the optimal yield (*red up arrow*) coinciding with the induction of P-starvation marker genes (*green down arrow*) in field-derived wheat roots shifting towards lower soil Olsen-P levels with improved P efficiency traits (Teng et al. 2013)

et al. 2009; Li et al. 2010; Yao et al. 2011) exposed to a short-term P starvation period; particularly transcriptome, proteome or metabolite profiling studies (Wang et al. 2002; Wasaki et al. 2003; Hammond et al. 2004; Calderón-Vázquez et al. 2008; Huang et al. 2008; Oono et al. 2011; Liang et al. 2010; Oono et al. 2013). Even if these profiling studies provided a useful tool to study the response mechanisms with regards to adaptation to nutrient stresses (Hammond et al. 2004; Nilsson et al. 2010), the majority used hydroponically grown plant material exposed to short-term P starvation. However, Hammond et al. (2011) used the transcriptional profiling technology to identify a predictive diagnostic gene set for detecting the physiological P_i status of potato under field conditions and at a range of P fertiliser application rates. This group of genes were determined by investigating the transcriptional P starvation responses in potato leaves grown hydroponically and were validated for exposure to various nutritional and abiotic stresses using the *Arabidopsis* orthologs (Hammond et al. 2011). This aspect is more focused on increasing the precision of fertiliser application but may also be a potential tool for genotypic screening PUE and PUE under agronomic conditions in the future.

The second approach deals with the over-expression of target genes resulting in partly contradictory observations due to the expression level of their role in the P sensing and regulating network (Rae et al. 2004; Zhou et al. 2008; Ren et al. 2012a, b; Tian et al. 2012; Guo et al. 2013; Wang et al. 2013).

The third approach uses quantitative trail loci (QTL) analysis to dissect the genetic basis of P efficiency and identify superior alleles or loci in different germplasm (Wissuwa et al. 2005; Zhu et al. 2005a, b; Su et al. 2006; Liang et al. 2010; Yang et al. 2011; Gamuyao et al. 2012) which should lead, if successful, to marker-assisted selection (MAS) in breeding for improved nutritional traits. All three approaches will be described and discussed in more detail within the sections below.

Root Morphology and Organic Acid Secretion

Due to the low mobility of P in the soil, the root architecture of crops in agricultural systems is strongly related to P distribution in the soil profile, determined by tillage, fertiliser and cultivation practices, which influence in turn the chemical dynamics of soil P and the rhizosphere (Niu et al. 2012). Phenotyping of root (architectural) parameters are difficult to evaluate as selection criteria as they are both time consuming and destructive (de Sousa et al. 2012), which makes the in vitro QTL analysis approach described below very attractive for breeders using MAS (Liang et al. 2010). Root morphology or primary root growth in maize seems not to be affected by P availability (Mollier and Pellerin 1999) and the extensive shoot-born root system and different root types of cereals (Hochholdinger and Zimmermann 2008) emphasises regulatory differences in crops compared to model plants such as *Arabidopsis*.

Nonetheless, genotypic differences in low P tolerance or high yield at low P availability were often associated with root growth properties or P uptake capability in crops (Hammond et al. 2009; Pariasca-Tanaka et al. 2009; Yao et al. 2011; Li et al. 2008a, b; Zhu and Lynch 2004; Zhu et al. 2005a, b; Gahoonia et al. 1996; 1997) showing that there is a large exploitable genetic variation in the root acquisition trait. There exists considerable genotypic variation in root hairs in barley and wheat cultivars showing that root hair length was strongly correlated to the rhizosphere P depletion ability (Gahoonia et al. 1997). A root hairless mutant in maize conferred significant grain yield loss (Hochholdinger et al. 2008). Similar observations were made when comparing different maize lines and their root hair length, plasticity and subsequent performance under low P (Zhu et al. 2010). A positive and negative regulatory role for transcription factors like PHR1 WRKY75 and BHLH32 has been suggested in *Arabidopsis* (Chen et al. 2007; Devaiah et al. 2007a, b; Bustos et al. 2010). Unfortunately, the underlying genetic mechanisms of germplasm variation for the root hair trait are not yet determined. MAS may facilitate root trait selection for breeding more P efficient cultivars, exemplified by studies showing that root morphology QTLs are likened to P acquisition efficiency in maize (Zhu et al. 2005a, b), wheat (Ren et al. 2012a, b) or soybean (Liang et al. 2010). Gene modification is a means of enhancing low P tolerance (Wang et al. 2013). For example, a ß-expansin gene in soybean, Gm-EXPB2, enhanced P uptake when it was over-expressed (Guo et al. 2011). Expansins are involved in cell wall extension (Zhao et al. 2012), including root hair formation (Yu et al. 2011) and are among up-regulated genes during P starvation (Calderón-Vázquez et al. 2008), suggesting a remodelling of the cell wall structure and integrity. Root morphology related genes, *Rtcs (rootless concerning crown and lateral seminal roots;* Hochholdinger and Zimmermann 2008), *Bk2 (brittle stalk-2*; Brady et al. 2007*)* and *Rth3 (root hairless 3*; Hochholdinger et al. 2008), were determined as being related to differential P responses and PAE capability in the seedling stage in two contrasting maize lines (de Sousa et al. 2012). However, even if observable only under low P conditions, root traits exhibited high heritability and a low coefficient of variation making them exploitable for breeding (de Sousa et al. 2012). Genes which were investigated belonged to a family of glycosylphophatidylinositol (GPI)-anchored proteins involved in root cell expansion, cell wall biosynthesis and root hair formation (de Sousa et al. 2012). Comparing the proteome of a low P tolerant and a sensitive oilseed rape genotype revealed that proteins related to lateral root formation such as auxin-responsive family proteins and sucrose-phosphate synthase-like proteins were up-regulated in roots and leaves (Yao et al. 2011) suggesting sources of tolerance lying in the P starvation signalling mechanism. Furthermore, Li et al. (2008a, b) revealed an increase of low-P tolerance when the phosphoprotein, phosphatase 2A isoform 4, was increased in abundance in the more tolerant genotype. This protein family is involved in auxin transport and reduced activity altered lateral root growth in *Arabidopsis* (Rashotte et al. 2001). Significant increases of CDC48 and other regulators of cell division and cell cycle, including Ran GTPase, MCM6 and importins, were assumed to be important factors mediating a better root development and accelerated cell proliferation in the meristem under P starvation (Li et al. 2008a, b).

Root system modification and organic acid secretion require a carbon supply which might be to the detriment of yield or growth but being nevertheless beneficial under conditions of low phosphorus availability (Zhu and Lynch 2004; Lynch and Ho 2005; Johnson et al. 1996; Yao et al. 2011). PUE in two different P efficient maize lines growing in nutrient solution culture was related to proteins that decreased citrate degradation, increased citrate synthesis and malate dehydrogenase activity in the roots (Li et al. 2008a). Furthermore, proteins related to carbon and energy metabolism were expressed to a higher extent in a low P-tolerant *Brassica napus* genotype compared to a low P-sensitive one (Yao et al. 2011). It has been hypothesised that transgenic plants that secret microbial phytases into the rhizosphere have potential for improved acquisition of organic P sources, but when grown in soil their growing performance matched the control plants (George et al. 2005; Richardson et al. 2000). Studies on bacterial citrate synthase genes in tobacco came to similar contradictory results (López-Bucio et al. 2000; Delhaize et al. 2001) and investigations of genotypic variation in root exuded wheat phosphatases activity could not relate their activity to the P content when plants were grown in soil (George et al. 2008). Nonetheless, Zhang et al. (2009) suggested that improved acquisition and therefore higher P uptake efficiency of two *Brassica napus* genotypes grow in soil was related to the ability to lower the pH or higher acid phosphatase activity in the rhizosphere. However, higher APase activity was observed in roots and particularly in shoot tissue in P scarce conditions in rice but without exhibiting genotypic differences (Yao et al. 2011). Over-expression of a wheat malate transporter Ta-ALMT1 in barley enhanced P uptake on acid soils in the short-term but not when the soil was limed (Delhaize et al. 2009). Over-expression of a root-associated purple phosphatase gene in rice, OsPAP10a, could promote better growth and a higher tiller number compared to the wild type under P sufficient conditions (Tian et al. 2012). In conclusion, results from *in vivo* studies appear contradictory when tested under soil conditions. However, genotypic variation of low P tolerance is related to root morphology and secretory traits, which may be exploitable.

P_i *Acquisition via Phosphate Transporters*

A strong induction of Pht1 transporters was reported in the majority of transcript profiling studies where plants were exposed to a short-term P starvation period and grown mainly in nutrient solution (Wang et al. 2002; Wasaki et al. 2003; Calderón-Vázquez et al. 2008; Huang et al. 2008; 2011) and in the field (Teng et al. 2013), which makes them obvious targets for genetic improvement. However, enhanced induction of TaPht1 transcripts might be either a short-term adaptation to local P depletion or unevenly distributed patterns of P availability. High persistent induction during severe long-term scarcity as an adaptation mechanism seems questionable and should be investigated on field-grown crops as performed by Teng et al. (2013). Enhancing P acquisition by over-expressing phosphate transporter genes has been reported in tobacco cell cultures (Mitsukawa et al. 1997) but could not be confirmed at the plant level (Rae et al. 2004). Promoter fusion to fluorescent proteins showed that high expression levels of root expressed phosphate transporter occur in trichoblast cells (Daram et al. 1998; Mudge et al. 2002; Schünmann et al. 2004) and remarkably in root tips and root hairs under low P conditions (Sánchez-Calderón et al. 2006). The increasing rate of P_i uptake during P deprivation via enhanced phosphate transporter expression presumably occurs as a result of increased V_{max}, rather than increased affinity (K_m), implying increasing high-affinity P transporter synthesis with similar kinetic properties (Raghothama 2005). As variability in P depletion profiles in the rhizosphere of wheat genotypes suggest genetic variability in root hair formation (Gahoonia et al. 1996; 1997), varietal expression differences with respect to preferentially root expressed Pht1 are very likely. Until now, there has been no evidence that genotypic variation exists that could be exploited in breeding (Rose and Wissuwa 2012): this should clearly be area for future research.

P Partitioning and Re-translocation Within the Crop

Preferential re-translocation of P to the roots was characteristic for more P-efficient rice genotypes (Wissuwa et al. 2005) or better remobilisation of P from shoot to root in oilseed rape (Hammond et al. 2009; Akhtar et al. 2008). Two rice genotypes, distinguishable in their ability to tolerate a P-limited environment, showed that genes like PEPC (phosphoenolpyruvate carboxylase), which are involved in the modified glycolysis bypassing ATP-requiring reactions, were up-regulated in the more tolerant but down-regulated in the more susceptible genotype (Li et al. 2010). Interestingly, over-expression of the bHLH transcription factor OsPTF1 enhanced low P tolerance and resulted in increasing expression of Glu-6-P translocator, H^+-ATPase, but also PEP carboxykinase in the shoot (Yi et al. 2005). A proteome study showed that among the higher over-accumulated proteins in the more P starvation tolerant genotype was an important enzyme of the pentose phosphate pathway, 6-phosphogluconate dehydrogenase (Li et al. 2008a).

The authors assumed that requirements of the sugar metabolism could be fulfilled more satisfyingly in the low P-tolerant genotype which was also represented by the larger proportion of sucrose in the total soluble sugar fraction (Li et al. 2008a). More abundant pyruvate phosphate dikinase, pyruvate kinase-like proteins and UDP-glucose pyrophosphorylase, which utilise PP_i (pyrophosphate) to produce ATP or UTP, in the low P-tolerant maize line could be assigned to a higher *in vivo* PUE (Li et al. 2008a, b). Another fundamental issue would be assessing the role or phosphate transporters involved in P loading into the grain (Rose and Wissuwa 2012).

Genes Involved in P Starvation Signalling Cascades

As discussed above, the transcription factor PHR1 seems to be a key regulator for downstream P-responsive genes through binding to a PHR1 specific binding sequence (P1BS) *cis*-element in model plants (Bustos et al. 2010; Rubio et al. 2001). In *Brassica napus*, the homologue BnPHR1 was predominantly expressed in the roots exposed to P limitation and over-expression enhanced remarkably the expression of the high-affinity transporter BnPT2 (Ren et al. 2012a, b). In wheat, three PHR1 homologues genes have been identified (Wang et al. 2013), which regulate genes such as TaPt2;1 (Tittarelli et al 2007; Guo et al. 2013) or TaIPS1 (Oono et al. 2013) that have been reported to contain the P1BS element. TaPht1-A1 transcriptionally activated the expression of the phosphate transporter TaPht1;2 in yeast cells (Wang et al. 2013). The promoter of the high-affinity transporter TaPht1;2 was more abundant in a P-efficient genotype than in an inefficient genotype (Miao et al. 2009). Furthermore, TaPHR1-A1 over-expression resulted in an up-regulation of P starvation response genes, stimulated lateral root branching, enhanced P uptake and P translocation and increased grain yield but not P distribution from shoot to the grains in pot and field trials under P deficient conditions (Wang et al. 2013). Pht1 transporter expression of TaPht1;2 in the roots and TaPht1;6 expression in the shoots increased under high and low P conditions, whereas other usually P starvation induced genes such as TaIPS1.2, TaPHO or TaSPX3 did not change their level of expression (Wang et al. 2013). These results indicated that TaPHR1 is an upstream regulator for Pht1 transporter but suggested other transcriptional factors being relevant for the induction of other P starvation induced genes as it is the case for OsPHR2 (Zhou et al. 2008). Oono et al. (2013) recently published a transcriptome study using *de novo* transcript assembly analysis, in order to investigate wheat seedlings, cv. Chinese spring, exposed to 10 days of P starvation. Genes of the phosphorylation category including protein kinases were among the up-regulated transcripts (Oono et al. 2013). Furthermore, genes belonging to oxidation-reduction processes, metabolic processes, carbohydrate metabolism, transcription process, lipid metabolism and transmembrane transport were induced, as well as AtWRKY6 and AtPHO1 homologues (Oono et al. 2013).

Rice-orthologous transcripts of PHR1, PHO2 and SIZ1 were detected but not all were highly responsive to P starvation (Oono et al. 2011, 2013). Significant was the induction of TaIPS1 homologous (Oono et al. 2013), suggesting that the IPS-mediated signalling cascade may also be functional as previously observed in model species including rice (Oono et al. 2011). This aspect is relevant when taking in account that genetic variation in PUE of barley exhibited a correlated expression of the low-affinity phosphate transporters, HvPht1;3 and HvPht1;6, with HvIPS1 expression (Huang et al. 2011). Higher PUE was also a consequence of higher root-shoot ratios under P limitation indicating an increase in carbohydrate partitioning (Huang et al. 2011). In model plants, IPS genes have been shown to be a miRNA399 antagonist and involved in the miR399-PHO2 regulatory loop

(Franco-Zorrilla et al. 2007; Doerner 2008). In wheat, TaIPS1 transcript levels were strongly repressed in roots and TaIPS2 transcript levels in shoots of P-deficient wheat by N deficiency (Li et al. 2008a, b) providing evidence of an influence on the signalling pathways of P homeostasis by the nitrogen nutritional status. Furthermore, nine wheat miRNAs were identified in addition to miRNA399 as responsive to P starvation in a variety-dependent manner (Zhao et al. 2013). TamiRNAs putatively target diverse gene families, which are down-regulated during P deficiency stress including transcriptional regulation, signal transduction, phytohormone and defence responses among several others (Zhao et al. 2013). Transgenic tomato lines over-expressing miRNA399 from *Arabidopsis* enhanced the secretion of acid phosphatases and protons in roots (Gao et al. 2010). A further example is the over-expression of another P starvation-induced transcription factor, OsPTF1 in rice, which increased tiller number, shoot biomass, panicle weight and P content under low Pi conditions (Yi et al. 2005). Genes which are regulated by OsPTF1 contain E-box and G-box elements but do not include high affinity transporters or acid phosphatases (Yi et al. 2005). In addition, total root length and root surface area may be increased resulting in higher P uptake rates (Yi et al. 2005). Both studies provide evidence of a promising method enhancing P uptake in crops although enhancing low P tolerance via this pathway needs more understanding of other involved proteins and factors.

Quantitative Trait Loci Identification

Until now, the indirect approach of QTL identification for P deficiency tolerance has been mostly exploited in rice and *Brassica* species, and approaches have focused on P acquisition parameters among different aspects and targets of P efficiency. To date, *Pup1* (*Phosphorus uptake 1*) is the only major QTL for P deficiency tolerance in rice coming from a landrace, which could actually be used by rice breeders for marker-assisted introgression into elite material (Wissuwa and Ae 2001, Wissuwa et al. 2002; Chin et al. 2010).

There are more known QTLs, often at very early growth stages (Su et al. 2006, 2009; Yang et al. 2010, 2011), related to yield components under low P conditions (Su et al. 2009; Chin et al. 2010; Ding et al. 2012; Gamuyao et al. 2012; Shi et al. 2013), to seed P concentrations (Ding et al. 2010; Zhao et al. 2008), to P uptake capability and morphological adaptation e.g. tiller number (Su et al. 2006; Wissuwa et al. 1998, 2002; Chin et al. 2010; Gamuyao et al. 2012) or root morphology (Zhu et al. 2005a, b, 2006; Liang et al. 2010; Yang et al 2011).

In wheat, a large number of QTLs on all chromosomes have been detected in a double haploid (DH) population derived from a P deficiency tolerant and low P-sensitive variety implying a polygenetic control of low P sensitivity (Su et al. 2006, 2009). There were three major loci associated with higher tiller number, shoot dry weight and shoot phosphate uptake under low P conditions suggesting that these alleles may be used for MAS (Su et al. 2006). Two of these loci are liked to genes associated with vernalization requirements including flowering time and shoot morphology (Su et al. 2006). Hammond et al. (2009) reported that QTLs associated with PUE measures were related to root traits in *B. oleracea* species which puts them back in the focus as potential targets for crop improvement. QTLs for root hair length and lateral root growth have been identified in maize under P deficiency, a trait which was observed in previous studies as associated with improved low P tolerance and may be controlled by many minor genes or loci with epistatic effects (Zhu et al. 2005a, b). An additional study on soybean possibly related root traits and P efficiency traits to three QTL clusters (Liang et al. 2010). As previously mentioned, a major storage form of phosphorus in grain crops is the non-desirable phytate, which diminishes nutritional quality and causes environmental problems. Despite being of strong interest for human health, QTL analysis focusing on seed traits or seed quality has not been performed extensively. Ding et al. (2012) observed an overlap of a P efficiency QTL with a QTL for seed P concentrations and in a RIL population (durum wheat x wild emmer wheat) and eight QTLs found for grain P content co-localised with QTLs for grain protein content (Peleg et al. 2009) which is correlated with both phytic acid and total P (Raboy et al. 2009). Zhu et al. (2005a, b) found seed phosphorus reserve-related QTLs in maize, which was only partly related to seed size. In rice grown with unlimited nutrients, Stangoulis et al. (2007) found a common QTL for phytate and P concentrations and interestingly, the phytate QTLs were distinct from those for micronutrients such as Fe, Zn or Mn.

So far, the underlying genes or their functional mechanisms, which are targeting a higher low P tolerance, remain still quite elusive (Pariasca-Tanaka et al. 2009; Chin et al. 2010; Shi et al. 2013). The region of a major QTL for low P tolerance in a rice cultivar, *Pup1 (Phosphate uptake 1)*, has been mapped and has been predicted to contains 60 genes. However, their mechanism for low P tolerance as not yet been identified (Ismail et al. 2007; Wissuwa et al. 2002). In one case, the receptor-like cytoplasmic Ser/Thr protein kinase PSTOL1 gene, present within the *Pup1* QTL region, seems to be involved in early crown root development and enhances yield and gene expression of such related to root growth when it was over-expressed (Gamuyao et al. 2012). Another approach for delivering putative candidate targets used a comparative mapping technique (*in silico* mapping) between the model plant *Arabidopsis* and the crop *Brassica napus* (Yang et al. 2010; Ding et al. 2012; Shi et al. 2013) which revealed yield associated QTLs being linked to genes which are involved in P homeostasis including P transport, transcriptional control, phospholipid and carbohydrate metabolism (Shi et al. 2013). Among these genes were glucose-6-phosphate transporters, BnSIZ1-A2, BnPHO1 or BnSQD2 and phosphate transporters (Shi et al. 2013) as well as BnIPS2 (Ding et al. 2012), which therefore confirm their potential roles as targets for crop improvement. Yang et al. (2010) observed a linkage of two functional gene-based markers (GBMs) potentially usable for MAS, BnIPS2 and BnGTP1, as well as orthologous genes for root development, auxin transport with two root morphology related QTLs.

In summary, recent approaches have proven to be a useful tool to dissect the genetic basis of P efficiency-related traits, and have detected a large number of

QTLs in several crop species. Nonetheless, its practical and effective application in MAS of breeding programmes has been rather limited up to now and would require higher precision, reliability and a proof of consistency across these known QTLs. Additionally, these findings emphasise the previously mentioned importance of post-transcriptional modification in the sensing and signalling network of P homeostasis, the low P tolerance, and the association of low P tolerance with root morphology.

Conclusions

Phosphorus is an essential macronutrient with crucial functions in plant macromolecular structure, energy metabolism and signal transduction, which can be a major constraint for high yield when it becomes limiting in crop production. There are economic, political and environmental reasons why P efficiency (PAE, PUE) and P fertiliser use in crops should be investigated for future crop improvement. Therefore, scientific interest has increased with the aim of finding the underlying molecular mechanisms for adaptation to low P accessibility and of identifying targets to archive high-yielding, low P-tolerant crops. Agronomic strategies for raising the amount of available fertiliser are constantly under assessment but the polygenetic basis of P deficiency tolerance is not yet understood. Field selection and screening for PAE/PUE traits, especially in relation to roots, are difficult to realise due to the complexity of soil P and agronomic practice, which may have significant impact on P availability and root properties. Another critical point is the shift from P acquisition to translocation processes when crops become generative and distribution in grain and seeds becomes predominant. Several approaches for investigating crops, such as the comparison of individual genotypes exposed to a short-term P starvation period, the over-expression of target genes and QTL analysis have resulted in contradictory observations. Nevertheless many potential target genes, which have been identified previously in model organisms using forward and reverse genetic approaches, are also be found in crops and are potentially exploitable. Phosphate transporters, several transcription factors, genes coding for proteins of the TCA cycle metabolism, phospholipid degradation, transfer and post-translational modifications are among the candidates who have been detected even if their role in the genetically diverse low P tolerance or PAE/PUE context seems complex and still rather elusive.

Acknowledgements Work at Rothamsted Research is supported via the 20:20 Wheat® Programme by the UK Biotechnology and Biological Sciences Research Council. The contribution of A. Gruen was supported by which received funding from the European Union Seventh Framework Programme (FP7/2007-2013) under grant agreement n° 264296.

References

Ai P, Sun S, Zhao J, Fan X, Xin W, Guo Q, Yu L, Shen Q, Wu P, Miller AJ, Xu A (2009) Two rice phosphate transporters, OsPht1;2 and OsPht1;6, have different functions and kinetic properties in uptake and translocation. Plant J 57:798–809

Akhtar MS, Oki Y, Adachi T (2008) Genetic variability in phosphorus acquisition and utilization efficiency from sparingly soluble P-sources by brassica cultivars under P-stress environment. J Agron Crop Sci 194:380–392

Alexova R, Millar AH (2013) Proteomics of phosphate use and deprivation in plants. Proteomics 13:609–623

Andersson MX, Stridh MH, Larsson KE, Liljenberg C, Sandelius AS (2003) Phosphate-deficient oat replaces a major portion of the plasma membrane phospholipids with the galactolipid digalactosyldiacylglycerol. FEBS Lett 537:128–132

Aung K, Lin SI, Wu CC, Huang YT, Su CL, Chiou TJ (2006) pho2, a phosphate overaccumulator, is caused by a nonsense mutation in a microRNA399 target gene. Plant Physiol 141:1000–1011

Bahl GS, Singh NT (1986) Phosphorus diffusion in soils in relation to some edaphic factors and its influence on P uptake by maize and wheat. J Agric Sci 107:335–341

Barber SA (1984) Phosphorus. In: Soil nutrient bioavailability – a mechanistic approach. Wiley, New York, pp 201–228

Bari R, Pant BD, Stitt M, Scheible WR (2006) PHO2, microRNA399, and PHR1 define a phosphate-signaling pathway in plants. Plant Physiol 141:988–999

Bates TR, Lynch JP (1996) Stimulation of root hair elongation in *Arabidopsis thaliana* by low phosphorus availability. Plant Cell Environ 19:529–538

Bates TR, Lynch JP (2001) Root hairs confer a competitive advantage under low phosphorus availability. Plant Soil 236:243–250

Batten GD, Khan MA (1987) Uptake and utilization of phosphorus and nitrogen by bread wheats grown under natural rainfall. Aust J Exp Agr 27:405–410

Batten GD (1992) A review of phosphorus efficiency in wheat. Plant Soil 146:163–168

Bayle V, Arrighi JF, Creff A, Nespoulous C, Vialaret J, Rossignol M, Gonzalez E, Paz-Ares J, Nussaume L (2011) *Arabidopsis thaliana* high-affinity phosphate transporters exhibit multiple levels of posttranslational regulation. Plant Cell 23:1523–1535

Bieleski RL (1968) Effect of phosphorus deficiency on levels of phosphorus compounds in Spirodela. Plant Physiol 43:1309–1316

Bieleski RL (1973) Phosphate pools, phosphate transport, and phosphate availability. Annu Rev Plant Physiol 24:225–252

Bollons HM, Barraclough PB (1997) Inorganic orthophosphate for diagnosing the phosphorus status of wheat plants. J Plant Nutr 20:641–655

Bollons HM, Barraclough PB (1999) Assessing the phosphorus status of winter wheat crops: inorganic orthophosphate in whole shoots. J Agric Sci 133:285–295

Brady SM, Song S, Dhugga KS, Rafalski JA, Benfey PN (2007) Combining expression and comparative evolutionary analysis. The COBRA gene family. Plant Physiol 143:172–187

Bucher M, Rausch C, Daram P (2001) Molecular and biochemical mechanisms of phosphate uptake into plants. J Plant Nutr Soil Sci 164:209–221

Bustos R, Castrillo G, Linhares F, Puga MI, Rubio V, Pérez- Pérez J, Solano R, Leyva A, Paz-Ares J (2010) Central regulatory system largely controls transcriptional activation and repression responses to phosphate starvation in *Arabidopsis*. PLoS Genet 6:e1001102

Byrne SL, Foito A, Hedley PE, Morris JA, Stewart D, Barth S (2011) Early response mechanisms of perennial ryegrass (*Lolium perenne*) to phosphorus deficiency. Ann Bot 107:243–254

Calderón-Vázquez C, Ilbarra-Laclette E, Caballero-Perez J, Herrera-Estrella L (2008) Transcript profiling of *Zea mays* roots reveals responses to phosphate deficiency and the plant-species-specific level. J Exp Bot 59:2479–2497

Calderón-Vázquez C, Sawers RJH, Herrera-Estrella L (2011) Phosphate deprivation in maize: genetics and genomics. Plant Physiol 156:1067–1077

Chen ZH, Nimmo GA, Jenkins GI, Nimmo HG (2007) BHLH32 modulates several biochemical and morphological processes that respond to Pi starvation in *Arabidopsis*. Biochem J 405:191–198

Chen YF, Li LQ, Xu Q, Kong YH, Wang H, Wu WH (2009a) The WRKY6 transcription factor modulates PHOSPHATE1 expression in response to low Pi stress in *Arabidopsis*. Plant Cell 21:3554–3566

Chen JY, Xu L, Cai YL, Xu J (2009b) Identification of QTLs for phosphorus utilization in maize (*Zea mays* L.) across P levels. Euphytica 167:245–252

Chen J, Liu Y, Ni J, Wang Y, Bai Y, Shi J, Gan J, Wu Z, Wu P (2011) Osphf1 regulates the plasma membrane localization of low- and high-affinity inorganic phosphate transporters and determines inorganic phosphate uptake and translocation in rice. Plant Physiol 157:269–278

Cheng L, Bucciarelli B, Liu J, Zinn K, Miller S, Patton-Vogt J, Allan D, Shen J, Vance CP (2011) White lupin cluster root acclimation to phosphorus deficiency and root hair development involve unique glycerophosphodiester phosphodiesterases. Plant Physiol 156:1131–1148

Chevalier F, Rossignol M (2011) Proteomic analysis of *Arabidopsis thaliana* ecotypes with contrasted root architecture in response to phosphate deficiency. J Plant Physiol 168:1185–1890

Chin JH, Lu X, Haefele SM, Gamuyao R, Ismail A, Wissuwa M, Heuer S (2010) Development and application of gene-based markers for the major rice QTL Phosphorus uptake 1. Theor Appl Genet 120:1073–1086

Chiou TJ, Lin S (2011) Signaling network in sensing phosphate availability in plants. Annu Rev Plant Biol 62:185–206

Chiou TJ, Aung K, Lin SI, Wu CC, Chiang SF, Sua CL (2006) Regulation of phosphate homeostasis by microRNA in *Arabidopsis*. Plant Cell 18:412–421

Chitwood DH, Timmermans MCP (2007) Target mimics modulate miRNAs. Nat Genet 39:935–936

Cordell D, Drangert J, White S (2009) The story of phosphorus: global food security and food for thought. Glob Environ Chang 12:292–305

Da Silva AE, Gabelman WH (1992) Screening maize inbred lines for tolerance to low-P stress conditions. Plant Soil 146:181–187

Daram P, Brunner S, Persson BL, Amrhein N, Bucher M (1998) Functional analysis and cell-specific expression of a phosphate transporter from tomato. Planta 206:225–233

Daram P, Brunner S, Rausch C, Steiner C, Amrheim N (1999) Pht2;1 encodes a low-affinity phosphate transporter from *Arabidopsis*. Plant Cell 11:2153–2166

Davies TGE, Ying J, Li ZS, Li J, Gordon-Weeks R (2002) Expression of analysis of putative high-affinity transporters in Chinese winter wheats. Plant Cell Environ 25:1325–1339

De Sousa SM, Clark RT, Mendes FF, de Oliveira AC, de Vasconcelos MJV, Parentoni SN, Kochian LV, Guimarães CT, Magalhães JV (2012) A role for root morphology and related candidate genes in P acquisition efficiency in maize. Funct Plant Biol 39:925–935

Delhaize E, Randall PJ (1995) Characterization of a phosphate-accumulator mutant of *Arabidopsis*-thaliana. Plant Physiol 107:207–213

Delhaize E, Hebb DM, Ryan PR (2001) Expression of a *Pseudomonas aeruginosa* citrate synthase gene in tobacco is not associated with either enhanced citrate accumulation or efflux. Plant Physiol 125:2059–2067

Delhaize E, Taylor P, Hocking PJ, Simpson RJ, Ryan PR, Richardson AE (2009) Transgenic barley (*Hordeum vulgare* L.) expressing the wheat aluminium resistance gene (TaALMT1) shows enhanced phosphorus nutrition and grain production when grown on an acid soil. Plant Biotech J 7:391–400

Devaiah BN, Karthikeyan AS, Raghothama KG (2007a) WRKY75 transcription factor is a modulator of phosphate acquisition and root development in *Arabidopsis*. Plant Physiol 143:1789–1801

Devaiah BN, Nagarajan VK, Raghothama KG (2007b) Phosphate homeostasis and root development in *Arabidopsis* are synchronized by the zinc finger transcription factor ZAT6. Plant Physiol 145:147–159

Dinkelacker B, Römheld V, Marschner H (1989) Citric acid excretion and precipitation of calcium citrate in the rhizosphere of white lupine. Plant Cell Environ 12:285–292

Ding G, Yang M, Hu Y, Liao Y, Shi L, Xu F, Meng J (2010) Quantitative trait loci affecting seed mineral concentrations in *Brassica napus* grown with contrasting phosphorus supplies. Ann Bot 105:1221–1234

Ding G, Zhao Z, Liao Y, Hu Y, Shi L, Long Y, Xu F (2012) Quantitative trait loci for seed yield and yield-related traits, and their responses to reduced phosphorus supply in *Brassica napus*. Ann Bot 109:747–759

Doerner P (2008) Phosphate starvation signaling: a threesome controls systemic P_i homeostasis. Curr Opin Plant Biol 11:536–540

Duan K, Yi K, Dang L, Huang H, Wu W, Wu P (2008) Characterization of a sub-family of *Arabidopsis* genes with the SPX domain reveals their diverse functions in plant tolerance to phosphorus starvation. Plant J 54:965–975

Duff SMG, Moorhead GBG, Lefebre DD, Plaxton WC (1989) Phosphate starvation inducible 'bypasses' of adenylate and phosphate dependent glycolytic enzymes in *Brassica nigra* suspension cells. Plant Physiol 90:1275–1278

Egle K, Manske G, Römer W, Vlek PLG (1999) Improved phosphorus efficiency of three new wheat genotypes from CIMMYT in comparison with an older Mexican variety. J Plant Nutr Soil Sci 162:353–358

Essigmann B, Güler S, Narang RA, Linke D, Benning C (1998) Phosphate availability affects the thylakoid lipid composition and the expression of SQD1, a gene required for sulfolipid biosynthesis in Arabidopsis thaliana. Proc Natl Acad Sci U S A 95:1950–1955

Finck A (1991) Düngermenge nach Pflanzenanalyse. In: Baumeister W (ed) Düngung. Ulmer, Stuttgart, pp 96–101

Fitter AH (2006) What is the link between carbon and phosphorus fluxes in arbuscular mycorrhizas? A null hypothesis for symbiotic function. New Phytol 172:3–6

Food and Agriculture Organization of The United Nations: current world fertilizer trends and outlook to 2015. Rome 2011

Franco-Zorrilla JM, Valli A, Todesco M, Mateos I, Puga IM, Rubio-Somoza I, Leyva A, Weigel D, Garcia JA, Paz-Ares J (2007) Target mimicry provides a new mechanism for regulation of microRNA activity. Nat Genet 39:1033–1037

Fujii H, Chiou TJ, Lin SI, Aung K, Zhu JK (2005) A miRNA involved in phosphate-starvation response in *Arabidopsis*. Curr Biol 15:2038–2043

Furihata T, Suzuki M, Sakurai H (1992) Kinetic characterization of two phosphate uptake systems with different affinities in suspension-cultured *Catharanthus roseus* protoplasts. Plant Cell Physiol 33:1151–1157

Gahoonia TS, Nielsen NE (1998) Direct evidence on participation of root hairs in phosphorus (^{32}P) uptake from soil. Plant Soil 198:147–152

Gahoonia TS, Care D, Nielsen NE (1996) Variation in acquisition of soil phosphorus among wheat and barley genotypes. Plant Soil 178:223–230

Gahoonia TS, Care D, Nielsen NE (1997) Root hairs and phosphorus acquisition of wheat and barley cultivars. Plant Soil 1991:181–188

Gahoonia TS, Nielsen NE, Lyshede OB (1999) Phosphorus (P) acquisition of cereal cultivars in the field at three levels of P fertilization. Plant Soil 211:269–281

Gahoonia TS, Nielsen NE, Joshi PA, Jahoor A (2001) A root hairless barley mutant for elucidation genetic of root hairs and phosphorus uptake. Plant Soil 235:211–219

Gamuyao R, Chin JH, Pariasca-Tanaka J, Pesaresi P, Catausan S, Dalid C, Slamet-Loedin I, Tecson-Mendoza EM, Wissuwa M, Heuer S (2012) The protein kinase Pstol1 from traditional rice confers tolerance of phosphorus deficiency. Nature 488:535–539

Gao N, Su Y, Mi J, Shen W, Shi W (2010) Transgenic tomato overexpressing ath-miR399d has enhanced phosphorus accumulation through increased acid phosphatase and proton secretion as well as phosphate transporters. Plant Soil 334:123–136

Gaude N, Tippmann H, Flemetakis E, Katinakis P, Udvardi M, Dörmann P (2004) The galactolipid digalactosyldiacylglycerol accumulates in the peribacteroid membrane of nitrogen-fixing nodules of soybean and lotus. J Biol Chem 279:34624–34630

Gaude N, Bortfeld S, Duensing N, Lohse M, Krajinski F (2012) Arbuscule-containing and non-colonized cortical cells of mycorrhizal roots undergo extensive and specific reprogramming during arbuscular mycorrhizal development. Plant J 69:510–528

Gaxiola RA, Edwards M, Elser JJ (2001) A transgenic approach to enhance phosphorus use efficiency in crops as part of a comprehensive strategy for sustainable agriculture. Chemosphere 84:840–845

George TS, Richardson AE (2008) Potential and limitations to improving crops for enhanced phosphorus utilization. In: White PJ, Hammond JP (eds) The ecophysiology of plant-phosphorus interactions, vol 7. Springer, Dordrecht, pp 247–270

George TS, Richardson AE, Hadobas PA, Simpson RJ (2004) Characterization of transgenic *Trifolium subterraneum* L. which expresses *phyA* and releases extracellular phytase: growth and P nutrition in laboratory media and soil. Plant Cell Environ 27:1351–1361

George TS, Simpson RJ, Hadobas PA, Richardson AE (2005) Expression of a fungal phytase gene in *Nicotiana tabacum* improves phosphorus nutrition of plants grown in amended soils. Plant Biotechnol J 3:129–140

George TS, Gregory PJ, Hocking P, Richardson AE (2008) Variation on root-associated phosphatase activities in wheat contributes to the utilization of organic P substrates *in vitro*, but does not explain differences in the P-nutrition of plants when grown in soils. Environ Exp Bot 64:2239–2249

Gerke J, Römer W, Jungk A (1994) The excretion of citric and malic-acid by proteoid roots of lupinus-albus l – effects on soil solution concentrations of phosphate, iron, and aluminum in the proteoid rhizosphere in samples of an oxisol and a luvisol. Z Pflanz Bodenkunde 157:289–294

Glassop D, Smith SE, Smith FW (2005) Cereal phosphate transporters associated with the mycorrhizal pathway of phosphate uptake in roots. Planta 222:688–698

Gomez-Roldan V, Fermas S, Brewer PB, Puech-Pagès V, Dun EA, Pillot JP, Letisse F, Matusova R, Danoun S, Portais JC, Bouwmeester H, Bécard G, Beveridge CA, Rameau C, Rochange SF (2008) Strigolactone inhibition of shoot branching. Nature 455:189–194

González E, Solano R, Rubio V, Leyva A, Paz-Ares J (2005) Phosphate transporter traffic facilitator1 is a plant-specific sec12-related protein that enables the endoplasmic reticulum exit of a high-affinity phosphate transporter in *Arabidopsis*. Plant Cell 17:3500–3512

Gregory PJ, George TS (2011) Feeding nine billion: the challenge to sustainable crop production. J Exp Bot 62:5233–5239

Gregory AL, Hurley BA, Tran HT, Valentine AJ, She YM, Knowles VL, Plaxton WC (2009) In vivo regulatory phosphorylation of the phosphoenolpyruvate carboxylase AtPPC1 in phosphate-starved *Arabidopsis thaliana*. Biochem J 420:57–65

Gu M, Xu K, Chen A, Zhu Y, Tang G, Xu G (2010) Expression analysis suggests potential roles of micro RNAs for phosphate and *arbuscular mycorrhizal* signaling in *Solanum lycopersicum*. Physiol Plant 138:226–237

Gunes A, Inal A, Alpaslan M, Cakmak I (2006) Genotypic variation in phosphorus efficiency between wheat cultivars grown under greenhouse and field conditions. Soil Sci Plant Nutr 52:470–478

Guo B, Jin Y, Wussler C, Blancaflor EB, Moter CM, Versaw WK (2008) Functional analysis of the *Arabidopsis* PHT4 family of intracellular phosphate transporters. New Phytol 177:889–898

Guo W, Zhao J, Li X, Qin L, Yan X, Hong Liao H (2011) A soybean b-expansin gene GmEXPB2 intrinsically involved in root system architecture responses to abiotic stresses. Plant J 66:541–552

Guo C, Zhao X, Liu X, Zhang L, Gu J, Li X, Lu W, Xiao K (2013) Function of wheat phosphate transporter gene TaPHT2;1 in Pi translocation and plant growth regulation under replete and limited Pi supply conditions. Planta 237:1163–1178

Güimil S, Chang HS, Zhu T, Sesma A, Osbourn A, Roux C, Ioannidis V, Oakeley EJ, Docquier M, Descombes P, Briggs SP, Paszkowski U (2005) Comparative transcriptomics of rice reveals an ancient pattern of response to microbial colonization. Proc Natl Acad Sci U S A 102:8066–8070

Gutjahr C, Banba M, Croset V, An K, Miyao A, An G, Hirochika H, Imaizumi-Anraku H, Paszkowski U (2008) Arbuscular mycorrhiza–specific signaling in rice transcends the common symbiosis signaling pathway. Plant Cell 20:2989–3005

Hamburger D, Rezzonico E, Petetot JM, Somerville C, Poirier Y (2002) Identification and characterization of the *Arabidopsisphol* gene involved in phosphate loading to the xylem. Plant Cell 14:889–920

Hammond JP, White PJ (2008) Sucrose transport in the phloem: integrating root responses to phosphorus starvation. J Exp Bot 59:93–109

Hammond JP, Bennet MJ, Bowen HC, Broadley MR, Eastwood DC, May ST, Rahn CR, Swarup R, Woolaway KE, White PJ (2003) Changes in gene expression in *Arabidopsis* shoots during phosphate starvation and the potential for developing smart plants. Plant Physiol 132:578–596

Hammond JP, Broadley MR, White PJ (2004) Genetic responses to phosphorus deficiency. Ann Bot 94:323–332

Hammond JP, Broadley MR, White PJ, King GJ, Bowen HC, Hayden R, Meacham MC, Mead A, Overs T, Spracklen WP, Greenwood DJ (2009) Shoot yield drives phosphorus use efficiency in *Brassica oleracea* and correlates with root architecture traits. J Exp Bot 60:1953–1968

Hammond JP, Broadley MR, Bowen HC, Spracklen WP, Hayden RM, White PJ (2011) Gene expression changes in phosphorus deficient potato (*Solanum tuberosum* L.) leaves and the potential for diagnostic gene expression markers. PLoS One 6:e24606

Harrison MJ, Dewbre GR, Liu J (2002) A phosphate transporter from *Medicago truncatula* involved in the acquisition of phosphate released by arbuscular mycorrhizal fungi. Plant Cell 14:2413–2429

Hayes JE, Zhu YG, Mimura T, Reid RJ (2004) An assessment of the usefulness of solution culture in screening for phosphorus efficiency in wheat. Plant Soil 261:91–97

Hermans C, Hammond JP, White PJ, Verbruggen N (2006) How do plants respond to nutrient shortage by biomass allocation? Trends Plant Sci 11:610–617

Hensel LL, Grbic V, Baumgarten DA, Bleecker AB (1993) Developmental and age-related processes that influence the longevity and senescence of photosynthetic tissues in Arabidopsis. Plant Cell 5:553–564

Hinsinger P (2001) Bioavailability of soil inorganic P in the rhizosphere as affected by root-induced chemical changes: a review. Plant Soil 237:173–195

Hochholdinger F, Zimmermann R (2008) Conserved and diverse mechanisms in root development. Curr Opin Plant Biol 11:70–74

Hochholdinger F, Wen TJ, Zimmermann R, Chimot-Marolle P, Silva ODE, Bruce W, Lamkey KR, Wienand U, Schnable PS (2008) The maize (*Zea mays* L.) roothairless3 gene encodes a putative GPI-anchored, monocot specific, COBRA-like protein that significantly affects grain yield. Plant J 54:888–898

Holford ICR (1997) Soil phosphorus: its measurement, and its uptake by plants. Aust J Soil Res 35:227–239

Hong JJ, Park YS, Bravo A, Bhattarai KK, Daniels DA, Harrison MJ (2012) Diversity of morphology and function in arbuscular mycorrhizal symbioses in *Brachypodium distachyon*. Planta 236:851–865

Hou XL, Wu P, Jiao FC, Jia QJ, Chen HM, Yu J, Song XW, Yi KK (2005) Regulation of the expression of OsIPS1 and OsIPS2 in rice via systemic and local Pi signalling and hormones. Plant Cell Environ 28:353–364

Hsieh LC, Lin SI, Shih ACC, Chen JW, Lin WY, Tseng CY, Li WH, Chi TJ (2009) Uncovering small RNA-mediated responses to phosphate deficiency in *Arabidopsis* by deep sequencing. Plant Physiol 151:2120–2132

Huang CY, Roessner U, Eickmeier I, Gene Y, Callahan DL, Sirley N, Langridge P, Bacie A (2008) Metabolite profiling reveals distinct changes in carbon and nitrogen metabolism in phosphate-deficient barley plants (*Hordeum vulgare* L.). Plant Cell Physiol 49:691–703

Huang CY, Shirley N, Genc Y, Shi B, Langridg P (2011) Phosphate utilization efficiency correlates with expression of low-affinity phosphate transporters and noncoding RNA, IPS1, in barley. Plant Physiol 156:1217–1229

Hürlimann HC, Pinson B, Stadler-Waibel M, Zeeman SC, Freimoser FM (2009) The SPX domain of the yeast low-affinity phosphate transporter Pho90 regulates transport activity. EMBO Rep 10:1003–1008

Ismail AM, Heuer S, Thomson MJ, Wissuwa M (2007) Genetic and genomic approaches to develop rice germplasm for problem soils. Plant Mol Biol 65:547–570

Jain A, Poling MD, Karthikeyan AS, Blakeslee JJ, Peer WA, Titapiwatanakun B, Murphy AS, Raghothama KG (2007) Differential effects of sucrose and auxin on localized phosphate deficiency-induced modulation of different traits of root system architecture in *Arabidopsis*. Plant Physiol 144:232–247

Jakobsen I, Abbott LK, Robson AD (1992) External hyphae of vesicular-arbuscular mycorrhizal fungi associated with *Trifolium subterraneum* L. 1. Spread of hyphae and phosphorus inflow into roots. New Phytol 120:371–380

Jones GPD, Jessop RS, Blair GJ (1992) Alternative methods for the selection of phosphorus efficiency in wheat. Field Crops Res 30:29–40

Johnson JF, Vance CP, Allan DL (1996) Phosphorus deficiency in *Lupinus albus*. Plant Physiol 112:31–41

Jungk A (2001) Root hairs and the acquisition of plant nutrients from soil. J Plant Nutr Soil Sci 164:121–129

Kanda H, Kasukabe Y, Fujita H, Washino T, Tachibana S (1994) Effect of low root temperature on ribonucleic-acid concentrations in figleaf gourd and cucumber roots differing in tolerance to chilling temperature. J Jpn Soc Hortic Sci 63:611–618

Karthikeyan AS, Varadarajan DK, Jain A, Held MA, Carpita NC, Raghothama KG (2007) Phosphate starvation responses are mediated by sugar signaling in *Arabidopsis*. Planta 225:907–918

Kirkby EA, Johnston AEJ (2008) Soil and fertilizer phosphorus in relation to crop nutrition. In: White PJ, Hammond JP (eds) The ecophysiology of plant-phosphorus interactions, vol 7. Springer Science+Business Media B.V., Dordrecht, pp 177–223

Koide RT, Kabir Z (2000) Extraradical hyphae of the mycorrhizal fungus *Glomus intraradices* can hydrolyse organic phosphate. New Phytol 148:511–517

Koyama H, Kawamura A, Kihara T, Hara T, Takita E, Shibata D (2000) Overexpression of mitochondrial citrate synthase in *Arabidopsis thaliana* improved growth on a phosphorus-limited soil. Plant Cell Physiol 41:1030–1037

Lan P, Li W, Schmidt W (2012) Complementary proteome and transcriptome profiling in phosphate-deficient *Arabidopsis* roots reveals multiple levels of gene regulation. Mol Cell Proteomics 11:1156–1166

Lauer MJ, Blevins DG, Sierzputowska-Gracz H (1989) ^{31}P-nuclear magnetic resonance determination of phosphate compartmentation in leaves of reproductive soybeans (*Glycine max* L.) as affected by phosphate nutrition. Plant Physiol 89:1331–1336

Lei M, Liu Y, Zhang B, Zhao Y, Wang X, Zhou Y, Raghothama KG, Liu D (2011) Genetic and genomic evidence that sucrose is a global regulator of plant responses to phosphate starvation in *Arabidopsis*. Plant Physiol 156:1116–1130

Li K, Xu C, Li Z, Zhang K, Yang A, Zhang J (2008a) Comparative proteome analyses of phosphorus responses in maize (*Zea mays* L.) roots of wild-type and a low-P-tolerant mutant reveal root characteristics associated with phosphorus efficiency. Plant J 55:927–939

Li Y, Tong Y, Bin L, Zhao H, Zhang X, Li Z (2008b) Expression of TaIPS genes in wheat seedlings with nitrogen and phosphorous starvation. Acta Bot Boreal Occident Sin 27:1303–1307

Li L, Liu C, Lian X (2010) Gene expression profiles in rice roots under low phosphorus stress. Plant Mol Biol 72:423–432

Liang Q, Cheng X, Mei M, Yan X, Liao H (2010) QTL analysis of root traits as related to phosphorus efficiency in soybean. Ann Bot 106:223–234

Liao M, Hocking PJ, Dong B, Delhaize E, Richardson AE, Ryan PR (2008) Variation in early phosphorus-uptake efficiency among wheat genotypes grown on two contrasting Australian soils. Aust J Agric Res 59:157–166

Lin WY, Lin SI, Chiou TJ (2009) Molecular regulators of phosphate homeostasis in plants. J Exp Bot 60:1427–1438

Liu C, Muchal US, Uthappa M, Kononowicz AK, Raghotama KG (1998) Tomato phosphate transporter genes are differentially regulated in plant tissues by phosphorus. Plant Physiol 116:91–99

Liu F, Wang Z, Ren H, Shen C, Li Y, Ling HQ, Wu C, Lian X, Wu P (2010) OsSPX1 suppresses the function of OsPHR2 in the regulation of expression of OsPT2 and phosphate homeostasis in shoots of rice. Plant J 62:508–517

Liu F, Chang X-J, Ye Y, Xie W-B, Wu P, Lian X-M (2011) Comprehensive sequence and whole-life-cycle expression profile analysis of the phosphate transporter gene family in rice. Mol Plant 4:1105–1122

Liu TY, Huang TK, Tseng CY, Lai YS, Lin SI, Lin WY, Chen JW, Chiou T (2012) PHO2-dependent degradation of PHO1 modulates phosphate homeostasis in Arabidopsis. Plant Cell 24:2168–2183

Lloyd JC, Zakhleniuk OV (2004) Responses of primary and secondary metabolism to sugar accumulation revealed by microarray expression analysis of the *Arabidopsis* mutant, pho3. J Exp Bot 55:1221–1230

López-Bucio J, de la Vega OM, Guevara-García A, Herrera-Estrella L (2000) Enhanced phosphorus uptake in transgenic tobacco plants that overproduce citrate. Nat Biotechnol 18:450–453

Lott JNA, Bojarski M, Kolasa J, Batten GD, Campbell LC (2009) A review of the phosphorus content of dry cereal and legume crops of the word. Int J Agric Resour Govern Ecol 8:351–370

Lundmark M, Kørner CJ, Nielsen TH (2010) Global analysis of microRNA in *Arabidopsis* in response to phosphate starvation as studied by locked nucleic acid-based microarrays. Physiol Plant 140:57–68

Lynch J (1995) Root architecture and plant productivity. Plant Physiol 109:7–13

Lynch J, Brown KM (2001) Topsoil foraging – an architectural adaptation of plants to low phosphorus availability. Plant Soil 237:225–237

Lynch JP, Ho MD (2005) Rhizoeconomics: carbon costs of phosphorus acquisition. Plant Soil 269:45–56

MacDonald GK, Bennet EM, Potter PA, Ramakutty N (2011) Agronomic phosphorus imbalances across the word's croplands. Proc Natl Acad Sci U S A 108:3086–3091

Manske GGB, Ortiz-Monasterio JI, van Ginkel M, González RM, Fischer RA, Rajaram S, Vlek PLG (2001) Importance of P uptake efficiency versus P utilization for wheat yield in acid and calcareous soils in Mexico. Eur J Agron 14:261–274

Marschner P (2012) Phosphorus. In: Marschner P (ed) Marschner's mineral nutrition of higher plant, 3rd edn. Academic Press/Elsevier Ltd, London

McDowell RW (2012) Minimising phosphorus losses from the soil matrix. Curr Opin Biotechnol 23:860–865

Mengel K, Kirkby EA (2001) Principles of plant nutrition, 5th edn. Kluwer Academic Publishers, Dordrecht, p 849

Miao J, Sun J, Liu D, Li B, Zhang A, Li Z (2009) Characterization of the promoter of phosphate transporter TaPHT1.2 differentially expressed in wheat varieties. J Genet Genomics 36:455–466

Michael B, Zink F, Lantzsch HJ (1980) Effect of phosphate application of phytin-P and other phosphate fractions in developing wheat grains. Z Planzenernähr Bodenk 143:369–376

Mimura T (2001) Physiological control of phosphate uptake and phosphate homeostasis in plant cells. Aust J Plant Physiol 28:653–658

Mimura T, Sakano K, Shimmen T (1996) Studies on the distribution, re-translocation and homeostasis of inorganic phosphate in barley leaves. Plant Cell Environ 19:311–320

Misson J, Raghothama KG, Jain A, Jouhet J, Block MA, Bligny R, Ortet P, Creff A, Somerville S, Rolland N, Doumas P, Nacry P, Herrerra-Estrella L, Nussaume L, Thibaud MC (2005) A genome-wide transcriptional analysis using Arabidopsis thaliana Affymetrix gene chips determined plant responses to phosphate deprivation. Proc Natl Acad Sci U S A 102:11934–11939

Mitsukawa N, Okumura S, Shirano Y, Sato S, Kato T, Harashima S, Shibata D (1997) Overexpression of an *Arabidopsis thaliana* high-affinity phosphate transporter gene in tobacco cultured cells enhances cell growth under phosphate-limited conditions. Proc Natl Acad Sci U S A 94:7098–7102

Miura K, Rus A, Sharkhuu A, Yokoi S, Karthikeyan AS, Raghothama KG, Baek D, Koo YD, Jin JB, Bressan RA, Yun DJ, Hasegawa PM (2005) The *Arabidopsis* SUMO E3 ligase SIZ1 controls phosphate deficiency responses. Proc Natl Acad Sci U S A 102:7760–7765

Miura M, Lee J, Gong Q, Ma S, Jin JB, Yoo CY, Miura T, Sato A, Bohnert HJ, Hasegawa PM (2011) SIZ1 regulation of phosphate starvation-induced root architecture remodeling involves the control of auxin accumulation. Plant Physiol 155:1000–1012

Mollier A, Pellerin S (1999) Maize root system growth and development as influenced by phosphorus deficiency. J Exp Bot 50:487–497

Morcuende R, Bari R, Gibon Y, Zhen W, Pant BD, Bläsing O, Usadel B, Czechowski T, Udvardi MK, Stitt M, Scheible W-R (2007) Genome-wide reprogramming of metabolism and regulatory networks of *Arabidopsis* in response to phosphorus. Plant Cell Environ 30:85–112

Muchhal US, Pardo JM, Raghothama KG (1996) Phosphate transporters from the higher plant *Arabidopsis thaliana*. Proc Natl Acad Sci U S A 93:10519–10523

Muchhal US, Raghothama KG (1999) Transcriptional regulation of plant phosphate transporters. Proc Natl Acad Sci U S A 96:5868–5872

Mudge SR, Rae AL, Diatloff E, Smith FW (2002) Expression analysis suggests novel roles for members of the Pht1 family of phosphate transporters in *Arabidopsis*. Plant J 31:341–353

Müller R, Nilsson L, Krintel C, Nielsen TH (2004) Gene expression during recovery from phosphate starvation in roots and shoots of *Arabidopsis thaliana*. Physiol Plant 122:233–243

Müller R, Nilsson L, Nielsen LK, Nielsen TH (2005) Interaction between phosphate starvation signalling and hexokinase-independent sugar sensing in *Arabidopsis* leaves. Physiol Plant 124:81–90

Nagy R, Vasconcelos MJV, Zhao S, McElver J, Bruce W, Amrhein N, Raghothama KG, Bucher M (2006) Differential regulation of five Pht1 phosphate transporters from maize (*Zea mays* L.). Plant Biol 8:186–197

Neumann G, Martinoia E (2002) Cluster roots – an underground adaptation for survival in extreme environments. TRENDS Plant Sci 7:162–167

Neumann G, Römheld V (1999) Root excretion of carboxylic acids and protons in phosphorus-deficient plants. Plant Soil 211:121–130

Nielsen TH, Krapp A, Röper-Schwarz U, Stitt M (1998) The sugar-mediated regulation of genes encoding the small subunit of Rubisco and the regulatory subunit of ADP glucose pyrophosphorylase is modified by phosphate and nitrogen. Plant Cell Environ 12:443–454

Nilsson L, Müller R, Nielsen TH (2007) Increased expression of the MYB-related transcription factor, PHR1, leads to enhanced phosphate uptake in Arabidopsis thaliana. Plant Cell Environ 30:1499–1512

Nilsson L, Müller R, Nielsen TH (2010) Dissecting the plant trascriptome and the regulatory response to tomato deprivation. Physiol Plant 139:129–143

Nilsson L, Lundmark M, Jensen PE, Nielsen TH (2012) The *Arabidopsis* transcription factor PHR1 is essential for adaptation to high light and retaining functional photosynthesis during phosphate starvation. Physiol Plant 144:35–47

Niu YF, Chai RS, Jin GL, Wang H, Tang CX, Zhang YS (2012) Responses of root architecture development to low phosphorus availability: a review. Ann Bot 112:391–408

Oono Y, Kawahara Y, Kanamori H, Mizuno H, Yamagata H, Yamamoto M, Hosokawa S, Ikawa H, Akahane I, Zhu Z, Wu J, Itoh T, Matsumoto T (2011) mRNA-seq reveals a comprehensive transcriptome profile of rice under phosphate stress. Rice 4:50–65

Oono Y, Kobayashi F, Kawahara Y, Yazawa T, Handa H, Itoh T, Matsumoto T (2013) Characterisation of the wheat (*Triticum aestivum* L.) transcriptome by de novo assembly for the discovery of phosphate starvation-responsive genes: gene expression in Pi-stressed wheat. BMC Genomics 14:77

Osborne LD, Rengel Z (2002) Screening cereals for genotypic variation in efficiency of phosphorus uptake and utilisation. Aust J Agric Res 53:295–303

Ozturk L, Eker S, Torun B, Cakmak I (2005) Variation in phosphorus efficiency among 73 bread and durum wheat genotypes grown in a phosphorus-deficient calcareous soil. Plant Soil 269:69–80

Pao SS, Paulsen IT, Saier MH (1998) Major facilitator superfamily. Microbiol Mol Biol Rev 63:1–134

Pariasca-Tanaka J, Satoh K, Rose T, Mauleon R, Wissuwa M (2009) Stress response versus stress tolerance: a transcriptome analysis of two rice lines contrasting in tolerance to phosphorus deficiency. Rice 2:167–185

Paszkowski U, Kroken S, Roux C, Briggs SP (2002) Rice phosphate transporters include an evolutionarily divergent gene specifically activated in arbuscular mycorrhizal symbiosis. Proc Natl Acad Sci U S A 99:13324–13329

Peleg Z, Cakmak I, Ozturk L, Yazici A, Jun Y, Budak H, Korol AB, Fahim TA, Saranga Y (2009) Quantitative trait loci conferring grain mineral nutrient concentrations in durum wheat x wild emmer wheat RIL population. Theor Appl Genet 119:353–369

Pérez-Torres CA, López-Bucio J, Cruz-Ramírez A, Ibarra-Laclette E, Dharmasiri S, Estelle M, Herrera-Estrella L (2008) Phosphate availability alters lateral root development in *Arabidopsis* by modulating auxin sensitivity via a mechanism involving the TIR1 auxin receptor. Plant Cell 20:3258–3272

Plaxton W, Tran HT (2011) Metabolic adaptation of phosphate-starved plants. Plant Physiol 156:1006–1015

Poirier Y, Thoma S, Somerville C, Schiefelbein J (1991) A mutant of *Arabidopsis* deficient in xylem loading of phosphate. Plant Physiol 97:1087–1093

Preuss C, Huang CY, Gilliham M, Tyerman SD (2010) Channel-like characteristics of the low-affinity barley phosphate transporter Pht1;6 when expressed in *xenopus* oocytes. Plant Physiol 152:1431–1441

Raboy V (2009) Approaches and challenges to engineering seed phytate and total phosphorus. Plant Sci 177:281–296

Rae AL, Cybinski DH, Jarmey JM, Smith FW (2003) Characterization of two phosphate transporters from barley; evidence for diverse function and kinetic properties among members of the Pht1 family. Plant Mol Biol 53:27–36

Rae AL, Jarmey JM, Mudge SR, Smith FW (2004) Over-expression of a high-affinity transporter in transgenic barley plants does not enhance phosphate uptake rates. Funct Plant Biol 31:141–148

Raghothama KG (1999) Phosphate acquisition. Annu Rev Plant Physiol Plant Mol Biol 50:665–693

Raghothama KG (2005) Phosphorous. In: Broadley MR, White PJ (eds) Plant nutritional genomics. Blackwell Publishing Ltd., Oxford, pp 112–126

Rashotte AM, DeLong A, Muday GK (2001) Genetic and chemical reductions in protein phosphatase activity alter auxin transport, gravity response, and lateral root growth. Plant Cell 13:1683–1697

Rausch C, Bucher M (2002) Molecular mechanisms of phosphate transport in plants. Planta 216:23–37

Raven JA (2008) Phosphorus and the future. In: White PJ, Hammond JP (eds) The ecophysiology of plant-phosphorus interactions. Springer, Dordrecht, pp 271–283

Raven JA (2012) Protein turnover and plant RNA and phosphorus requirements in relation to nitrogen fixation. Plant Sci 188:25–35

Rebafka FP, Bationo A, Marschner H (1993) Phosphorus seed coating increases phosphorus uptake, early growth and yield of pearl-millet (pennisetum-glaucum (l) r br) grown on an acid sandy soil in niger, west-africa. Fertil Res 35:151–160

Ren F, Guo QQ, Chang LL, Chen L, Zhao CZ, Zhong H, Li XB (2012a) *Brassica napus PHR1* gene encoding a myb-like protein functions in response to phosphate starvation. PLoS One 7:e44005

Ren Y, He X, Liu D, Li J, Zhao X, Li B, Tong Y, Zhang A, Li Z (2012b) Major quantitative trait loci for seminal root morphology of wheat seedlings. Mol Breed 30:139–148

Reymond M, Svistoonof FS, Loudet O, Nussaume L, Desnos T (2006) Identification of QTL controlling root growth response to phosphate starvation in *Arabidopsis thaliana*. Plant Cell Environ 29:115–125

Richardson AE, Hadobas PA, Hayes JE (2000) Acid phosphomonoesterase and phytase activities of wheat (*Triticum aestivum* L.) roots and utilization of organic phosphorus substrates by seedings grown in sterile culture. Plant Cell Environ 23:397–405

Richardson AE (1994) Soil microorganisms and phosphorus availability. In: Pankhurst CE, Doube BM, Gupta VVS, Grace PR (eds) Soil biota: management in sustainable farming systems. CSIRO Australia, Melbourne, pp 50–62

Römer R, Schilling G (1986) Phosphorus requirements of the wheat plant in various stages of its life cycle. Plant Soil 91:221–229

Rose TJ, Wissuwa M (2012) Rethinking internal phosphorus utilization efficiency: a new approach is needed to improve PUE in grain crops. In: Sparks DL (ed) Advances in agronomy. Elsevier, London, pp 185–218

Rose TJ, Rengel Z, Ma Q, Bowden JW (2007) Differential accumulation patterns of phosphorus and potassium by canola cultivars compared to wheat. J Plant Nutr Soil Sci 170:404–411

Rose TJ, Rose MT, Pariasca-Tanaka J, Heuer S, Wissuwa M (2011) The frustration with utilization: why have improvements in internal phosphorus utilization efficiency in crops remained so elusive? Front Plant Sci 2:73. doi:10.3389/fpls.2011.00073

Rouached H, Arpat AB, Poirier Y (2010) Regulation of phosphate starvation responses in plants: signaling players and cross-talks. Mol Plant 3:288–299

Rouached H (2011) Multilevel coordination of phosphate and sulfate homeostasis in plants. Plant Signal Behav 6:952–955

Rubio V, Linhares F, Solano R, Martín AC, Iglesias J, Leyva A, Paz-Ares J (2001) A conserved MYB transcription factor involved in phosphate starvation signaling both in vascular plants and in unicellular algae. Gene Dev 15:2122–2133

Sánchez-Calderón L, López-Bucio J, Chacón-López A, Cruz-Ramírez A, Nieto-Jacobo F, Dubrovsky JG, Herrera-Estrella L (2005) Phosphate starvation induces a determinate developmental program in the roots of *Arabidopsis thaliana*. Plant Cell Physiol 46:174–184

Sánchez-Calderón L, López-Bucio J, Chacón-López A, Gutiérrez-Ortega A, Hernández-Abreu E, Herrera-Estrella L (2006) Characterization of low phosphorus insensitive mutants reveals a crosstalk between low phosphorus-induced determinate root development and the activation of genes involved in the adaptation of *Arabidopsis* to phosphorus deficiency. Plant Pysiol 140:879–889

Sattelmacher B, Horst WJ, Becker HC (1994) Factors that contribute to genetic variation for nutrient efficiency of crop plants. Z Pflanzen Bodenk 157:215–224

Schachtman DP, Shin R (2007) Nutrient sensing and signaling: NPKS. Annu Rev Plant Biol 58:47–69

Schachtman DP, Reid RJ, Ayling SM (1998) Phosphorus uptake by plants: from soil to cell. Plant Physiol 116:447–453

Schünmann PHD, Richardson AE, Smith FW, Delhaize E (2004) Characterization of promoter expression patterns derived from the Pht1 phosphate transporter genes of barley (*Hordeum vulgare* L.). J Exp Bot 55:855–865

Secco D, Baumann A, Poirier Y (2010) Characterization of the rice PHO1 gene family reveals a key role for OSPHO1;2 in phosphate homeostasis and the evolution of a distinct clade in dicotyledons. Plant Physiol 152:1693–1704

Secco D, Wang C, Arpat BA, Wang Z, Poirier Y, Tyerman SD, Wu P, Shou H, Whelan J (2012) The emerging importance of the SPX domain-containing proteins in phosphate homeostasis. New Phytol 193:842–851

Shi T, Li R, Zhao Z, Ding G, Long Y, Meng J, Xu F, Shi L (2013) QTL for yield traits and their association with functional genes in response to phosphorus deficiency in *Brassica napus*. PLoS One 8:e54559

Shin H, Shin HS, Chen R, Harrison MJ (2006) Loss of At4 function impacts phosphate distribution between the roots and the shoots during phosphate starvation. Plant J 45:712–726

Smith FW, Cybinski DH, Rae AL (1999) Regulation of expression of genes encoding phosphate transporters in barley roots. In: Gissel-Nielsen G, Jensen A (eds) Plant nutrition-molecular biology and genetics. Kluwer Academic Publisher, Dordrecht, pp 145–150

Solaiman Z, Marschner P, Wang D, Rengel Z (2007) Growth, P uptake and rhizosphere properties of wheat and canola genotypes in an alkaline soil with low P availability. Biol Fertil Soils 44:143–153

Stangoulis JCR, Huynh BL, Welch RM, Choi EY, Graham RD (2007) Quantitative trait loci for phytate in rice grain and their relationship with grain micronutrient content. Euphytica 154:289–294

Stefanovic A, Ribot C, Rouached H, Wang Y, Chong J, Belbahri L, Delessert S, Poirier Y (2007) Members of the PHO1 gene family show limited functional redundancy in phosphate transfer to the shoot, and are regulated by phosphate deficiency via distinct pathways. Plant J 50:982–994

Strong WM, Best EK, Cooper JE (1997) Phosphate fertilizer residues in wheat-growing soils of the western downs. Qld Aust J Soil Res 35:341–354

Su J, Xiao Y, Li M, Liu Q, Li B, Tong Y, Jia J, Li Z (2006) Mapping QTLs for phosphorus-deficiency tolerance at wheat seedling stage. Plant Soil 281:25–36

Su J, Zheng Q, Li HW, Li B, Jing RL, Tong YP, Li ZS (2009) Detection of QTLs for phosphorus use efficiency in relation to agronomic performance of wheat grown under phosphorus sufficient and limited conditions. Plant Sci 176:824–836

Suzuki Y, Kihara-Doi T, Kawazu T, Miyake C, Makino A (2010) Differences in rubisco content and its synthesis in leaves at different positions in Eucalyptus globulus seedlings. Plant Cell Environ 33:1314–1323

Svistoonoff S, Creff A, Reymond M, Sigoillot-Claude C, Ricaud L, Blanchet A, Nussaume L, Desnos T (2007) Root tip contact with low-phosphate media reprograms plant root architecture. Nat Genet 39:792–796

Syers JK, Johnston AE, Curtin D (2008) Efficiency of soil and fertilizer phosphorus use – reconciling changing concepts of soil phosphorus behaviour with agronomic information. FAO Fertil Plant Nutr Bull 18:27–44

Tarafdar JC, Marschner H (1994) Phosphatase activity in the rhizosphere and hydrosphere of a mycorrhizal wheat supplied with inorganic and organic phosphorus. Soil Biol Biochem 26:387–395

Teng W, Deng Y, Chen YP, Xu XF, Chen RY, Lv Y, Zhao YY, Zhao XQ, He X, Li B, Tong YP, Zhang FS, Li ZS (2013) Characterization of root response to phosphorus supply from morphology to gene analysis in field-grown wheat. J Exp Bot 64:1403–1411

Theodorou ME, Plaxton WC (1993) Metabolic adaptations of plant respiration to nutritional phosphate deprivation. Plant Physiol 101:339–344

Tesfaye M, Temple SJ, Allan DL, Vance CP, Samac DA (2001) Overexpression of malate dehydrogenase in transgenic alfalfa enhances organic acid synthesis and confers tolerance to aluminum. Plant Physiol 127:1836–1844

Tian J, Wang C, Zhang Q, He X, Whelan J, Shou H (2012) Overexpression of *OsPAP10a*, a root-associated acid phosphatase, increased extracellular organic phosphorus utilization in rice. J Integr Plant Biol 54:631–639

Ticconi CA, Delatorre CA, Lahner B, Salt DE, Abel S (2004) *Arabidopsis* pdr2 reveals a phosphate-sensitive checkpoint in root development. Plant J 37:801–814

Tiessen H (2008) Phosphorus in a global environment. In: White PJ, Hammond JP (eds) The ecophysiology of plant-phosphorus interactions, vol 7. Springer, Dordrecht, pp 1–7

Tittarelli A, Milla L, Vargas F, Morales A, Neupert C, Meisel LA, Salvo-G H, Peñaloza E, Muñoz G, Corcuera LJ, Silva H (2007) Isolation and comparative analysis of the wheat TaPT2 promoter: identification in silico of new putative regulatory motifs conserved between monocots and dicots. J Exp Bot 58:2573–2582

Uhde-Stone C, Zinn KE, Ramirez-Yanez M, Li A, Vance CP (2003) Nylon filter arrays reveal differential gene expression in proteoid roots of white lupin in response to phosphorus deficiency. Plant Physiol 131:1064–1079

Ullrich-Eberius CI, Novacky A, Fischer E, Lüttge U (1981) Relationship between energy-dependent phosphate uptake and the electrical membrane potential in *Lemna gibba* G1. Plant Physiol 67:797–801

Van Mooy BAS, Rocap G, Fredricks HF, Evans CT, Devol AH (2006) Sulfolipids dramatically decrease phosphorus demand by picocyanobacteria in oligotrophic marine environments. Proc Natl Acad Sci U S A 103:8607–8612

Van Mooy BAS, Fredricks HF, Pedler BE, Dyhrman ST, Karl DM, Koblizek M, Lomas MW, Mincer TJ, Moore LR, Moutin T, Rappe MS, Webb EA (2009) Phytoplankton in the ocean use non-phosphorus lipids in response to phosphorus scarcity. Nature 458:69–72

Vance CP, Uhde-Stone C, Allan DL (2003) Phosphorus acquisition and use: critical adaptations by plants for securing a non-renewable resource. New Phytol 157:423–447

Veneklaas EJ, Lambers H, Bragg J, Finnegan PM, Lovelock CE, Plaxton WC, Price CA, Scheible WR, Shane MW, White PJ, Raven JA (2012) Opportunities for improving phosphorus-use efficiency in crop plants. New Phytol 195:306–320

Versaw WK, Harrison MJ (2002) A chloroplast phosphate transporter PHT2;1 influences allocation of phosphate within the plant and phosphate-starvation responses. Plant Cell 14:1751–1766

Wang YH, Garvin DF, Kochian LV (2002) Rapid induction of regulatory and transporter genes in response to phosphorus, potassium, and iron deficiencies in tomato roots. Evidence for cross talk and root/rhizosphere-mediated signals. Plant Physiol 130:1361–1370

Wang Y, Ribot C, Rezzonico E, Poirier Y (2004) Structure and expression profile of the *Arabidopsis* PHO1 gene family indicates a broad role in inorganic phosphate homeostasis. Plant Physiol 135:400–411

Wang C, Ying S, Huang H, Li K, Wu P, Shou H (2009a) Involvement of OsSPX1 in phosphate homeostasis in rice. Plant J 57:895–904

Wang Z, Hu H, Huang H, Duan K, Wu ZA, Wu P (2009b) Regulation of OsSPX1 and OsSPX3 on expression of *OsSPX* domain genes and Pi-starvation signaling in rice. J Integr Plant Biol 51:663–674

Wang X, Shen J, Liao H (2010a) Acquisition or utilization, which is more critical for enhancing phosphorus efficiency in modern crops? Plant Sci 179:302–306

Wang L, Chen F, Zhang F, Mi G (2010b) Two strategies for achieving higher yield under phosphorus deficiency in winter wheat grown in field conditions. Field Crops Res 118:36–42

Wang J, Sun J, Miao J, Guo J, Shi Z, He M, Chen Y, Zhao X, Li B, Han FP, Tong Y, Li Z (2013) A wheat phosphate starvation response regulator Ta-PHR1 is involved in phosphate signalling and increases grain yield in wheat. Ann Bot. doi:10.1093/aob/mct080

Wasaki J, Yonetani R, Kuroda S, Shinano T, Yazaki J, Fujii F, Shimbo K, Yamamoto K, Sakata K, Sasaki T, Kishimtot N, Kikuchi S, Yamagishi MY, Osaki M (2003) Transcriptomic analysis of metabolic changes by phosphorus stress in rice roots. Plant Cell Environ 26:1515–1523

White PJ, Veneklaas EJ (2012) Nature and nurture: the importance of seed phosphorus content. Plant Soil 357:1–8

White PJ, Broadley MR, Gregory PJ (2012) Managing the nutrition of plants and people. Appl Environ Soil Sci 2012: Article ID 104826

Williamson LC, Ribrioux SPCP, Fitter AH, Leyser HMO (2001) Phosphate availability regulates root system architecture in *Arabidopsis*. Plant Physiol 126:875–882

Wissuwa M, Yano M, Ae N (1998) Mapping of QTLs for phosphorus-deficiency tolerance in rice (*Oryza sativa* L.). Theor Appl Genet 97:777–783

Wissuwa M, Ae N (2001) Further characterization of two QTLs that increase phosphorus uptake of rice (Oryza sativa L.) under phosphorus deficiency. Plant Soil 237:275–286

Wissuwa M, Wegner J, Ae N, Yano M (2002) Substitution mapping of Pup1: a major QTL increasing phosphorus uptake of rice from a phosphorus-deficient soil. Theor Appl Genet 105:890–897

Wissuwa M, Gamat G, Ismail AM (2005) Is root growth under phosphorus deficiency affected by source or sink limitations? J Exp Bot 56:1943–1950

Wissuwa M, Mazzola M, Picard C (2009) Novel approaches in plant breeding for rhizosphere-related traits. Plant Soil 321:409–430

Wu P, Wang XM (2008) Role of OsPHR2 on phosphorus homeostasis and root hairs development in rice (*Oryza sativa* L.). Plant Signal Behav 3:674–675
Yang YS, Paszkowski U (2011) Phosphate import at the arbuscule: just a nutrient? MPMI 24:1296–1299
Yang M, Ding G, Shi L, Feng J, Xu F, Meng J (2010) Quantitative trait loci for root morphology in response to low phosphorus stress *in Brassica napus*. Theor Appl Genet 121:181–193
Yang M, Ding G, Shi L, Xu F, Meng J (2011) Detection of QTL for phosphorus efficiency at vegetative stage in *Brassica napus*. Plant Soil 339:97–111
Yang SY, Grønlund M, Jakobsen I, Suter-Grotemeyer M, Rentsch D, Miyao A, Hirochika H, Kumar CS, Sundaresan V, Salamin N, Catausan S, Mattes N, Heuer S, Paszkowskia U (2012) Nonredundant regulation of rice arbuscular mycorrhizal symbiosis by two members of the phosphate transporter1 gene family. Plant Cell 24:4236–4251
Yao Y, Sun H, Xu F, Zhang X, Liu S (2011) Comparative proteome analysis of metabolic changes by low phosphorus stress in two *Brassica napus* genotypes. Planta 233:523–537
Yi K, Wu Z, Zhou J, Du L, Guo L, Wu Y, Wu P (2005) OsPTF1, a novel transcription factor involved in tolerance to phosphate starvation in rice. Plant Physiol 138:2087–2096
Yu Z, Kang B, He X, Lv S, Bai Y, Ding W, Chen M, Cho H, Wu P (2011) Root hair-specific expansins modulate root hair elongation in rice. Plant J 66:725–734
Zakhleniuk OV, Raines CA, Lloyd JC (2001) pho3: a phosphorus-deficient mutant pf *Arabidopsis thaliana* (L.) Heynh. Planta 212:529–534
Zeng HQ, Zhu YY, Huang SQ, Yang ZM (2010) Analysis of phosphorus-deficient responsive miRNAs and cis-elements from soybean (*Glycine max* L.). J Plant Physiol 167:1289–1297
Zhang H, Forde BG (2000) Regulation of *Arabidopsis* root development by nitrate availability. J Exp Bot 51:51–59
Zhang H, Huang Y, Ye X, Shi L, Xu F (2009) Genotypic differences in phosphorus acquisition and the rhizosphere properties of *Brassica napus* in response to low phosphorus stress. Plant Soil 320:91–102
Zhao J, Jamar DCL, Lou P, Wang Y, Wu J, Wang X, Bonnema G, Koornneef M, Vreugdenhil D (2008) Quantitative trait loci analysis of phytate and phosphate concentrations in seeds and leaves of *Brassica rapa*. Plant Cell Environ 31:887–900
Zhao MR, Han YY, Feng YN, Li F, Wang W (2012) Expansins are involved in cell growth mediated by abscisic acid and indole-3-acetic acid under drought stress in wheat. Plant Cell Rep 31:671–685
Zhao X, Liu X, Guo C, Gu J, Xiao K (2013) Identification and characterization of microRNAs from wheat (*Triticum aestivum* L.) under phosphorus deprivation. J Plant Biochem Biotechnol 22:113–123
Zhou J, Jiao FC, Wu Z, Li Y, Wang X, He X, Zhong W, Wu P (2008) OsPHR2 is involved in phosphate-starvation signaling and excessive phosphate accumulation in shoots of plants. Plant Physiol 146:1673–1686
Zhu J, Lynch JP (2004) The contribution of lateral rooting to phosphorus acquisition efficiency in maize (Zea mays) seedlings. Funct Plant Biol 31:949–958
Zhu YG, Smith SE (2001) Seed phosphorus (P) content affects growth, and P uptake of wheat plants and their association with arbuscular mycorrhizal (AM) fungi. Plant Soil 231:105–112
Zhu J, Shawn M, Kaepple SM, Lynch JP (2005a) Mapping of QTLs for lateral root branching and length in maize (*Zea mays* L.) under differential phosphorus supply. Theor Appl Genet 111:688–695
Zhu J, Shawn M, Kaepple SM, Lynch JP (2005b) Mapping of QTL controlling root hair length in maize (*Zea mays* L.) under phosphorus deficiency. Plant Soil 270:299–310
Zhu J, Mickelson SM, Kaeppler SM, Lynch JP (2006) Detection of quantitative trait loci for seminal root traits in maize (*Zea mays* L.) seedlings grown under differential phosphorus levels. Theor Appl Genet 113:1–10
Zhu J, Zhang C, Lynch JP (2010) The utility of phenotypic plasticity of root hair length for phosphorus acquisition. Funct Plant Biol 37:313–322

Chapter 5
Micronutrient Use Efficiency – Cell Biology of Iron and Its Metabolic Interactions in Plants

Ilaria Forieri and Ruediger Hell

Abstract Iron (Fe) is an intriguing nutrient due to its dual nature. Its redox properties make it essential for different vital processes in plant cells. But an excess of Fe can be toxic as it catalyses the formation of reactive oxygen species. Therefore Fe homeostasis must be tightly regulated. Different mechanisms contribute to the regulation, including the control of uptake, the intracellular chelation by different molecules and the partitioning into the organelles and storage locations. Despite its high abundance in soil, Fe solubility is extremely low. Fe availability represents a significant constraint to plant growth and plants have developed distinct strategies to ensure Fe solubilisation and uptake. The Fe-S clusters in the electron transport chain of mitochondria and chloroplasts represent an important sink of Fe. Recent observations suggest that a co-regulation exists between Fe and sulfur metabolism. This is most likely the outcome of the high demand for Fe and S required for the biosynthesis of Fe-S clusters. In the following chapter the uptake strategies and their regulation mechanisms will be introduced. Moreover, different aspects of the regulation of Fe homeostasis in the cell will be presented, including the partitioning in the organelles. In the last section different evidences towards the interaction between Fe and S metabolism will be discussed.

Keywords Iron • Micronutrients • Homeostasis • Regulation • Iron transporters • Deficiency • Partitioning

Introduction: Fe Importance for Plant Nutrition

Iron (Fe) is a chemical element with atomic number 26, which belongs, together with manganese, cobalt, nickel, copper and zinc, to the metals of the first transition series in the periodic table. Fe is the most abundant element found on planet Earth, as it constitutes a significant part of the inner and outer core. It is the fourth most

I. Forieri • R. Hell (✉)
Department for Plant Molecular Biology, Centre for Organismal Studies (COS), University of Heidelberg, Heidelberg, Germany
e-mail: ruediger.hell@cos.uni-heidelberg.de

M.J. Hawkesford et al. (eds.), *Nutrient Use Efficiency in Plants*, Plant Ecophysiology 10, DOI 10.1007/978-3-319-10635-9_5

common element in the crust, after oxygen, silicon and aluminium. As a transition metal, Fe can easily accept and donate electrons. Thus its oxidation state can vary in a broad range between −2 and +6, but the most common under current atmospheric conditions are +2 (ferrous iron) and +3 (ferric iron). This redox property of Fe and its capability to form complexes with different ligands make this element indispensable for different biological processes in all the living organisms. Indeed several proteins (Fe-proteins) found in different organisms rely on Fe as a cofactor for proper functioning. Fe is also essential for plant metabolism, where it participates in vital cellular functions such as photosynthesis, respiration and chlorophyll biosynthesis.

Despite its essential role for life, an excess of free Fe can be detrimental to the cell because it can react with oxygen catalysing the formation of reactive oxygen species (ROS) such as superoxide ($O_2^{\cdot-}$) and hydroxyl radical ($OH^{\cdot}$) via the Fenton reaction (reviewed by Hell and Stephan 2003):

$$Fe^{3+} + O_2^{\cdot-} \rightarrow Fe^{2+} + O_2$$

$$Fe^{2+} + H_2O_2 \rightarrow Fe^{3+} + OH^- + OH^{\cdot}$$

$$\text{Resulting in}: O_2^{\cdot-} + H_2O_2 \rightarrow O_2 + OH^- + OH^{\cdot}$$

These radicals constitute a severe danger for the cell particularly the hydroxyl radical, which is very reactive and can indiscriminately oxidise DNA, polyunsaturated fatty acids in lipids (lipid peroxidation), amino acids in proteins and sugars.

To avoid potentially toxic reactions, protein-bound Fe is found incorporated into structures such as heme or coordinated with sulfur (S) to form Fe-S cluster. Heme contains a Fe atom in the centre of a large heterocyclic organic ring, the porphyrin, made of four pyrrolic groups joined by methine bridges. Among heme proteins, a fundamental role is played by hemoglobin and myoglobin in vertebrates as they contain a Fe atom which binds to oxygen. The most common heme proteins in plants are cytochromes, which participate in the electron transport process of mitochondria and chloroplasts. Other heme proteins are catalase and peroxidases that are involved in scavenging of ROS.

Fe-S clusters are versatile and ubiquitous cofactors of many different enzymes that participate in respiration, photosynthesis, DNA repair and replication, sulfur and nitrogen assimilation and ribosome biosynthesis (Balk and Pilon 2011). They are formed from Fe atoms and sulfur in the form of acid-labile sulfide and are bound to proteins via the sulfhydryl groups of cysteine residues. Different forms are found in plants, the most common are the 2Fe-2S and 4Fe-4S bound to four cysteine (Cys) residues. Other types include the 2Fe-2S Rieske-type cluster coordinated by 2 Cys and 2 His residues and 3Fe-4S liganded by 3 Cys (reviewed by Couturier et al. 2013).

Toxicity of free Fe ions may be avoided by chelation by different compounds such as the non-proteinogenic amino acid nicotianamine, citrate and the storage via ferritin proteins.

Ferrous Fe is relatively soluble but it is easily oxidised to ferric Fe by atmospheric oxygen. The solubility of Fe^{3+} is highly influenced by the soil pH. Indeed in alkaline conditions Fe^{3+} hydrolyses water producing $Fe(OH)_3$ that polymerises and precipitates together with inorganic anions. Free Fe^{3+} is soluble up to 10^{-6} M at pH 3.3, but in aerated soil at neutral-basic pH, the concentration of free Fe^{3+} and Fe^{2+} is estimated to be less than 10^{-15} M (Marschner 1995). This is much lower than the optimal concentration needed by plants that require between 10^{-4} and 10^{-8} M Fe^{3+}. This solubility problem strongly impacts Fe availability that represents a severe constraint for plant development and yield. In particular, Fe is considered the third most limiting plant nutrient after nitrogen and phosphorus. Hence, even though Fe is quite abundant in soil, Fe deficiency represents a major problem for worldwide agriculture in calcareous-alkaline soil, which comprises 30 % of all arable land. Upon Fe deficiency, plants display typical symptoms such as leaf chlorosis and reduced growth. Therefore the quest for Fe use efficient crop plants is a goal of plant breeding and biotechnology.

Plants are always dealing with the dual nature of Fe and have developed sophisticated mechanisms to ensure adequate Fe acquisition from the soil and at the same time to make it available for biological processes in the cell avoiding toxic reactions.

Fe content in crop plants, which constitute a widely utilised staple food, significantly influences Fe assimilation by the human population. Fe deficiency is one of the most diffuse nutritional problems in the world, with around 30 % of the world population affected according to World Health Organization (http://www.who.int/nutrition/topics/ida/en/). Therefore, understanding the mechanisms used by plants to cope with changing Fe availability is a prerequisite not only for improving crop yield but also for a positive impact on human nutrition.

In this chapter, the strategies adopted by plants for Fe uptake will be reviewed, with focus on mechanisms of regulation. The partitioning of Fe in the cell and the interaction with other nutrients such as sulfur will also be presented.

Fe Deficiency and Plant Responses

Higher plants use distinct strategies to ensure Fe solubilisation and uptake. In the 1980s, Römheld and Marschner (1986) divided plants into two groups according to their Fe uptake mechanisms. Dicotyledonous and non-graminaceous monocotyledonous plants belong to the Strategy I or reduction strategy group, whereas Poaceae to the Strategy II or chelation strategy group.

Strategy I is based on (1) soil acidification to increase Fe solubility, (2) reduction of Fe^{3+} to Fe^{2+} in the rhizosphere and (3) uptake of Fe^{2+} across the root plasma membrane (see Fig. 5.1). This strategy was first characterised in *Lycopersicum esculentum* (tomato) and *Pisum sativum* (pea) as model crop plants. Recently, most of the studies in this respect have focused on *Arabidopsis thaliana*, which represents a powerful tool for cell biology investigations. The genes responsible for the

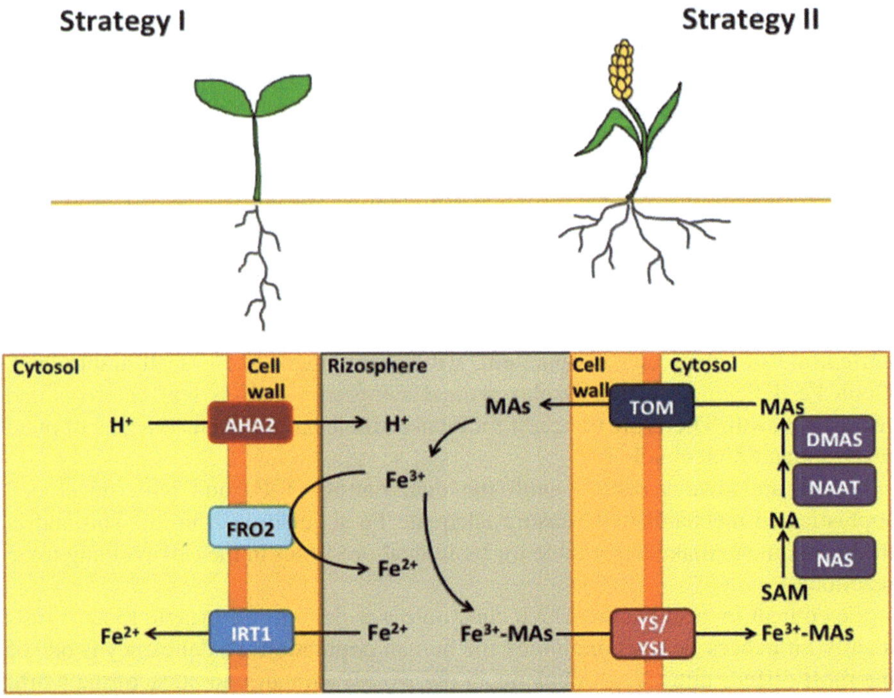

Fig. 5.1 Fe deficiency responses in plants. Strategy I (non-graminaceous plants) and Strategy II (graminaceous plants) are presented. In the *rectangles* the key enzymes of the two strategies are shown. Abbreviations: *AHA2 Arabidopsis* H^+-ATPase, *DMAS* deoxymugineic acid synthase, *FRO2* ferric chelate reductase 2, *IRT1* iron regulated transporter 1, *MAs* mugineic acids, *NAAT* nicotianamine aminotransferase, *NAS* nicotianamine synthase, *TOM* transporter of mugineic acids, *YS* yellow stripe, *YSL* yellow stripe like. YSL refers to orthologs of YS in plants other than maize

different steps of the strategy I have been identified and cloned and Fe deficiency results in an up-regulation of their expression. Firstly, the H^+-ATPase family (HA) excretes protons into the rhizosphere to increase Fe solubility (Palmgren 2001). In *Arabidopsis* the *HA2* gene particularly is induced in Fe deficiency (Santi and Schmidt 2009). The reduction of Fe^{3+} is catalysed by the ferric-chelate reductase oxidase 2 (FRO2) in *Arabidopsis* (Robinson et al. 1999) and by FRO1 in pea (Waters et al. 2002). FRO proteins are integral membrane proteins that belong to a superfamily of flavocytochromes and can transfer electrons from cytosolic NADPH to FAD across the plasma membrane (Robinson et al. 1999). *FRO2* was isolated as allelic to the *frd1* mutants in *Arabidopsis* (Yi and Guerinot 1996). These mutants are not able to induce the Fe chelate reductase activity, although they are still able to acidify the rhizosphere upon Fe deficiency. Moreover, these mutants cannot translocate radiolabeled Fe from root to the shoot when Fe is provided as chelated Fe^{3+}. Altogether these results shown that FRO activity is uncoupled from the HA activity and that Fe^{3+} reduction to Fe^{2+} is a prerequisite for the transport.

After reduction, the uptake of Fe^{2+} in the root epidermal cells is performed by the iron-regulated transporter 1 (IRT1) in *Arabidopsis* (Vert et al. 2003). Orthologs of IRT1 were cloned in both pea and tomato (Cohen et al. 1998; Eckhardt et al. 2001). The *Arabidopsis* knock-out mutant *irt1* is lethal unless plants are watered with an excess of Fe (Vert et al. 2002). Another transporter, IRT2, is also induced in roots when exposed to Fe shortage (Vert et al. 2001) but its knock-out mutant is not affected under normal Fe conditions. The attempt to complement the *irt1* mutant with *IRT2* driven by a constitutive promoter did not rescue the phenotype, showing that the two transporters have different roles in Fe uptake (Varotto et al. 2002). Both transporters belong to the zinc-regulated transporter iron-regulated transporter like protein family (ZIP) that takes its name from the first transporter that was identified, the zinc regulated transporter ZRT. Indeed IRT1 and IRT2 are not highly specific for Fe and can mediate the import of a broad spectrum of metal species including zinc Zn^{2+}, manganese Mn^{2+} and cadmium Cd^{2+}. The *Arabidopsis thaliana* genome encodes for 16 ZIP proteins (Mäser et al. 2001) and they function in the uptake of different bivalent metal ions.

IRT1 is regulated at different levels. Its expression promptly responded after the plants were transferred to Fe deficiency (Connolly et al. 2002). In particular, the mRNA accumulated after 24 h and the protein level peaked after 72 h. After Fe resupply the IRT1 protein was already almost undetectable after 12 h, indicating that its expression is tightly regulated. An over-expressing line of *IRT1* showed accumulation of the protein only under Fe deficiency, indicating a fine regulation of Fe homeostasis at the uptake level. Indeed IRT1 protein is rapidly degraded in response to changing Fe conditions and this degradation is mediated by ubiquitination (Kerkeb et al. 2008; Barberon et al. 2011). Recently the specific E3 ubiquitin ligase IRT1 DEGRADATION FACTOR 1 (IDF1) was identified in a screen of insertional mutants (Shin et al. 2013).

The main feature of Strategy II is the excretion of chelating compounds such as mugineic acids (MAs) in the root rhizosphere that chelate Fe^{3+}, to enhance solubility and allow for mobilisation of Fe^{3+} (Fig. 5.1). The name mugineic acid is derived from the Japanese word komugi for wheat from which these compounds had first been isolated. The MAs biosynthetic pathway is conserved among the Poaceae family, which comprises many of the most important food plants: rice, wheat and maize, and starts from *S*-adenosyl-L-methionine (SAM). Three molecules of SAM are converted into nicotianamine (NA) in one reaction by nicotianamine synthase (NAS). NA is a precursor for MAs in strategy II plants but additionally functions as Fe chelator in different plant organs, and seems to be essential not only for the uptake of Fe from the soil, but also for the cell-to-cell and long distance transport within the plant (Schuler et al. 2012). Studies on NA started with the analysis of the tomato mutant *chloronerva*. This mutant is NA-free and shows retarded growth and intercostal chlorosis of young leaves. Map-based cloning revealed that chloronerva is a single copy gene in tomato and encodes for NAS (Ling et al. 1999). The NAS genes have been cloned from other plant species such as barley, rice and *Arabidopsis*, showing that NA carries out functions in strategy I and II species (reviewed by Hell and Stephan 2003; Klatte et al. 2009). The expression of most NAS genes is strongly induced upon Fe deficiency. NA is then further processed by NA aminotransferase (NAAT) and deoxymugineic acid synthase (DMAS) to form 2′-deoxymugineic acid (DMA). DMA is the starting point for the synthesis of all the other chemical forms of MAs (Nakanishi et al. 2000). The secretion of MAs is diurnally regulated, with a peak in the morning (Cakmak et al. 1998). The transporter of mugineic acid family phytosiderophores 1 (TOM) has been identified in rice and barley as responsible for the secretion of MAs (Nozoye et al. 2011). After secretion, MAs can bind to Fe^{3+} and the MA-Fe^{3+} complexes are taken up by the root YELLOW STRIPE 1 (YS1) and YELLOW STRIPE 1-like transporters (YSL1) (Curie et al. 2009; Inoue et al. 2009). The study of these transporters started with the analysis of the maize mutant *yellow stripe 1*. This mutant shows leaf chlorosis and fails to take up phytosiderophores from the soil (Von Wiren et al. 1994). The *YELLOW STRIPE 1* gene was then mapped and cloned (Curie et al. 2001) and found to encode for a membrane transporter that mediates the uptake of phytosiderophores bound to Fe^{3+}. The biochemical function is directly shown by the ability of YS1 to restore the growth of a yeast strain deficient in Fe uptake, when phytosiderophores are present in the media. Its mRNA accumulates significantly under Fe deficiency in both root and shoot. The latter was surprising as phytosiderophores were not expected to be transported in green tissues as they are only present in the root (Curie et al. 2001). Indeed, although the release of phytosiderophores is a prerogative of graminaceous plants, YSL transporters are found also in non-graminaceous taxa (Chu et al. 2010, 2013). In particular, *Arabidopsis thaliana* has eight YSL genes (Mäser et al. 2001). In addition, YSL transporters have been found to transport NA-Fe^{3+} complexes (reviewed by Chu et al. 2013), adding to the fundamental role of NA as Fe chelator in both strategy I and strategy II plants.

Interestingly, rice additionally possesses an iron transporter, OsIRT1, an ortholog of AtIRT1. However, rice roots show a very low ferric-chelate reductase activity, suggesting a role of OsIRT1 in the direct uptake of Fe^{2+} in anaerobic growth conditions that is typical for this crop (Ishimaru et al. 2006).

Both graminaceous and non-graminaceous plants possess other divalent metal transporters, which can facilitate Fe assimilation. The NRAMP (natural resistance-associated macrophage protein) family transporters have been first found in mammals and have then been cloned from *Arabidopsis thaliana* (Curie et al. 2000) There are six genes encoding for NRAMP proteins in *Arabidopsis* (Mäser et al. 2001). They can mediate the uptake of several divalent metals, including Fe^{2+}, zinc Zn^{2+}, manganese Mn^{2+}, nickel Ni^{2+} and cadmium Cd^{2+}. The expression of three of these transporters, NRAMP1, NRAMP2 and NRAMP4, is induced by Fe deficiency in roots and leaves. In particular, NRAMP1 is thought to mediate the uptake of Fe and other essential nutrients, such as manganese from the soil (Curie et al. 2000; Cailliatte et al. 2010). The other two transporters NRAMP3 and NRAMP4 are involved in the Fe distribution to developing seeds in low Fe conditions (Lanquar et al. 2005).

Regulation of the Strategies

The regulation of Fe deficiency responses is very complex and requires the coordination of several regulatory elements. The presence of different pathway and feedback signals constitute an important aspect in the regulation.

Several transcription factors that are involved in the regulation of the Fe uptake machinery have been already identified in different Strategy I plants (Fig. 5.2a). The key element/regulator in this respect was first identified in the tomato *fer* mutant. Map-based cloning revealed that *FER* encodes for a basic helix-loop-helix (bHLH) transcription factor (Ling et al. 2002). *Arabidopsis* possesses an ortholog of *FER*, which has been named *FIT* (*FER-like iron deficiency-induced transcription factor*, also named before *FIT1/FRU/AtbHLH029*; Bauer et al. 2007). *FIT* expression is repressed upon full Fe supply, whereas the expression is highly induced in Fe deficiency. FIT positively regulates the expression of different Fe-responsive genes, including *IRT1* and *FRO2*. The *fit* mutant shows leaf chlorosis and decreased Fe content and fails to induce the typical Strategy I responses (Colangelo and Guerinot 2004). Moreover, the mutant dies at the seedling stage unless watered with additional Fe. *FRO2* mRNA level is severely downregulated in the mutant and the FRO activity cannot be detected. The transcript level of *IRT1* is decreased but still detectable in *fit* plants whereas the protein IRT1 is not present. These results suggest that FIT can control the Strategy I responses at different level, regulating the gene expression but also the turnover of IRT1 protein.

FIT can also interact directly with other bHLH factors, such as bHLH038 and bHLH039. This interaction is thought to serve in modulating the plant response to Fe deficiency (Yuan et al. 2008). These two factors belong together with bHLH100 and bHLH101 to a specific sub-group of bHLH and their expression is strongly induced upon Fe deficiency (Wang et al. 2007). They are also functioning independently from FIT to mediate Fe deficiency responses (Sivitz et al. 2012).

FIT can also directly interact with ETHYLENE INSENSITIVE 3 and ETHYLENE INSENSITIVE 3–LIKE1 (Lingam et al. 2011). This interaction provides the molecular link between ethylene and the responses to Fe starvation, which was elusive before. Ethylene is known to be a positive regulator of the induction of different Fe responsive genes. The ethylene downstream transcription factors EIN3 and EIL1 are required for FIT accumulation and therefore thought to inhibit its proteasomal degradation, thus enhancing the plant responses to Fe deficiency (Lingam et al. 2011).

Microarray analysis aimed at finding new regulatory candidates identified the bHLH transcription factor POPEYE (PYE) (Long et al. 2010). PYE is upregulated specifically in the cells of the root perycicle upon Fe deficiency. The mutant *pye* displays severely impaired growth under – Fe condition; therefore PYE seems to play a fundamental role in the roots of plants exposed to Fe deficiency. Moreover, PYE is proposed to negatively regulate a cluster of Fe-responsive genes, amongst these *NAS4* and *FRO3*. PYE can directly interact with PYE homologues, such as IAA-Leu Resistant3 (ILR3) and bHLH115. ILR3 in turn interacts with another

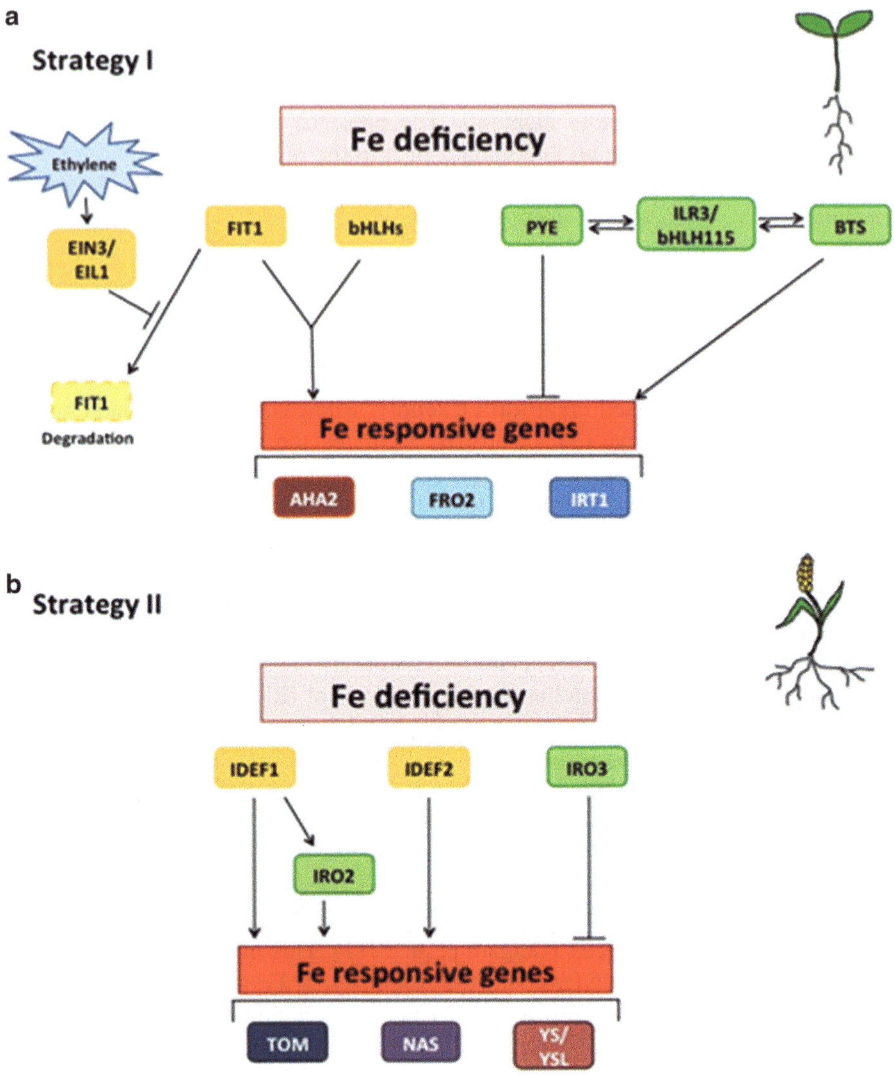

Fig. 5.2 Regulation of Fe deficiency responses in (**a**) Strategy I and (**b**) Strategy II plants. *Rectangles* indicate important regulatory transcription factors of the two Strategies and downstream Fe responsive genes. *Arrows* indicate positive or negative regulation. Abbreviations: *bHLH* basic helix-loop-helix, *FIT* FER-like iron deficiency induced, *EIN3*/*EIL1* ETHYLENE INSENSITIVE 3/ETHYLENE INSENSITIVE 3–LIKE, *PYE* POPEYE, ILR3, *BTS* BRUTUS, *IDEF* iron deficiency responsive element-binding factor, *IRO* iron-related transcription factor

regulatory protein named BRUTUS (BTS). BTS possesses three different domains, one with putative E3 ligase activity, one for transcriptional regulation and one for Fe binding. Unlike *pye*, *bts* mutants appear more resistant to Fe deprivation and show a better growth in – Fe conditions, with longer roots and greener shoots. A direct interaction between PYE and BTS has not been reported, but interestingly BTS interacts with the PYE interactors ILR3 and bHLH115. It is therefore speculated that this interaction participates in the regulation of Fe deficiency responses in the root. The induction of PYE under Fe limiting conditions might serve to regulate Fe homeostasis in the plant. Additionally BTS, the antagonist of PYE, might help in this regulation controlling PYE activity (Long et al. 2010).

Other regulatory elements have been identified in Strategy II plants (Fig. 5.2b). The analysis was based on stepwise promoter analysis of the barley *IDS2* gene in tobacco and allowed the identification of two key regulators of Fe deficiency responses, the *cis*-acting iron deficiency responsive element 1 (IDE1) and IDE2 (Kobayashi et al. 2003). IDE1 and IDE2 were the first discovered *cis*-acting elements related to nutrient deficiency. From sequence alignment of the promoters of several Fe responsive genes it emerged that these *cis*-elements are quite conserved among different plant species. Indeed they have been found in several genes, *e.g. HvNAAT*, *HvNAS*, *OsNAS2*, *OsNAS3*, *OsIRT1*, *AtIRT1* and *AtFRO2*. IDE1 and IDE2 can interact with two rice transcription factors IDE-binding factor 1 (IDEF1) and IDEF2 (Kobayashi et al. 2007; Ogo et al. 2008). These two factors are members of the ABI3/VP1 (ABSCISIC ACID INSENSITIVE 3/VIVIPAROUS 1) family and NAC (NO APICAL MERISTEM, *Arabidopsis* transcription activation factor and CUP SHAPED COTYLEDON) family, respectively. IDEF1 and IDEF2 are constitutively expressed in vegetative tissues and can regulate two different sets of genes (Kobayashi et al. 2009). IDE1 regulates most of the Fe related genes in normal Fe conditions and during the early responses to Fe deficiency. Interestingly, IDEF1 can switch its target genes in the late stages of Fe deficiency. IDEF2 instead maintains the same target genes during the responses to Fe deficiency and it is known to positively regulate the expression of *OsYSL2* (Kobayashi et al. 2010). Therefore, IDEF2 is also involved in the correct partitioning of Fe between roots and shoot.

Many regulators from graminaceous plants have been identified by a microarray analysis approach. The most extensively studied candidate is *OsIRO2*, which encodes a bHLH transcription factor (Ogo et al. 2011). Its expression is positively regulated by IDEF1 in Fe deficiency. OsIRO2 can in turn positively regulate different Strategy II genes, such as *OsNAS1*, *OsNAS2*, *OsNAAT1*, *TOM1* and *OsYSL15*. Another bHLH transcription factor, *OsIRO3*, is present in rice and its expression is induced by Fe deficiency (Zheng et al. 2010). It seems to be a negative regulator of several genes related to Fe deficiency responses.

Intriguingly, sequence comparison with *Arabidopsis* transcription factors has shown that OsIRO2 is similar to AtbHLH038, 039, 100 and 101, whereas IRO3 is similar to PYE (Ogo et al. 2006). Thus far, no correspondent of FIT has been found in graminaceous plants and no orthologue of IDEF1 and IDEF2 has been found in non-graminaceous plants. Therefore, it seems that the regulatory mechanisms are only partially conserved between Strategy I and Strategy II plants.

Apart from the positive regulator ethylene, other signaling molecules and plant hormones participate in the regulation of Fe deficiency responses. Among these nitric oxide (NO), carbon dioxide and auxin also contribute to the induction of several Fe-responsive genes. In contrast, cytokinin and jasmonic acid can negatively regulate the expression of different Fe genes such as *IRT1*, *FRO2* and *FIT* (reviewed by Kobayashi and Nishizawa 2012).

While the regulation of the Fe deficiency responses has been elucidated, the Fe sensing mechanism in the root remains unidentified. Recently the IDEF1 transcription factor has been found to bind directly to Fe and other divalent metals via its proline-rich domains and histidine-asparagine residues (Kobayashi et al. 2012). Thus, this transcription factor might be one factor for the sensing of the actual Fe situation in the cell and thus the Fe availability.

Fe Homeostasis in the Cell

After uptake in the root, further steps are required in order to allocate Fe in the rest of the plant. Fe must first be transported from the root epidermis through the root tissues to be loaded into the xylem. Due to its low solubility, a symplastic transport is assumed, but little is known about the mechanism and possible carrier (Morrissey and Guerinot 2009). Once it reaches shoot tissues, other mechanisms must be involved for the unloading and the transport into the different cellular compartments.

The solubility problem requires that Fe must always be in a chelated form during transport within the plant. Another important reason for the chelation is the potential toxicity of free ionic Fe that can catalyse the formation of reactive oxygen species (ROS) causing cell damage. Citrate, NA and MAs are the known predominant Fe chelators. In particular, citrate plays a fundamental role in chelating Fe in the xylem. Indeed, citrate-Fe(III) complexes have been found in the xylem sap of tomato plants (Rellan-Alvarez et al. 2010). FERRIC REDUCTASE DEFECTIVE 3 (FRD3) is an *Arabidopsis* multidrug and toxin efflux (MATE) transporter that plays a fundamental role in balancing Fe homeostasis. Its orthologue, OsFRDL1, has been found in rice. Both transporters mediate the efflux of citrate into the xylem. The *frd3* mutant displays chlorotic and dwarf phenotype, constitutive up-regulation of the Fe deficiency genes and of FRO activity. A high level of Fe is found in the roots of the mutant, due to inefficient Fe translocation to the shoot, emphasising the importance of citrate for Fe transport from root to shoot (Durrett et al. 2007; Rogers and Guerinot 2002). As FRD3 and FRDL1 efflux citrate in the Fe-free form, other transporters must be involved in the transport of Fe into the xylem. The *Arabidopsis* ferroportin 1/iron regulated 1 (AtFPN/AtREG1) could be involved in this process. Although direct evidence is still lacking, the localisation, the promoter activity and the mutant phenotype make this transporter a promising candidate (Morrissey et al. 2009).

Other transporters are involved in unloading the xylem into phloem. Members of the YSL family are widely expressed in different tissues also of non-graminaceous plants, suggesting a role in Fe translocation from xylem to phloem besides the uptake of MA-Fe complexes from the soil. Indeed YSL transporters have been found to mediate the transport of NA-Fe complexes (Curie et al. 2009). The *Arabidopsis* YSL1 and YSL2 proteins were found to localise to the plasma membrane and to function in yeast complementation assay (Chu et al. 2010). They are active in leaves and in flowers and are therefore required for the fertility and the development of seeds and for the distribution of Fe to the seeds. The two transporters and *AtYSL3* are quite closely related but they have distinct functions in the plant, as neither *YSL1* nor *YSL2* under control of *YSL3* promoter could complement the double mutant *ysl1ysl3*. YSL4 and YSL6 were found to localise to the chloroplast (Divol et al. 2013), to the tonoplast and to internal membranes (Chu et al. 2013). They are thought to mediate the release of Fe from the chloroplast in case of Fe overload, thus controlling Fe homeostasis.

NA certainly represents the principal Fe chelator in the cell for different reasons (reviewed by Hell and Stephan 2003). It can form complexes with both Fe^{3+} and Fe^{2+} at neutral and basic pH, NA-Fe complexes are unlikely to react with oxygen in the Fenton reaction, NA is found in all plant tissues and also all plant species its concentration positively correlates with the root areas of Fe uptake. NA is also involved in loading the seeds with Fe (Klatte et al. 2009). It is, however, also able to bind and transport other transition metals such as zinc (Haydon et al. 2012).

Partitioning of Fe in the Organelles

The partitioning of Fe to the organelles must be tightly regulated, due to the high requirement for the biosynthesis of Fe-S clusters in both chloroplasts and mitochondria. In addition, synthesis of cytosolic Fe-S depends on provision of a precursor from the mitochondria (Balk and Pilon 2011). The import of Fe into the chloroplast also represents an important strategy to store Fe in a non-toxic and available form. Indeed the largest amount of Fe in plant cells is found in the chloroplast, where 80–90 % of Fe is accumulated (Marschner 1995). The members of the ferritin family (FER) play a fundamental role to prevent oxidative damage in case of Fe overload. Ferritins are spherical protein complexes formed by 24 subunits. They can internalise Fe atoms in their central cavity and can release them when needed (Briat et al. 2010). Animal ferritins are regulated mostly at the translational level, while phytoferritins are mainly subjected to transcriptional regulation.

In *Arabidopsis*, there are four ferritin isoforms (FER1, FER2, FER3 and FER4). A loss-of-function approach was used to investigate the role of this protein in different plant tissues (Ravet et al. 2009). The analysis showed that plants lacking ferritins were more sensitive to excess of Fe, with reduced growth and defects in flower development. Moreover, loss-of-function mutant plants presented differential regulation of genes related to Fe uptake and higher level of ROS and consequently higher activity of detoxifying enzymes. Electron microscopy has shown that plant ferritins localise to the plastids, mainly to non-photosynthetic ones such as proplastids, etioplasts and amyloplasts (Seckback 1982). The loss-of-function approach provided more evidence that ferritins are not actually required for the proper formation of the photosynthetic chloroplast or for the functioning of the photosynthetic apparatus. Indeed, ferritins seem to play a fundamental role in the protection against oxidative stress (Ravet et al. 2009).

Other studies have attempted to further elucidate the localisation of ferritins in plant cells.

According to Zancani et al. (2004) ferritins can also localise to the mitochondria. Indeed, according to bioinformatics analyses, the *Arabidopsis* AtFER4 is the isoform most likely to be targeted to the mitochondria. A study was conducted based on the knock-out mutant *atfer4*. An antibody against FER was applied to protein fractions of isolated mitochondria and a ferritin signal was found in mitochondria isolated from wild type plants subjected to high Fe supply. The signal was not present in the fraction isolated from the mutant plants. This mitochondrial isoform seems to be of great importance for balancing Fe homeostasis in heterotrophic tissues, as shown by work on suspension cell cultures (Tarantino et al. 2010). Petit et al. (2001) identified a *cis*-element in the region of maize ferritin gene *ZmFER1* and in its orthologue from *Arabidopsis AtFER1*. This regulator named iron-dependent regulatory sequence (IDRS) is able to repress the transcription of the gene under low Fe conditions. IDRS also has additional functions in *Arabidopsis*, where it triggers the expression of *AtFER1* under dark-induced senescence but not in age-dependent senescence and in seedlings (Tarantino et al. 2003). This suggests that more regulatory elements must be involved in the regulation of *AtFER1* expression under such conditions.

Time for coffee (TIC) has been found in a luciferase-based genetic screen of the *AtFER1* promoter (Duc et al. 2009). TIC has been previously described as a nuclear component of the circadian clock. Mutants of TIC are chlorotic unless supplied with exogenous iron and are hypersensitive to iron during the early stages of development. Thus TIC is a central regulator of *AtFER1* as it represses its expression in low Fe conditions, in a way that is independent from IDRS. The *tic* mutants also fail to repress other genes induced by Fe overload under low Fe, pointing out that TIC-dependent pathways are fundamental for the response to Fe overload.

Another Fe binding protein, which has been reported to be involved in the protection against photo-oxidative damage is frataxin. This protein has been hypothesised to participate in the mitochondrial biosynthesis of Fe-S cluster acting as Fe donor. Its importance for plant cell has been demonstrated by analysis of T-DNA insertion mutants in *Arabidopsis*. Indeed frataxin knock-out mutants are lethal, while the knockdown ones are viable but accumulate high levels of ROS and induce the expression of genes encoding for ROS scavenging proteins (Busi et al. 2006).

The import of Fe into the chloroplast is linked to ferritin function. The permease in chloroplast 1 (PIC1) was first identified as member of the chloroplast inner membrane translocon complex TIC (Teng et al. 2006). This transporter emerged as a possible candidate in a bioinformatic screening for the Fe importer in the chloroplast proteome, due to its biochemical characteristics such as hydrophobicity, basic isoelectric point and predicted transmembrane domains, and was therefore renamed permease in chloroplast 1 PIC1 (Duy et al. 2007). The loss-of-function mutants of PIC1 display chlorotic phenotype and severely impaired growth. Moreover, the mutation causes severe problems in chloroplast development and leads to disturbed metal homeostasis in leaves.

FRO7, one of the members of the *Arabidopsis* FRO family, localises to the inner chloroplast membrane and it is thought to be essential for seedling development. The growth of the fro7 mutant is significantly impaired in alkaline conditions and in media lacking sugar. Moreover, chloroplasts isolated from the mutant exhibit significantly lower FRO activity and Fe content compared to the wild type (Jeong et al. 2008). These results suggested an important role of FRO7 as a chloroplast Fe transporter during photosynthesis and development.

In addition, the mitochondria contribute to cellular Fe homeostasis. Indeed a mutation in the STARIK/ATM3 gene that encodes for a mitochondria ABC transporter leads to chlorosis and reduced growth. The mitochondria of the *starik* mutant accumulate more non-heme and non-protein bound Fe, as the biosynthesis of Fe-S clusters in the mitochondria is linked to the intracellular Fe by this transporter (Kushnir et al. 2001; Bernard et al. 2009). A mitochondria iron transporter (MIT) has been identified in a screening of T-DNA (transfer DNA) of rice in Fe deficiency conditions (Bashir et al. 2011). The homozygous knock-out mutant *mit* is lethal, highlighting the importance of this transporter for plant growth. In contrast, the heterozygous mutant is viable but severely impaired in growth, exhibiting an accumulation of Fe in the shoot with less Fe in the mitochondria.

The vacuole also functions to store Fe and avoid toxicity. The vacuolar iron transporter (VIT) mediates the transport of Fe from the cytosol into the vacuole and plays a fundamental role in seed and seedling development, as shown by the analysis of the *vit1-1* mutant (Kim et al. 2006). In contrast, NRAMP3 and NRAMP4 are influx transporters that function to export Fe from the vacuole into the cytosol during seed germination (Lanquar et al. 2005). Their contribution is essential to provide Fe during development until the seedlings can start to take up Fe from the environment.

Interaction of Fe and S Metabolism (Uptake, Fe-S Clusters)

Uptake and homeostasis of Fe is known to be linked to other metals such as zinc (Lin et al. 2009; Deinlein et al. 2012). More recently the connection between Fe and sulfur (S) metabolism is being recognised, mainly because Fe is required together with S for the biosynthesis of the Fe-S clusters.

The thylakoids in chloroplasts harbour ferredoxin, photosystem I (PSI) and cytochrome b_6f complex, which belong to the photosynthetic electron transport chain. In the stroma, other Fe-S proteins are found; among these there are nitrite reductase and two key enzymes for sulfur metabolism, sulfite reductase and APR. In mitochondria, major Fe-S proteins are Complex I, II and III of the respiratory chain and aconitase. Other Fe-S proteins are found in the nucleus and function in DNA replication (Balk and Pilon 2011) and damage repair (Liu et al. 2003).

The coordination of the plant's demand for Fe-S clusters supports the hypothesis of a co-evolution between the Fe and the S metabolisms and the development of interaction traits between them.

Chelated Fe and reduced S in form of cysteine represent the substrates of the biosynthetic pathway. The Fe donor molecule is not known yet, whereas it is known that sulfur is mobilised from cysteine by a cysteine desulfurase. The protein frataxin has received attention for its putative role as Fe donor in the mitochondria assembly pathway (Busi et al. 2006).

As sulfur is present as acid-labile sulfide (S^{-2}) in the Fe-S cluster, two additional electrons are needed to reduce elemental sulfur S^0 to S^{-2} (Lill 2009). In the first step of the pathway, the Fe-S cluster is assembled on scaffold proteins. In the second step, the cluster is transferred to the specific apoprotein, which provides free amino acidic residues to bind it. Additional carrier proteins are involved in this step. The assembly machineries have been characterised in plants: chloroplasts contain the ISC (iron-sulfur cluster) biosynthetic pathway, whilst mitochondria contain the SUF-like (sulfur mobilization) pathway and in the cytoplasm the CIA (cytosolic iron-sulfur cluster assembly) pathway (reviewed by Balk and Pilon 2011; Couturier et al. 2013). The CIA is dependent on the mitochondria SUF assembly machinery, which provided the sulfide-containing compound that is used for the biosynthesis of the cluster. The mitochondrial ABC transporter, STARIK/ATM3, is thought to be involved in this process. Indeed the mutant *atm3* shows severely impaired activity of cytosolic aconitase while the mitochondria and plastidic isoforms are unaffected. Furthermore, *atm3* does not accumulate Fe in the mitochondria and the general Fe homeostasis is not affected (Bernard et al. 2009).

Upon nutrient deficiency, a complex reprogramming of cell metabolism occurs, in order to maintain viability. The strong requirement of Fe and S for the biosynthesis of Fe-S clusters in the organelles might constitute a feedback signal for the co-regulation of the assimilation pathways. Indeed the existence of such signals has been recently proposed for Fe and S metabolism (Vigani et al. 2013; Chan et al. 2013). The plant responses triggered by Fe or S deficiency have been well characterised. The consequences of the combined shortage of these nutrients however have only seldom been investigated, and might impact particularly on Fe-S cluster assembly. Significant interactions between Fe uptake mechanisms and external sulfate supply have been reported. It has been shown that S deficiency limits the capacity to cope with Fe shortage in tomato plants, preventing the expression of Fe chelate reductase *FRO1* and reducing the activity of Fe^{2+} transporter (Zuchi et al. 2009). Other studies using barley plants reported a positive correlation between S supply in the growth media and the plant capability of coping with Fe deficiency. Indeed phytosiderophores represent another important junction between Fe and S metabolism, as they are derived from the S-containing amino acid methionine. The release rate of phytosiderophores was diminished in barley plants upon sulfate deficiency, due to a decrease of the methionine level. After sulfate resupply, plants increased the release of phytosiderophores when exposed to Fe shortage (Astolfi et al. 2010, 2012). Recently the effect of Fe deficiency on sulfur metabolism has been analyzed in durum wheat (Ciaffi et al. 2013). Wheat plants grown under sufficient S supply showed an up-regulation of certain S deficiency responses when exposed to Fe deficiency. The expression of the high affinity sulfate transporters was increased in the root, as well as of several genes of the S metabolic pathway.

Recently we found that also in *Arabidopsis thaliana* the expression of the two key genes for the uptake of Fe (*IRT1*) and of S (*SULTR1;1*) correlates with the supply of both Fe and S in the growth media. The expression is differentially regulated in case of double nutrient shortage (Forieri et al. 2013). We suggested that Fe-S cluster availability might function in sensing and signalling of combined Fe and S deficiencies. Altogether, these analyses strongly support the existence of a co-regulation between the metabolic pathways, as the limitation of one nutrient influences the uptake of the other one. Such a co-regulation is very likely to be the outcome of a complex remodelling of the whole plant metabolism upon nutrient limitation as known for the prolonged deficiency of the single nutrients (Schuler et al. 2011; Nikiforova et al. 2003). Hence, we propose different signals that might contribute to this co-regulation, such as the sensing of the Fe and S concentrations in the root rhizosphere or within root cells, ROS, metabolism intermediates, Fe-S cluster assembly machineries or Fe-S proteins.

Conclusions

Fe is one of the most fascinating elements for life functions due to its redox properties and is of great importance for human nutrition. Crop plants are the direct or indirect source of Fe in our food, and research on the dicot model plant *Arabidopsis thaliana* and also on rice has provided tremendous advances in our understanding of plant Fe homeostasis in the past years. In particular, the primary uptake processes into the root are now based on molecular evidence for the genes involved in Strategy I (reduction based) and Strategy II (chelation based). The allocation of Fe from the rhizodermis to xylem and phloem for supply of young and growing tissues and recirculation to roots is, however, much less well understood. Finally, transport processes inside cells are beginning to be unraveled, explaining how iron homeostasis is mediated between cytosol, plastids, mitochondria and the vacuole. The key genes involved in these processes also represent possible candidates in the search for Fe use efficient plants.

A crucial process in homeostatic control is the chelation of the almost insoluble and redox active free Fe ions. Fe chelated by nicotianamine is carried across plasmalemma and endomembranes and also for long distance transport. Future work needs to address the mechanisms of donation of Fe from chelators to acceptor molecules such as heme, Fe-S clusters and proteins. In addition, the regulatory networks of transcription factors that

(continued)

function in the sensing of Fe deficiency, adaptation of root morphology and coordination with uptake of other nutrients are only beginning to be discovered. Recent observations indicate a co-regulation of Fe homeostasis with the uptake and metabolism of S, most likely triggered by the demand for Fe-S clusters in the electron transport chains of plastids and mitochondria. Detailed understanding of Fe homeostasis is prerequisite for the generation of Fe efficient plants and enhanced Fe contents in food.

References

Astolfi S, Zuchi S, Hubberten H-M, Pinton R, Hoefgen R (2010) Supply of sulphur to S-deficient young barley seedlings restores their capability to cope with iron shortage. J Exp Bot 61:799–806

Astolfi S, Zuchi S, Neumann G, Cesco S, Di Toppi LS, Pinton R (2012) Response of barley plants to Fe deficiency and Cd contamination as affected by S starvation. J Exp Bot 63:1241–1250

Balk J, Pilon M (2011) Ancient and essential: the assembly of iron–sulfur clusters in plants. Trends Plant Sci 16:218–226

Barberon M, Zelazny E, Robert S, Conejero G, Curie C, Friml J, Vert G (2011) Monoubiquitin-dependent endocytosis of the iron-regulated transporter 1 (IRT1) transporter controls iron uptake in plants. Proc Natl Acad Sci U S A 108:E450–E458

Bashir K, Ishimaru Y, Shimo H, Nagasaka S, Fujimoto M, Takanashi H, Tsutsumi N, An G, Nakanishi H, Nishizawa NK (2011) The rice mitochondrial iron transporter is essential for plant growth. Nat Commun 2:322

Bauer P, Ling HQ, Guerinot ML (2007) FIT, the FER-like iron deficiency induced transcription factor in *Arabidopsis*. Plant Physiol Biochem 45:260–261

Bernard DG, Cheng Y, Zhao Y, Balk J (2009) An allelic mutant series of ATM3 reveals its key role in the biogenesis of cytosolic iron-sulfur proteins in *Arabidopsis*. Plant Physiol 151:590–602

Briat JF, Duc C, Ravet K, Gaymard F (2010) Ferritins and iron storage in plants. Biochim Biophys Acta 1800:806–814

Busi MV, Maliandi MV, Valdez H, Clemente M, Zabaleta EJ, Araya A, Gomez-Casati DF (2006) Deficiency of *Arabidopsis thaliana* frataxin alters activity of mitochondrial Fe-S proteins and induces oxidative stress. Plant J 48:873–882

Cailliatte R, Schikora A, Briat JF, Mari S, Curie C (2010) High-affinity manganese uptake by the metal transporter NRAMP1 is essential for *Arabidopsis* growth in low manganese conditions. Plant Cell 22:904–917

Cakmak I, Erenoglu B, Gülüt K, Derici R, Römheld V (1998) Light-mediated release of phytosiderophores in wheat and barley under iron or zinc deficiency. Plant Soil 202:309–315

Chan KX, Wirtz M, Phua SY, Estavillo GM, Pogson BJ (2013) Balancing metabolites in drought: the sulfur assimilation conundrum. Trends Plant Sci 18:18–29

Chu HH, Chiecko J, Punshon T, Lanzirotti A, Lahner B, Salt DE, Walker EL (2010) Successful reproduction requires the function of *Arabidopsis* Yellow Stripe-Like1 and Yellow Stripe-Like3 metal-nicotianamine transporters in both vegetative and reproductive structures. Plant Physiol 154:197–210

Chu H-H, Conte SS, Chan Rodriguez D, Vasques K, Punshon T, Salt DE, Walker EL (2013) *Arabidopsis thaliana* Yellow Stripe1-Like4 and Yellow Stripe1-Like6 localize to internal cellular membranes and are involved in metal ion homeostasis. Front Plant Sci 4. doi:10.3389/fpls.2013.00283

Ciaffi M, Paolacci AR, Celletti S, Catarcione G, Kopriva S, Astolfi S (2013) Transcriptional and physiological changes in the S assimilation pathway due to single or combined S and Fe deprivation in durum wheat (*Triticum durum* L.) seedlings. J Exp Bot 64:1663–1675
Cohen CK, Fox TC, Garvin DF, Kochian LV (1998) The role of iron-deficiency stress responses in stimulating heavy-metal transport in plants. Plant Physiol 116:1063–1072
Colangelo EP, Guerinot ML (2004) The essential basic helix-loop-helix protein FIT1 is required for the iron deficiency response. Plant Cell 16:3400–3412
Connolly EL, Fett JP, Guerinot ML (2002) Expression of the IRT1 metal transporter is controlled by metals at the levels of transcript and protein accumulation. Plant Cell 14:1347–1357
Couturier J, Touraine B, Briat J-F, Gaymard F, Rouhier N (2013) The iron-sulfur cluster assembly machineries in plants: current knowledge and open questions. Front Plant Sci 4. doi:10.3389/fpls.2013.00259
Curie C, Alonso JM, Le Jean M, Ecker JR, Briat JF (2000) Involvement of NRAMP1 from *Arabidopsis* thaliana in iron transport. Biochem J 347(Pt 3):749–755
Curie C, Panaviene Z, Loulergue C, Dellaporta SL, Briat JF, Walker EL (2001) Maize yellow stripe1 encodes a membrane protein directly involved in Fe(III) uptake. Nature 409:346–349
Curie C, Cassin G, Couch D, Divol F, Higuchi K, Le Jean M, Misson J, Schikora A, Czernic P, Mari S (2009) Metal movement within the plant: contribution of nicotianamine and yellow stripe 1-like transporters. Ann Bot 103:1–11
Deinlein U, Weber M, Schmidt H, Rensch S, Trampczynska A, Hansen TH, Husted S, Schjoerring JK, Talke IN, Kramer U, Clemens S (2012) Elevated nicotianamine levels in *Arabidopsis halleri* roots play a key role in zinc hyperaccumulation. Plant Cell 24:708–723
Divol F, Couch D, Conejero G, Roschzttardtz H, Mari S, Curie C (2013) The *Arabidopsis* Yellow Stripe LIKE4 and 6 transporters control iron release from the chloroplast. Plant Cell 25:1040–1055
Duc C, Cellier F, Lobreaux S, Briat JF, Gaymard F (2009) Regulation of iron homeostasis in *Arabidopsis thaliana* by the clock regulator time for coffee. J Biol Chem 284:36271–36281
Durrett TP, Gassmann W, Rogers EE (2007) The FRD3-mediated efflux of citrate into the root vasculature is necessary for efficient iron translocation. Plant Physiol 144:197–205
Duy D, Wanner G, Meda AR, Von Wiren N, Soll J, Philippar K (2007) PIC1, an ancient permease in *Arabidopsis* chloroplasts, mediates iron transport. Plant Cell 19:986–1006
Eckhardt U, Mas Marques A, Buckhout TJ (2001) Two iron-regulated cation transporters from tomato complement metal uptake-deficient yeast mutants. Plant Mol Biol 45:437–448
Forieri I, Wirtz M, Hell R (2013) Towards new perspectives on the interaction of iron and sulfur metabolism in plants. Front Plant Sci 4. doi:10.3389/fpls.2013.00357
Haydon MJ, Kawachi M, Wirtz M, Hillmer S, Hell R, Kramer U (2012) Vacuolar nicotianamine has critical and distinct roles under iron deficiency and for zinc sequestration in *Arabidopsis*. Plant Cell 24:724–737
Hell R, Stephan U (2003) Iron uptake, trafficking and homeostasis in plants. Planta 216:541–551
Inoue H, Kobayashi T, Nozoye T, Takahashi M, Kakei Y, Suzuki K, Nakazono M, Nakanishi H, Mori S, Nishizawa NK (2009) Rice OsYSL15 is an iron-regulated iron (III)-deoxymugineic acid transporter expressed in the roots and is essential for iron uptake in early growth of the seedlings. J Biol Chem 284:3470–3479
Ishimaru Y, Suzuki M, Tsukamoto T, Suzuki K, Nakazono M, Kobayashi T, Wada Y, Watanabe S, Matsuhashi S, Takahashi M, Nakanishi H, Mori S, Nishizawa NK (2006) Rice plants take up iron as an Fe^{3+}-phytosiderophore and as Fe^{2+}. Plant J 45:335–346
Jeong J, Cohu C, Kerkeb L, Pilon M, Connolly EL, Guerinot ML (2008) Chloroplast Fe(III) chelate reductase activity is essential for seedling viability under iron limiting conditions. Proc Natl Acad Sci U S A 105:10619–10624
Kerkeb L, Mukherjee I, Chatterjee I, Lahner B, Salt DE, Connolly EL (2008) Iron-induced turnover of the *Arabidopsis* IRON-REGULATED TRANSPORTER1 metal transporter requires lysine residues. Plant Physiol 146:1964–1973

Kim SA, Punshon T, Lanzirotti A, Li L, Alonso JM, Ecker JR, Kaplan J, Guerinot ML (2006) Localization of iron in *Arabidopsis* seed requires the vacuolar membrane transporter VIT1. Science 314:1295–1298

Klatte M, Schuler M, Wirtz M, Fink-Straube C, Hell R, Bauer P (2009) The analysis of *Arabidopsis* nicotianamine synthase mutants reveals functions for nicotianamine in seed iron loading and iron deficiency responses. Plant Physiol 150:257–271

Kobayashi T, Nishizawa NK (2012) Iron uptake, translocation and regulation in higher plants. Annu Rev Plant Biol 63:131–152

Kobayashi T, Nakayama Y, Itai RN, Nakanishi H, Yoshihara T, Mori S, Nishizawa NK (2003) Identification of novel cis-acting elements, IDE1 and IDE2, of the barley IDS2 gene promoter conferring iron-deficiency-inducible, root-specific expression in heterogeneous tobacco plants. Plant J 36:780–793

Kobayashi T, Ogo Y, Itai RN, Nakanishi H, Takahashi M, Mori S, Nishizawa NK (2007) The transcription factor IDEF1 regulates the response to and tolerance of iron deficiency in plants. Proc Natl Acad Sci U S A 104:19150–19155

Kobayashi T, Itai RN, Ogo Y, Kakei Y, Nakanishi H, Takahashi M, Nishizawa NK (2009) The rice transcription factor IDEF1 is essential for the early response to iron deficiency and induces vegetative expression of late embryogenesis abundant genes. Plant J 60:948–961

Kobayashi T, Ogo Y, Aung MS, Nozoye T, Itai RN, Nakanishi H, Yamakawa T, Nishizawa NK (2010) The spatial expression and regulation of transcription factors IDEF1 and IDEF2. Ann Bot 105:1109–1117

Kobayashi T, Itai RN, Aung MS, Senoura T, Nakanishi H, Nishizawa NK (2012) The rice transcription factor IDEF1 directly binds to iron and other divalent metals for sensing cellular iron status. Plant J 69:81–91

Kushnir S, Babiychuk E, Storozhenko S, Davey MW, Papenbrock J, De Rycke R, Engler G, Stephan UW, Lange H, Kispal G, Lill R, Van Montagu M (2001) A mutation of the mitochondrial ABC transporter Sta1 leads to dwarfism and chlorosis in the *Arabidopsis* mutant starik. Plant Cell 13:89–100

Lanquar V, Lelievre F, Bolte S, Hames C, Alcon C, Neumann D, Vansuyt G, Curie C, Schroder A, Kramer U, Barbier-Brygoo H, Thomine S (2005) Mobilization of vacuolar iron by AtNRAMP3 and AtNRAMP4 is essential for seed germination on low iron. EMBO J 24:4041–4051

Lill R (2009) Function and biogenesis of iron-sulphur proteins. Nature 460:831–838

Lin YF, Liang HM, Yang SY, Boch A, Clemens S, Chen CC, Wu JF, Huang JL, Yeh KC (2009) *Arabidopsis* IRT3 is a zinc-regulated and plasma membrane localized zinc/iron transporter. New Phytol 182:392–404

Ling HQ, Koch G, Baumlein H, Ganal MW (1999) Map-based cloning of chloronerva, a gene involved in iron uptake of higher plants encoding nicotianamine synthase. Proc Natl Acad Sci U S A 96:7098–7103

Ling HQ, Bauer P, Bereczky Z, Keller B, Ganal M (2002) The tomato fer gene encoding a bHLH protein controls iron-uptake responses in roots. Proc Natl Acad Sci U S A 99:13938–13943

Lingam S, Mohrbacher J, Brumbarova T, Potuschak T, Fink-Straube C, Blondet E, Genschik P, Bauer P (2011) Interaction between the bHLH transcription factor FIT and ETHYLENE INSENSITIVE3/ETHYLENE INSENSITIVE3-LIKE1 reveals molecular linkage between the regulation of iron acquisition and ethylene signaling in *Arabidopsis*. Plant Cell 23:1815–1829

Liu Z, Hong SW, Escobar M, Vierling E, Mitchell DL, Mount DW, Hall JD (2003) *Arabidopsis* UVH6, a homolog of human XPD and yeast RAD3 DNA repair genes, functions in DNA repair and is essential for plant growth. Plant Physiol 132:1405–1414

Long TA, Tsukagoshi H, Busch W, Lahner B, Salt DE, Benfey PN (2010) The bHLH transcription factor POPEYE regulates response to iron deficiency in *Arabidopsis* roots. Plant Cell 22:2219–2236

Marschner H (1995) Mineral nutrition of higher plants. Academic/Harcourt Brace & Company, London, pp 313–324

Mäser P, Thomine S, Schroeder JI, Ward JM, Hirschi K, Sze H, Talke IN, Amtmann A, Maathuis FJ, Sanders D, Harper JF, Tchieu J, Gribskov M, Persans MW, Salt DE, Kim SA, Guerinot ML (2001) Phylogenetic relationships within cation transporter families of *Arabidopsis*. Plant Physiol 126:1646–1667

Morrissey J, Guerinot ML (2009) Iron uptake and transport in plants: the good, the bad, and the ionome. Chem Rev 109:4553–4567

Morrissey J, Baxter IR, Lee J, Li L, Lahner B, Grotz N, Kaplan J, Salt DE, Guerinot ML (2009) The ferroportin metal efflux proteins function in iron and cobalt homeostasis in *Arabidopsis*. Plant Cell 21:3326–3338

Nakanishi H, Yamaguchi H, Sasakuma T, Nishizawa NK, Mori S (2000) Two dioxygenase genes, Ids3 and Ids2, from *Hordeum vulgare* are involved in the biosynthesis of mugineic acid family phytosiderophores. Plant Mol Biol 44:199–207

Nikiforova V, Freitag J, Kempa S, Adamik M, Hesse H, Hoefgen R (2003) Transcriptome analysis of sulfur depletion in *Arabidopsis thaliana*: interlacing of biosynthetic pathways provides response specificity. Plant J 33:633–650

Nozoye T, Nagasaka S, Kobayashi T, Takahashi M, Sato Y, Sato Y, Uozumi N, Nakanishi H, Nishizawa NK (2011) Phytosiderophore efflux transporters are crucial for iron acquisition in graminaceous plants. J Biol Chem 286:5446–5454

Ogo Y, Itai RN, Nakanishi H, Inoue H, Kobayashi T, Suzuki M, Takahashi M, Mori S, Nishizawa NK (2006) Isolation and characterization of IRO2, a novel iron-regulated bHLH transcription factor in graminaceous plants. J Exp Bot 57:2867–2878

Ogo Y, Kobayashi T, Nakanishi Itai R, Nakanishi H, Kakei Y, Takahashi M, Toki S, Mori S, Nishizawa NK (2008) A novel NAC transcription factor, IDEF2, that recognizes the iron deficiency-responsive element 2 regulates the genes involved in iron homeostasis in plants. J Biol Chem 283:13407–13417

Ogo Y, Itai RN, Kobayashi T, Aung MS, Nakanishi H, Nishizawa NK (2011) OsIRO2 is responsible for iron utilization in rice and improves growth and yield in calcareous soil. Plant Mol Biol 75:593–605

Palmgren MG (2001) Plant plasma membrane H+-ATPases: powerhouses for nutrient uptake. Annu Rev Plant Physiol Plant Mol Biol 52:817–845

Petit JM, Van Wuytswinkel O, Briat JF, Lobreaux S (2001) Characterization of an iron-dependent regulatory sequence involved in the transcriptional control of AtFer1 and ZmFer1 plant ferritin genes by iron. J Biol Chem 276:5584–5590

Ravet K, Touraine B, Boucherez J, Briat JF, Gaymard F, Cellier F (2009) Ferritins control interaction between iron homeostasis and oxidative stress in *Arabidopsis*. Plant J 57:400–412

Rellan-Alvarez R, Giner-Martinez-Sierra J, Orduna J, Orera I, Rodriguez-Castrillon JA, Garcia-Alonso JI, Abadia J, Alvarez-Fernandez A (2010) Identification of a tri-iron (III), tri-citrate complex in the xylem sap of iron-deficient tomato resupplied with iron: new insights into plant iron long-distance transport. Plant Cell Physiol 51:91–102

Robinson NJ, Procter CM, Connolly EL, Guerinot ML (1999) A ferric-chelate reductase for iron uptake from soils. Nature 397:694–697

Rogers EE, Guerinot ML (2002) FRD3, a member of the multidrug and toxin efflux family, controls iron deficiency responses in *Arabidopsis*. Plant Cell 14:1787–1799

Romheld V, Marschner H (1986) Evidence for a specific uptake system for iron phytosiderophores in roots of grasses. Plant Physiol 80:175–180

Santi S, Schmidt W (2009) Dissecting iron deficiency-induced proton extrusion in *Arabidopsis* roots. New Phytol 183:1072–1084

Schuler M, Keller A, Backes C, Philippar K, Lenhof H-P, Bauer P (2011) Transcriptome analysis by GeneTrail revealed regulation of functional categories in response to alterations of iron homeostasis in *Arabidopsis thaliana*. BMC Plant Biol 11:87

Schuler M, Rellan-Alvarez R, Fink-Straube C, Abadia J, Bauer P (2012) Nicotianamine functions in the phloem-based transport of iron to sink organs, in pollen development and pollen tube growth in *Arabidopsis*. Plant Cell 24:2380–2400

Seckback J (1982) Ferreting out the secrets of plant ferritin – a review. J Plant Nutr 5:369–394
Shin LJ, Lo JC, Chen GH, Callis J, Fu H, Yeh KC (2013) IRT1 DEGRADATION FACTOR1, a RING E3 ubiquitin ligase, regulates the degradation of IRON-REGULATED TRANSPORTER1 in *Arabidopsis*. Plant Cell 25:3039–3051
Sivitz AB, Hermand V, Curie C, Vert G (2012) *Arabidopsis* bHLH100 and bHLH101 control iron homeostasis via a FIT-independent pathway. PLoS One 7:e44843
Tarantino D, Petit JM, Lobreaux S, Briat JF, Soave C, Murgia I (2003) Differential involvement of the IDRS cis-element in the developmental and environmental regulation of the AtFer1 ferritin gene from *Arabidopsis*. Planta 217:709–716
Tarantino D, Santo N, Morandini P, Casagrande F, Braun H-P, Heinemeyer J, Vigani G, Soave C, Murgia I (2010) AtFer4 ferritin is a determinant of iron homeostasis in *Arabidopsis thaliana* heterotrophic cells. J Plant Physiol 167:1598–1605
Teng YS, Su YS, Chen LJ, Lee YJ, Hwang I, Li HM (2006) Tic21 is an essential translocon component for protein translocation across the chloroplast inner envelope membrane. Plant Cell 18:2247–2257
Varotto C, Maiwald D, Pesaresi P, Jahns P, Salamini F, Leister D (2002) The metal ion transporter IRT1 is necessary for iron homeostasis and efficient photosynthesis in *Arabidopsis thaliana*. Plant J 31:589–599
Vert G, Briat JF, Curie C (2001) *Arabidopsis* IRT2 gene encodes a root-periphery iron transporter. Plant J 26:181–189
Vert G, Grotz N, Dedaldechamp F, Gaymard F, Guerinot ML, Briat JF, Curie C (2002) IRT1, an *Arabidopsis* transporter essential for iron uptake from the soil and for plant growth. Plant Cell 14:1223–1233
Vert GA, Briat JF, Curie C (2003) Dual regulation of the *Arabidopsis* high-affinity root iron uptake system by local and long-distance signals. Plant Physiol 132:796–804
Vigani G, Zocchi G, Bashir K, Philippar K, Briat JF (2013) Signals from chloroplasts and mitochondria for iron homeostasis regulation. Trends Plant Sci 18:305–311
Von Wiren N, Mori S, Marschner H, Romheld V (1994) Iron inefficiency in maize mutant ys1 (*Zea mays* L. cv Yellow-Stripe) is caused by a defect in uptake of iron phytosiderophores. Plant Physiol 106:71–77
Wang HY, Klatte M, Jakoby M, Baumlein H, Weisshaar B, Bauer P (2007) Iron deficiency-mediated stress regulation of four subgroup Ib BHLH genes in *Arabidopsis thaliana*. Planta 226:897–908
Waters BM, Blevins DG, Eide DJ (2002) Characterization of FRO1, a pea ferric-chelate reductase involved in root iron acquisition. Plant Physiol 129:85–94
Yi Y, Guerinot ML (1996) Genetic evidence that induction of root Fe(III) chelate reductase activity is necessary for iron uptake under iron deficiency. Plant J 10:835–844
Yuan Y, Wu H, Wang N, Li J, Zhao W, Du J, Wang D, Ling HQ (2008) FIT interacts with AtbHLH38 and AtbHLH39 in regulating iron uptake gene expression for iron homeostasis in *Arabidopsis*. Cell Res 18:385–397
Zancani M, Peresson C, Biroccio A, Federici G, Urbani A, Murgia I, Soave C, Micali F, Vianello A, Macrì F (2004) Evidence for the presence of ferritin in plant mitochondria. Eur J Biochem 271:3657–3664
Zheng L, Ying Y, Wang L, Wang F, Whelan J, Shou H (2010) Identification of a novel iron regulated basic helix-loop-helix protein involved in Fe homeostasis in *Oryza sativa*. BMC Plant Biol 10:166
Zuchi S, Cesco S, Varanini Z, Pinton R, Astolfi S (2009) Sulphur deprivation limits Fe-deficiency responses in tomato plants. Planta 230:85–94

Chapter 6
Boron: A Promising Nutrient for Increasing Growth and Yield of Plants

Himanshu Bariya, Snehal Bagtharia, and Ashish Patel

Abstract Boron (B) is a vital nutrient for plant growth and metabolism. Lack of B in plant tissues causes reductions in crop yields, whilst an excess supply of B may also seriously damage plant tissues and sometimes leads to plant death. Appropriate amounts of B in plants are crucial for normal growth and it significantly increases seed germination and seedling growth. Moreover B has positive effects on the uptake and utilization of other nutrients at the whole plant level and it may improve nutrient use efficiency (NUE) and nutrient demand and supply (NDS). NUE mainly reflects efficiency of extraction of mineral nutrients from soil along with their integration and recycling, whereas NDS nutrient shows how efficiently plant can fulfil the demand and supply rate of required nutrient at different stages and conditions of plant life cycle. Despite a substantial existing literature, the understanding of B interactions with other nutrients remains unclear.

Keywords Boron • NUE (nitrogen use efficiency) • Soil interactions • Deficiency • Reproductive growth • Photosynthesis • Nutrient interactions

Introduction: Boron in the Soil and in the Plant

Boron belongs to the metalloid elements having properties of both metals and non-metals (Marschner 1995). There is very low abundance of B in nature (Kot 2009) but it is broadly distributed in all the layers of the soil. B abundance range in rocks averages about 10–20 mg B kg^{-1}. In sea water it can range from 1 to 10 mg B kg^{-1} and as far as river is concern the B concentration is about 1/350 that of sea water (Power and Woods 1997).

Warington (1923) established the requirement of B for plant growth and functioning, and many recent reports suggest an essentiality of B for all vascular plants.

H. Bariya • A. Patel (✉)
Department of Biotechnology, Hemchandra North Gujarat University,
Patan 384265, Gujarat, India
e-mail: uni.ashish@gmail.com

S. Bagtharia
Gujarat State Biotechnology Mission (GSBTM), DST – Government of Gujarat,
Gandhinagar 382017, Gujarat, India

M.J. Hawkesford et al. (eds.), *Nutrient Use Efficiency in Plants*,
Plant Ecophysiology 10, DOI 10.1007/978-3-319-10635-9_6

B deficiency or toxicity affects various metabolic and physiological processes (Blevins and Lukaszewski 1998; Bolaños et al. 2004). Less than 10 mg kg^{-1} B in the soil is considered to be a B deficient soil (Woods 1994). Moreover, most of this B is in a bound form in rocks and is not readily available to plants. Boric acid is the most common form of B liberated during weathering of rocks (Nable et al. 1997) and is easily absorbed by plant roots, but this represents only 10 % of the total B in the soil (Power and Woods 1997).

Soil pH, texture, temperature, and organic matter affect soil B availability; among these the soil pH is one of the most significant characteristics for B availability (Goldberg 1997). Boric acid is a very weak acid and when the pH is below 7, it appears in its undissociated form; at alkaline pH, boric acid dissociates to form the borate anion:

$$B(OH)_3 + H_2O \Longleftrightarrow B(OH)_4^- + H^+ \qquad (pKa\ 9.25)$$

Therefore, at common soil pH values (5.5–7.5), B exists mainly as soluble uncharged boric acid ($B(OH)_3$), and in this form B is absorbed by plant roots (Hu and Brown 1997; Power and Woods 1997; Sathya et al. 2013). In flooded conditions, B is easily solubilised and leached resulting in a deficiency of B for growing plants.

A plant requirement for B varies between species and is dependent on the conditions in which they grow. The B requirement for one plant species may be toxic or deficient for other species (Blevins and Lukaszewski 1998). There are three main groups of plants based on B requirements, one consists of graminaceous species, a second are monocotyledons and a third are dicotyledons and lactifers, having minimum, moderate and high B demands, respectively (Blevins and Lukaszewski 1998; Goldbach et al. 2001). It is necessary to study mechanisms of B uptake and transport, as well as translocation in plant systems, to optimise agricultural production. For example, B deficiency results in major disorders, which may decrease plant growth in soil with excess water. Molecular mechanisms involved in B transport from the soil to root cells and xylem have two different routes. Transport of B is dependent mainly on its availability and it is mostly transported into the plant from the soil by passive diffusion, which is possible when adequate B is available. When B availability is low or under deficiency conditions, uptake is facilitated by active transport, which requires energy.

B absorbed by roots must be delivered to the xylem for further transport around the plant. When there is sufficient B availability, absorbed B moves by passive diffusion involving MIPs channels (Dannel et al. 2002; Miwa and Fujiwara 2010). In limited or deficient B conditions, transport of B towards the xylem is facilitated via a specific B transporter (BOR), which is an energy dependent transport process (Fig. 6.1). Takano et al. (2002) had identified such a transporter of B under limiting conditions, known as BOR1 in *Arabidopsis*. Subsequently, BOR1-like genes have been reported in *Eucalyptus* (Domingues et al. 2005) and in rice (Nakagawa et al. 2007). Expression of two genes, NIP5;1 and BOR1, are decreased by transcriptional and post-translational regulation respectively, under sufficient B supply (Miwa and Fujiwara 2010; Yang et al. 2013).

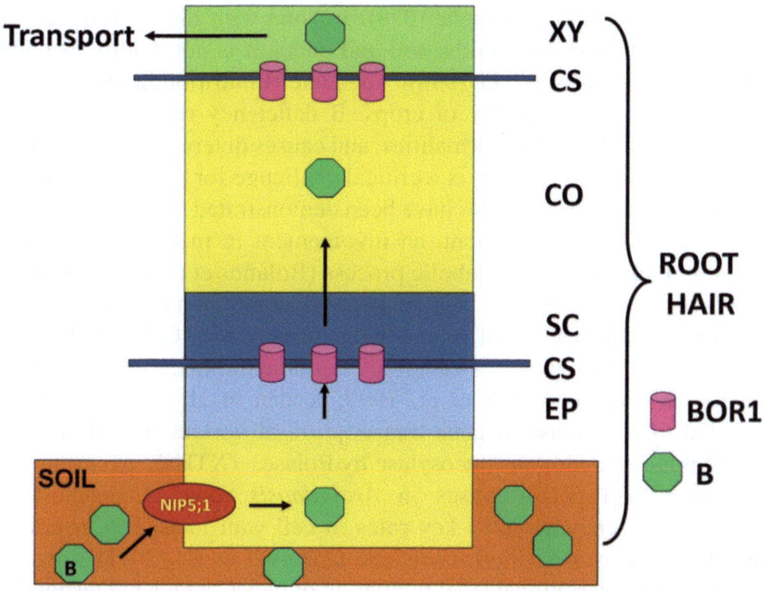

Fig. 6.1 Schematic diagram of boron transport from soil to xylem via root tissue. In sufficient B condition NIP5;1 imports B from the soil to epidermal, cortex and endodermal cells, and BOR 1 exports B from stelar cell (xylem loading)

After loading in the xylem, B is transported to the shoot through the transpiration stream (Wimmer and Eichert 2013; Bogiani et al. 2014). Transport of B through phloem was also reported, and such transport differs between species (Brown and Hu 1996; Brown and Shelp 1997; Ganie et al. 2013). In certain plants, B is transported and translocated to reproductive and vegetative tissue via the phloem (Matoh and Ochiai 2005). It is also suggested that there is a formation of a boron-diol complex involving a sugar alcohol, which acts as the transport molecule (Brown and Hu 1996; Hu et al. 1997). Transgenic tobacco and rice with enhanced sorbitol production had a higher ability to transport B through the phloem towards the plant shoot (Brown et al. 1999; Bellaloui et al. 2003). Plants that produce sugar alcohols like sorbitol and trehalose, have the ability of B transport via the phloem, whereas plants without any production of sugar alcohols have no such transport system (Stangoulis et al. 2001; Takano et al. 2001; Matoh and Ochiai 2005).

Role of Boron in Plant Functioning

Since the beginning of twentieth century, B has been considered to be an important micronutrient for plant growth, but there are very few records regarding its actual biochemical role. Deficiency of B is common all over the world, which has an important agronomic impact (Gupta 1979). In soils with high percolatory water, B is easily leached downward in the soil and hence it is not readily available for plants (Blevins and Lukaszewski 1998). Adequate B nutrition is required for high production as well as for quality of crops. B deficiency results in biochemical, metabolic and physiological abnormalities, and causes diverse disease symptoms in plants and hence adequate supply is a critical challenge for plant nutrition.

In the last 10 years, several roles have been demonstrated for B in plant function, including as a cell wall component, an involvement in membrane structure and integrity, and involvement in metabolic process (Bolaños et al. 2004). To date, one of the most accepted roles of B in plant physiological function is the formation of an ester with one apiose residue of rhamnogalacturonan II (RG II) in the cell wall (Kobayashi et al. 1996), which is essential for maintaining cell wall permeability and also for rigidity (Fleischer et al. 1999; Ryden et al. 2003). Moreover, B deficiency led to a decrease of gene transcription of various hydrolytic enzymes such as xyloglucan endotransglucosylase/hydrolases (XTHs), expansins, pectin methylesterases, and pectin lyases in *Arabidopsis* roots (Camacho-Cristóbal et al. 2002). These enzymes play key roles in cell wall loosening, necessary for cell elongation (Cosgrove 1999). Camacho-Cristobal et al. (2011) suggested an influence of B in transcriptional level regulation of genes, which are responsible for cell wall synthesis and its modification.

Many research reports suggested roles of B in plasma membrane transport processes, and in membrane integrity by cross-linking the membrane molecules containing hydroxlated ligands such as glycoproteins and glycolipids (Goldbach et al. 2001; Wimmer et al. 2009). The membrane potential in *Daucus carota* is changed under limited B (Blaser-Grill et al. 1989) and activity of the proton-pumping ATPase was reduced in *Helianthus annuus* roots (Ferrol and Donaire 1992). Similarly, it has been reported that B deficiency alters plasma membrane permeability for ions and other solutes (Wang et al. 1999; Carmen Rodriguez Hernandez et al. 2013). The impact of B on ion fluxes can be mediated by direct or indirect effects of B on plasma membrane-bound proton-pumping ATPase (Cara et al. 2002). The activity of the K^+ stimulated ATPase in B-deficient maize roots was considerably lower than in control plants (Pollard et al. 1977). These results indicated that the action of B could be associated with membrane components. It is still unclear whether B directly interacts with membrane proteins or indirectly modifies membrane properties with subsequent changes in enzymatic activities.

The literature indicates possible roles of B in several other metabolic functions. For instance, it has been shown that B deficiency causes qualitative and quantitative changes in phenol metabolism (Pandey and Archana 2012; Hajiboland et al. 2013). Additionally B deficiency affects nitrogen metabolism (Bolaños et al. 1994). B-deficient plants showed lower nitrate reductase activity and enhanced amounts of nitrate; these observations strongly suggest a role of B in the de novo synthesis of the nitrate reductase protein or facilitation of nitrate absorption (Ruiz et al. 1998).

Effect of B on Photosynthesis

Photosynthesis is a complex series of reactions that culminate in the reduction of carbon dioxide. The effect of B nutrition on photosynthetic processes has been rarely studied. Decreases in B supply reduce soluble proteins and chlorophyll in leaves, which are important constituents for Hill reactions and photosynthesis (Mukhopdhyaya et al. 2013).

There was no direct involvement of B on rate of photosynthesis, but from recent research it is evident that B has a positive effect on photosynthesis under normal (optimum) level. Moreover, there are reports of increases in chlorophyll pigment and carotenoids by foliar spraying of B leading to increases in photosynthetic rate (Thurzo et al. 2010). Ganie et al. (2013) also showed that net photosynthetic rate was increased due to increases in plant light harvesting pigments such as chlorophyll and carotene in the leaves.

B deficiency negatively affects photosynthesis by decreasing photosynthetic oxygen evolution rates and hence the efficiency of photosystem II (Kastori et al. 1995; El-Shintinawy 1999). Photosynthetic rate (P_n) drastically decreased in cotton plants when grown in B deficient soil (Zhao and Oosterhuis 2002). It was reported that inhibition of photosynthesis was a result of reduced Hill reactions and low intercellular CO_2 concentrations (Sharma and Ramchandra 1990). Some experimental evidence indicated close relationships between gas exchange parameters and B deficiency, implying that these parameters were possibly affected by external B supply and in turn prejudiced growth. Previous studies, under B deficiency, supported a change in photosynthetic enzyme activities that were undoubtedly involved in a decrease in P_n (Sharma and Ramchandra 1990). B deficiency was followed by reduction in the leaf stomatal conductance (g_s) and rate of photosynthesis (Huang et al. 2005; Han et al. 2008). Related declines in plant g_s at B deficiency were initiated by high oxidative damage in leaves (Huang et al. 2005). Moreover, low g_s also decreases E (Han et al. 2008). It was advocated that the presence of free hexoses can elicit regulation of the Calvin cycle and, hence, can obstruct P_n (Han et al. 2008). A diminution in pigments under B deficiency was shown for citrus (Han et al. 2008). The synthesis of unnecessary starch possibly disrupts chloroplast structure, leading to poorer CO_2 assimilation and decreased chlorophyll content (Han et al. 2008). It is likely that B deficiency caused changes in chloroplast structure, which eventually affected pigment content (Pandey and Pandey 2008). B deficiency may also affect photosynthetic responses by the modifications in the structure and function of chloroplast thylakoids.

Effect on Reproductive Growth

According to Loomis and Durst (1992) B is essential for generative growth. Boron is involved in metabolism of carbohydrates and phenolic acids, which are crucial for growth of pollen tubes. Decreases in crop propagative yield or seed/fruit quality in low B soils can be due to diminished reproductive development early or late in the flowering/fruiting cycle. It has often been seen that reproductive growth, mainly flowering, fruit and seed set and seed yield, is particularly sensitive to B deficiency compared to asexual growth (e.g. Woodbridge et al. 1971; Dear and Lipsett 1987; Noppakoonwong et al. 1997). Likewise, substantial yield decreases can arise without expression of indications of deficiency during prior somatic growth. B also plays an important role in synthesis and metabolism of nucleic acids (Hundt et al. 1970).

As described above, B is responsible for enhancing chlorophyll content and rate of photosynthesis (P_n), as well as inducing dry matter production in plants, and therefore may result both in enhancing flowering and also the transport of photosynthetic products to reproductive stages, ultimately leading to yield improvement (Du Ying Qiong et al. 1999).

Factors influencing the impact of a low external B supply on sexual reproduction in flowering plants are likely to include: the capacity of roots to obtain B from soil (Hu and Brown 1997); the mobility of B in the phloem (Brown and Shelp 1997); the relative sink size in floral parts for photosynthate; the capacity to redistribute B from vegetative tissues to reproductive organs; the rate of transpiration by floral organs; the functional necessity for B in reproductive tissues; and the distribution and richness of B-binding compounds in the apoplastic pathway between the vein endings and the most distal floral tissue. The B requirement for flowering is indicated by the sensitivity of pollen development to low B and the generally high concentrations of B that occur in reproductive parts of the flower. Under conditions of low external B supply, levels in the anthers and pistils do not decline to the low levels measured in leaves. B concentrations are higher in the stamen than in the pistil. The physiological roles for B in sexual reproduction have yet to be fully defined and there is a need for experimentation in this area.

Many of the studies that have been undertaken do not give definitive information as to whether plants were critically deficient in B at the time of flowering, or the B status of floral tissues was not determined at the time of impairment in cellular development/function, or cell structure and metabolism were examined long after the primary effect of B took place. An example is the observation that B deficiency results in male sterility, a condition that can be induced by deficiencies in other nutrients (e.g. Mn, Cu) or unfavourable environmental conditions (e.g. water deficit, high/low temperature). These observations reveal nothing about the processes that are being affected by low B supply and do not enable us to conclude whether the effect of low B supply on male sterility is a direct or indirect event. As root function is greatly impaired in severely B-deficient plants in containers and this can impact on whole plant physiology, the requirement of pollen development for B should be studied under conditions of controlled external supply.

Flowering Response

There is no evidence that B deficiency prevents initiation of or delays floral development. When plants are grown without B or are transferred into B-free nutrient solutions in pot experiments, the apical meristems may abort and therefore flowers do not develop, as shown for peach (*Prunus persica*; Kamali and Childers 1970). However, under field conditions, where B deficiency stresses occur more gradually, plants may have time to adapt to deficiency. For example, in peanut (*Arachis hypogaea*), the flowering period was extended in B-deficient plants (Harris and Brolman 1966), resulting in low-B plants producing as many flowers as B-adequate plants. In species where the flowers occur in compact inflorescences, and these are terminal on the stem (e.g. sunflower, wheat), low B has a greater impact on reproduction because the plant has less ability to modulate reproductive growth than species with axillary inflorescences and indeterminate growth. The first group of plants is more prone to pollen sterility under low B than the latter group of plants. Experiments on wheat by Li et al. (1978), suggested that absorbed B was transported from soil to floret and spikelet organs, which resulted in accumulation of B in seeds. Finally, low B can result in plants being functionally male sterile (e.g. wheat, Li et al. 1978; rice, Garg et al. 1979; barley, Simojoki 1972), although cases of female sterility have been reported (e.g. maize, Vaughan 1977; avocado, Coetzer and Robbertse 1987). The external B supply does not appear to alter the frequency of unisexual flowers, although some workers have attempted unsuccessfully to do this by applying B sprays (Singh 1994).

Effect of B on Nutrient Use Efficiency, Demand and Supply

Nutrient use efficiency (NUE) may be expressed as productivity of the plant per content of applied nutrients. Enhancement of NUE is vital for improvement of crop production in marginal lands with poor nutrient availability. NUE for plants is reliant on the ability to efficiently take up nutrients from the soil, but also on translocation, storage and usage within the plant, and on the environment (North et al. 2009). NUE is largely dependent on nutrient availability in the soil or applied medium. Nutrient demand and supply (NDS) is how efficiently the plant can fulfil the demand and supply rate for required nutrients at different stages and conditions of plant growth and development. Under limiting conditions of nutrients, plants showed decreases in nutrient uptake compared to sufficient nutrient conditions (preserving the nutrition for future demand).

Large variations in defining nutrient efficient plants and methods used in calculating nutrient use efficiency, makes it difficult to compare results of different studies. The effort to measure yield response to an applied nutrient is further confounded by other factors, such as variable soil fertility levels, climatic conditions, crop rotations, and changes in production practices that affect nutrient use efficiency (Stewart et al. 2005). In simple terms, efficiency is the ratio of output

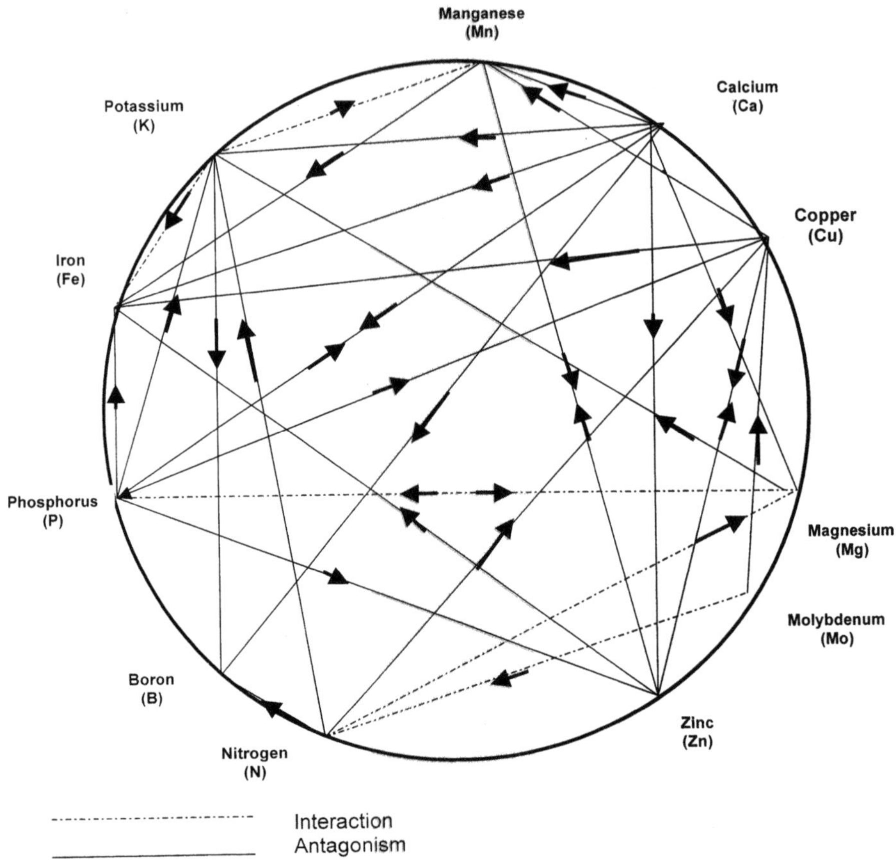

Fig. 6.2 Mulder's chart shows some of the interactions between plant nutrients. Interaction: A decrease in availability to the plant of a nutrient by the action of another nutrient (see direction of *arrow*). Stimulation: An increase in the need for a nutrient by the plant because of the increase in the level of another nutrient (*dotted line*)

(economic yield) to input (fertilisers) for a process or complex system (Crop Science Society of America 1992).

Plant nutrients rarely work in isolation. Interactions among nutrients are important because a deficiency of one restricts the uptake and use of another. Figure 6.2 shows interactions of major plant nutrients with each other (Khan Towhid Osman 2013). Numerous studies have demonstrated that interactions between B and other nutrients, primarily N, P and K, impact crop yields and nutrient efficiency.

For example, for experiments carried out in our research group on groundnut it was evident that uptake of almost all macro and micronutrients by straw and seeds showed a linear relationship ($p < 0.05$) among the different nutrients, which were absorbed by roots. B at a level of 1 kg ha^{-1} resulted in significant uptake of macro and micronutrients from soil to seed and straw. Data strongly support improvement of

NUE as well as NDS in groundnut at 1.0 kg ha^{-1} B supply. Other levels of B treatment did not significantly increase the uptake of nutrients in groundnut. The same trend of nutrient uptake by plants at different levels of applied B has also been reported by Nadia et al. (2006) and 200 ppm B sprayed on groundnut plant showed an increase in uptake of N, P, K and Fe, Mn and Zn. The highest value of yield and yield components were received from the plant treated with 200 ppm B (Nadia et al. 2006). Leaf venation, xylem stream, and transpiration are the primarily factors involved in the accumulation of B in leaves (Oertli and Richardson 1970; Shelp and Shattuck 1987).

These results are similar to the observations of McIlrath et al. (1960). It has been suggested that selected nutrients in the soil have antagonist and/or synergistic effect on the uptake of other nutrients by roots of developing plants (Malvi 2011). Moreover, B interactions, either synergistic and/or antagonistic, may affect plant nutrition under both deficiencies as well as in toxic conditions (Tariq and Mott 2007).

Deficiency, sufficiency and toxicity of B may exert an effect on mineral nutrient content of plants but such an interaction has not been well studied or reported. The results of many reports in this direction are conflicting, which may be due to different experimental systems with different crop plants and varieties (Lombin and Bates 1982; Mozafar 1989). B is directly or indirectly involved in many physiological and biochemical processes and may affect other plant nutrients (Bolaños et al. 2004). Therefore, one might expect relationships between B and other nutrient utilization to be very complex. Examples of effects of B on availability and uptake of plant nutrients other than B are described below.

Parks et al. (1944) were the first researchers who reported that with graded B levels, the concentrations of NH_4-N, NO_3-N, Org-N, P, K, Ca, Mg, Na, Zn, Cu, Fe, Mn, Mo and B were altered in the tomato leaflets as much as several-fold. In addition, they stated that B supply had specific effects, and the trends found were completely dissimilar with respect to different elements. In the absence of B, the concentrations of N, K, Ca, Mg, Na, Cu and Mn in tobacco leaves were increased and the concentrations of P, Fe and Al were decreased as compared to plants fed with a B adequate nutrient solution (Steinburg et al. 1955).

Baker and Cook (1959) reported that P, K and Mg were higher and Ca was lower in severely B deficient alfalfa plants compared to healthy plants, perhaps due to the dilution effect which occurred in the healthy plants. In increasing B conditions, the concentration of Cu, Fe, Mn, Mo, and B were increasing in perennial fodder grass, but the reverse trend occurred in the case of uptake and ash content for micronutrients accept B (Mcllarth et al. 1960). Cu and K content in grass showed highly significant positive correlations, while Ca and Mg contents showed negative correlations with B contents for 98 grasses at the flowering stage when grown in increasing B nutrition (Tolgyesi and Kozma 1974). Touchton and Boswell (1975) observed that P, K, Ca, Mg, Na, Zn, Cu, Fe, Mn, Mo and Al concentrations varied slightly with location, but were not affected by the method or rate of B application. Only the B concentration in tissues was significantly increased with regard to method rate and location. Increasing B nutrition enhanced phytotoxicity and some interactions among nutrients due to increased concentrations of Zn, Cu, Fe and Mn in the leaves, stem and roots of bush bean plants (Wallace et al. 1977). But contradictory results were reported by Leece (1978), who observed that with high levels of applied B, the concentrations of N, P, K, Ca, Zn, Cu, Fe and Mn (not Mg) in maize crop were depressed. The reverse results were obtained when no B was applied. Increasing B supply in soil resulted in the decrease of leaf N and P in tomato, suggesting B antagonism. The contrary was the case with a B effect on leaf K, Ca, Mg and Na (Aduayi 1978). Yadav and Manchanda (1979) noted that with an increase in the B content of soil, Ca and Mg concentration in wheat and Gramineae crops significantly decreased, whereas N, P and K contents were significantly increased.

Moreover, with differential supply of B in nutrient solution, the concentration of Fe, Mn, P and Ca in the shoots and roots of tomato increased, and B reduced the translocation of Mn, P and Fe whilst Ca remained unchanged (Alvarez-Tinaut et al. 1980). Addition of B in the nutrient solution decreased the absorption of N, P Ca, Mg and B, induced K accumulation, while Na remained unaffected in lamina stem and roots of *Cabernet sauvignon* wine plants (Downton and Hawker 1980). Gomez-Rodriguez et al. (1981) found a highly significant inverse correlation between B and Mn concentrations in leaves of sunflower, while Cu, Fe and Zn concentrations were not changed by different B levels in the nutrient solution. Marked reductions in Fe and Mn adsorption, but an increase in Zn uptake were recorded in bean plants grown in B deficient medium. The transport of Fe, Mn and Zn was increased in the trifoliate leaves, while that in shoots was reduced. It appears that, B is involved in the physiological processes controlling uptake and transport of nutrients like Mn, Fe and Zn (Dave and Kannan 1981).

Lombin and Bates (1982) found that with increasing B levels, the uptake of K, Mn, Zn, Cu, Mo and B was increased in alfalfa, peanut and soybean crops, but had no apparent effect on that uptake of Ca ad Mg in all crops. Similar detrimental effects of B on the uptake of Ca and Mg were reported by Singh and Singh (1983), who observed varying B level significantly increased the concentration of N, P, K, Na and B and decreased Ca and Mg concentrations in lentil plants. Applied B increased the N, P, K, Na and B content but decrease Ca and Mg contents of barley crops, whilst uptake of N, P, Na and B in grain and straw significantly increased, and K uptake remained unaffected (Singh and Singh 1983). Francois (1986) reported that with increasing B in the soil solution the concentration of B, P, K and Mg tended to increase in tomato leaves, whilst Ca and Na showed inconsistence trends. Studies on the chemical composition of radish, using sand culture techniques, indicated that Ca and P concentrations decreased significantly and K, Mg and Na remained unchanged with the increasing B levels (Francois 1986). Morsey and Taha (1986) reported that applied soil B and foliar application increased the concentration and uptake of N, P, K, Mn and B in both shoots and roots of sugar plants. Patel and Golakia (1986) demonstrated the effect of soil B on the uptake of N, P, K, Ca, Zn, Cu, Fe and Mn by a groundnut crop. Interestingly they outlined the mechanisms of action for some nutrients in relation to the B effect: for example, B increased an uptake and could be responsible for a favourable effect on nodulation. A positive effect of B and P uptake, which altered the permeability of plasma lemma at the root surface, resulted in increased P absorption. Uptake of K increased because of their mutual synergistic relationship, but Ca decreased due to antagonistic effect. Uptake of Fe and Cu were positively correlated, while Mn and Zn negatively correlated with applied B. The deficient state of B resulted in decreased the leaf N, P, Ca, Mg, Fe, Cu, Zn and B in tomato. On the other hand, excess B increased the concentration of nutrients with greater significance for K, Mg and Fe followed by Ca and Mn and in smaller quantity Cu and Zn (Carpena-Artes and Carpena-Ruiz 1987).

B toxicity had no consistent effects on the tissue concentration of P, K, Ca, Mg, Zn, Cu, Fe and Mn for five barley and six wheat cultivars grown in nutrient solution and no interactions were found among B nutrients and cultivars (Nable 1989). Higher levels of applied B significantly depressed the N and enhanced P and K contents in three cuttings of *Trifolium alexandrinum* (berseem; Pal et al. 1989). Singh et al. (1990) reported that the concentration of P, Mg and Zn in wheat increased and Ca, K, Cu, Fe and Mn decreased with increasing B in soil. On the other hand, an increasing supply of B significantly decreases the uptake of P, K, Ca, Mg and Mn, while that of Zn, Cu and Fe increased. They concluded that high levels of applied B had an antagonistic effect on the uptake of nutrients and this could be due to the toxic effect of B on root cells, resulting in impaired nutrient absorption processes. Alvarez-Tinaut (1990) found positive correlations between B and Fe and Cu contents of sunflower, suggesting that B could indirectly affect catalase activity via Fe and Cu. Positive correlations between Zn and B also indicated that B could indirectly affect the enzyme through modification of the Zn content. Concentrations, total uptake and ratios of certain nutrients in radish top and root change with differential B supply to nutrient solution (Tariq 1997). However, this study also suggested that changes occurring were mainly due to the B effect and partially due to antagonism between Ca and B. It is clear from the reported literature, that B interactions, either synergism or/and antagonism, can affect plant nutrition under both deficiency and toxicity conditions.

With increasing B supply in nutrient medium, leaf content of P became high (usually younger leaves shows higher P concentration) and there was a minor decrease in K, Ca and Mg compared to the average concentration for leaves (Furlani et al. 2003).

Yang and Gu (2004) demonstrated the effect of B on Al toxicity for soybean seedlings. They showed that high supply of B was found to induce Al toxicity by significantly increasing growth parameters including root length at 2 mM, and fresh weight at 5 mM Al for two different cultivars. Similar results have been described by Hossain and Hossain (2004), who confirmed the relationship of B with Al. The ratio between Ca and B in the plant is sometimes used to identify B deficiency. In one study, the supply of Ca and B to four maize cultivars considerably improved shoot dry matter production (Kanwal et al. 2008).

B is responsible for changes in other nutrients in soil-plant interactions (Tariq and Mott 2006). They also showed optimum productivity of radish plants at 0.5 mg l^{-1} B supply. Toxic effects confirmed by significant productivity decreases were found at increased levels of B supply. The amount of B, Zn and Cu in plants was increased and amount of Fe, Mn and Mo were reduced. Except B, the net uptake of all microelements decreased with increasing levels of B in the nutrient supply, and exerted close connection to the growth response of radish plants. Moreover, Zn/Cu ratio increased and ratio of Mn/Fe and Mn/Zn decreased, while Fe/Cu exerted unpredictable trend with increasing B levels. Inoculation with biofertilisers (*Rhizobium* strains) alone or combined with different levels of B increased significantly the uptake of N, P, K, Fe, Mn, Zn and B by shoot and seeds of peanut in both seasons as compared with the corresponding treatments without biofertilisers. The highest values of N, P, K, Fe and Mn uptake by straw and seeds of peanut plants were obtained by using (200 ppm of B + *Rhizobium* spp.) in two successive seasons, while the highest values of Zn and B uptake by straw and seeds in both seasons were obtained by using (300 ppm of B + inoculation with *Rhizobium*; Nadia et al. 2006). B interactions (synergism and/or antagonism) can affect plant nutrition under both deficient and toxic levels (Tariq and Mott 2007).

There is no significant effect on the residual Fe in the soil when using B fertiliser. Results suggest that Zn and B fertilisers had no role in the changes of residual Fe and Mn in the soil relative to the normal levels, and other factors are operative on the accumulation of residual Fe and Mn in the soil relative to its normal levels. The effect of Zn and B interaction on the residual Fe and Mn in the soil was insignificant (Aref 2010).

Our findings showed that different level of B applied to the groundnut plant affect uptake of nutrients in an irregular fashion. Interactions between nutrients and applied B indicted uptake of N, P, K, Mg, Mn, Zn and Fe are indirectly dependent on increasing supply of B resulting in increasing NUE and NDS, but at a certain level. Our recent observations suggested that at 1 kg ha^{-1} B level, NUE of groundnut plant was higher in terms of absorption of mineral nutrients compared to the 0.5 kg ha^{-1} B level. At the 2.0 kg ha^{-1} B level, groundnut varieties showed decreases in nutrients uptake capacity, resulting in decreases in NUE. Increases in supply of nutrients may affect nutrient uptake capacity of plants, which was confirmed when levels of absorbed macro- and micronutrients in groundnut plants at 2.0 kg ha^{-1} B supply were determined. At 0.5 and 1.0 kg ha^{-1} B, NDS in groundnut plants was higher. NDS was improved in both varieties of groundnut at 1.0 kg ha^{-1} B supply. Higher levels of B might be exerting antagonist effects on uptake of nutrients, which might be affecting NUE and NDS to groundnut plants. Observations of nutrient uptake capacity of groundnut plants in all three conditions (limiting, sufficient and toxic) strongly suggested that 1.0 kg ha^{-1} B improved NUE and NDS in groundnut plants. Despite substantial literature, the mechanism of B interaction with other nutrients still remains unclear and needs more investigation in terms of improvement in NUE and NDS. NUE is usually studied for only one nutrient and improving NUE for any one nutrient may affect the NUE of other nutrients; this is still a question of interest.

Concluding Remarks

Boron is a crucial element for normal development and growth of plants. This chapter highlights the positive effects of boron on crops and also describes the interaction of B with the uptake of other nutrients. It further highlights the importance of understanding the mechanisms of B action in plants and determining the molecular mechanisms of plant responses to toxicity and deficiency of B, to allow improvement of crops for tolerance to both conditions. Deficient or toxic amounts of B may both have adverse effect on plants and alter the uptake of other nutrients by direct or indirect interactions. Increasing supply of B may result in increasing nutrient use efficiency (NUE) and nutrient demand and supply (NDS). Alternatively higher levels of B may exert antagonist effects on uptake of nutrients, negatively affecting NUE and NDS. The mechanism underlying the B interaction with other nutrients still remains unclear and requires further investigation. NUE is studied typically for only one nutrient at a time and improving NUE for any one nutrient almost certainly affects the NUE of other nutrients. However, the interactions of B with other plant elements are complex and may exert antagonistic or synergistic effects, which may be specific for species, growth medium and different environments.

References

Aduayi EA (1978) Role of boron on growth components and elemental composition of Ife plum tomato. Commun Soil Sci Plant Anal 9:1–11

Alvarez-Tinaut MC, Leal A, Recalde-Martinez L (1980) Iron-manganese interaction and its relation to boron levels in tomato plants. Plant Soil 55:377–388

Alvarez-Tonaut MC (1990) Correlation between boron and other micronutrients. In: Behaviour, function and significance of boron in agriculture. Report on an International Workshop at St. John's College, Oxford, England. 23–25 July, 1990. Borax Consolidated Limited, London, SW 1P 1HT

Aref F (2010) Influence of zinc and boron interaction on residual available iron and manganese in the soil after corn harvest. Am Eur J Agric Environ Sci 8:767–772

Baker AS, Cook RL (1959) Green house studies on alfalfa with soil type, soil reaction and borax fertilization as variables. Agron J 51:1–4

Bellaloui N, Yadavc RC, Chern MS, Hu H, Gillen AM, Greve C, Dandekar AM, Ronald PC, Brown PH (2003) Transgenically enhanced sorbitol synthesis facilitates phloemboron mobility in rice. Physiol Plant 117:79–84

Blaser-Grill J, Knoppik D, Amberger A, Goldbach H (1989) Influence of boron on the membrane potential in *Elodea densa* and *Helianthus annuus* roots and H^+ extrusion of suspension cultured *Daucus carota* cells. Plant Physiol 90:280–284

Blevins DG, Lukaszewski KM (1998) Boron in plant structure and function. Annu Rev Plant Physiol Plant Mol Biol 49:481–500

Bogiani JC, Sampaio TF, Abreu-Junior C, Rosolen CA (2014) Boron uptake and translocation in some cotton cultivars. Plant Soil 375:241–253

Bolaños L, Esteban E, de Lorenzo C, Fernández-Pascual M, de Felipe MR, Garate A, Bonilla I (1994) Essentiality of boron for symbiotic dinitrogen fixation in pea (*Pisum sativum*) rhizobium nodules. Plant Physiol 104:85–90
Bolaños L, Lukaszewski K, Bonilla I, Blevins D (2004) Why boron? Plant Physiol Biochem 42:907–912
Brown PH, Hu H (1996) Phloem mobility of boron is species dependent: evidence of phloem mobility in sorbitol-rich species. Ann Bot 77:497–505
Brown PH, Shelp BJ (1997) Boron mobility in plants. Plant Soil 193:85–101
Brown PH, Bellaloui N, Hu H, Dandekar A (1999) Transgenically enhanced sorbitol synthesis facilitates phloem boron transport and increases tolerance of tobacco to boron deficiency. Plant Physiol 119:17–20
Camacho-Cristóbal JJ, Anzellotti D, González-Fontes A (2002) Changes in phenolic metabolism of tobacco plants during short-term boron deficiency. Plant Physiol Biochem 40:997–1002
Camacho-Cristóbal JJ, Rexach J, Herrera-Rodríguez MB, Navarro-Gochicoa MT, González-Fontes A (2011) Boron deficiency and transcript level changes. Plant Sci 181:85–89
Cara FA, Sánchez E, Ruiz JM, Romero L (2002) Is phenol oxidation responsible for the short-term effects of boron deficiency on plasma-membrane permeability and function in squash roots? Plant Physiol Biochem 40:853–858
Carmen Rodriguez- Hernandez M, Moreno DA, Carvajal M, Carmen–Martinez- Ballesta M (2013) Interactive effects of boron and NaCl stress on water and nutrient transport in two broccoli cultivars. Funct Plant Biol 40(7):739–748
Carpena-Artes O, Carpena-Ruiz RO (1987) Effect of boron in tomato plant. Leaf evaluations. Agrochimica 31:391–400
Coetzer LA, Robbertse PJ (1987) Pollination biology of Persea Americana Fuerte. S Afr Avocado Growers' Assoc Yearb 10:43–45
Cosgrove DJ (1999) Enzymes and other agents that enhance cell wall extensibility. Annu Rev Plant Physiol Plant Mol Biol 50:391–417
Crop Science Society of America (CSSA) (1992) Glossary of crop science terms. CSSA, Madison
Dannel F, Pfeffer H, Römheld V (2002) Update on boron in higher plant. Uptake, primary translocation and compartmentation. Plant Biol 4:193–204
Dave IC, Kannan S (1981) Influence of boron deficiency on micronutrients absorption by *Phaseoulus vulgaris* and protein contents in cotyledons. Acta Physiol Plant 3:27–32
Dear B, Lipsett J (1987) The effect of boron supply on the growth and seed production of subterranean clover (*Trifloium subtereneum* L). Aust J Agric Res 38:537–546
Domingues DS, Leite SMM, Farro APC, Coscrato VE, Mori ES, Furtado EL, Wilcken CF, Velini ED, Guerrini IA, Maia IG, Marino CL (2005) Boron transport in Eucalyptus. 2. Identification in silico of a putative boron transporter for xylem loading in eucalypt. Genet Mol Biol 28:625–629
Downton WJS, Hawkar JS (1980) Interaction of boron and chloride on growth and mineral composition of cabernet sauvignon vines. Am J Enol Vitic 31:277–282
Du Ying Q, Rong LX, Hua HJ, Zhiyao H (1999) Influence of B and/or Mo application on the growth and yield of peanut. Chin J Oil Crop Sci 21:61–66
El-Shintinawy F (1999) Structural and functional damage caused by boron deficiency in sunflower leaves. Photosynthetica 36:565–573
Ferrol N, Donaire JP (1992) Effect of boron on plasma membrane proton extrusion and redox activity in sunflower cells. Plant Sci 86:41–47
Fleischer A, O'Neill MA, Ehwald R (1999) The pore size of non-graminaceous plant cell walls is rapidly decreased by borate ester cross-linking of the pectic polysaccharide rhamnogalacturonan II. Plant Physiol 121:829–838
Francosis LE (1986) Effect of excess boron on broccoli, cauliflower and radish. Am Soc Hortic Sci 111:494–498

Furlani ÂMC, Carvalho CP, de Freitas JG, Verdial MF (2003) Wheat cultivar tolerance to boron deficiency and toxicity in nutrient solution. Sci Agric 60:359–370

Ganie MZ, Akhter F, Bhat MA, Malik AR, Mihd Junaid J, Shah MA, Bhat AH, Bhat TA (2013) Boron- a critical nutrient element for plant growth and productivity with reference to temperate fruits. Curr Sci 104:76–85

Garg OK, Sharma AN, Kona GRSS (1979) Effect of boron on the pollen vitality and yield of rice plants (*Oryza sativa* L. var. Jaya). Plant Soil 52:575–578

Goldbach HE, Yu Q, Wingender R, Schulz M, Wimmer M, Findeklee P, Baluska F (2001) Rapid response reactions of roots to boron deprivation. J Plant Nutr Soil Sci 164:173–181

Goldberg S (1997) Reactions of boron with soil. Plant Soil 193:35–48

Gomez-Rodriguez MV, Gomez-Ortega M, Alvarez-Tinaut MC (1981) Boron, copper, iron, manganese and zinc content in leaves of flowering sunflower plants (*Helianthus annuus* L.) grown with different boron supplies. Plant Soil 62:461–464

Gupta U (1979) Boron nutrition of crops. Adv Agron 31:273–307

Hajiboland R, Rad B, Bastani S (2013) Phenolic metabolism in boron deficient tea (*Camellia sinensis* L. O. Kuntze) plants. Acta Biol Hung 64:196–206

Han S, Chen L, Jiang H, Smith BR, Yang L, Xie C (2008) Boron deficiency decreases growth and photosynthesis, and increases starch and hexoses in leaves of citrus seedlings. J Plant Physiol 165:1331–1341

Harris HC, Brolmann JB (1966) Comparison of calcium and boron deficiencies of the peanut II. seed quality in relation to histology and viability. Agron J 58:578–582

Hossain AKMZ, Hossain MA (2004) Effects of aluminum and boron supply on growth of seedlings among 15 cultivars of wheat (*Triticum aestivum* L.) grown in Bangladesh. Soil Sci Plant Nutr 50:189–195

Hu HN, Brown PH (1997) Absorption of boron by plant roots. Plant Soil 193:49–58

Hu H, Penn SG, Lebrilla CB, Brown PH (1997) Isolation and characterization of soluble B complexes in higher plants. The mechanism of phloem mobility of boron. Plant Physiol 113:649–655

Huang L, Ye Z, Bell WR, Dell B (2005) Boron nutrition and chilling tolerance of warm climate crop species. Ann Bot 96:755–767

Hundt I, Schilling G, Fisher F, Bergmann W (1970) Investigations on the influence of the micronutrient boron on nucleic acid metabolism. Albrecht-Thaer-Arch 14:825–837

Kamali AR, Childers NF (1970) Growth and fruiting of peach in sand culture as affected by boron and fritted form of trace elements. J Am Soc Hortic Sci 95:652–656

Kanwal S, Rahmatullah Aziz T, Maqsood MA, Abbas N (2008) Critical ratio of calcium and boron in maize shoot for optimum growth. J Plant Nutr 31:1535–1542

Kastori R, Plesnicar M, Pankovic D, Sakac Z (1995) Photosynthesis, chlorophyll fluorescence and soluble carbohydrates in sunflower leaves as affected by boron deficiency. J Plant Nutr 18:1751–1763

Khan Towhid Osman (2013) Plant nutrients and soil fertility management. In: Soils: principles, properties and management. Springer, Dordrecht/Heidelberg/New York, pp 129–159

Kobayashi M, Matoh T, Azuma J (1996) Two chains of rhamnogalacturonan II are crosslinked by borate-diol ester bonds in higher plant cell walls. Plant Physiol 110:1017–1020

Kot FS (2009) Boron sources, speciation and its potential impact on health. Rev Environ Sci Biotechnol 8:3–28

Leece DR (1978) Effects of boron on the physiological activity of zinc in maize. Aust J Agric Res 29:739–749

Li B, Li H, Kui WH, Chao MC, Jern WS, Li HP, Chu WJ, Wang L (1978) Studies on cause of sterility of wheat. J Northeast Agric Coll 3:1–19

Lombin GL, Bates TE (1982) Comparative responses of peanut, alfalfa and soybeans to varying rates of boron and manganese on two calcareous Ontario soils. Can J Soil Sci 62:1–9

Loomis WD, Durst RW (1992) Chemistry and biology of boron. Biofactors 3:229–239

Malvi UR (2011) Interaction of micronutrients with major nutrients with special reference to potassium. Karnataka J Agric Sci 24:106–109

Marschner H (1995) Mineral nutrition of higher plants, 2nd edn. Academic, San Diego

Matoh T, Ochiai K (2005) Distribution and partitioning of newly taken-up boron in sunflower. Plant Soil 278:351–360

McIlrath WJ, Debruyn JA, Skok J (1960) Influence of boron supply on the micronutrients content of setaria shoots. Soil Sci 89:117–121

Miwa K, Fujiwara T (2010) Boron transport in plants: co-ordinated regulation of transporters. Ann Bot 105:1103–1108

Morsey MA, Taha EM (1986) Effect of boron, manganese and their combination on sugar beet under El-Minia conditions. 2: concentration and uptake of N, P, K, B and Mn. Ann Agric Sci Ain Shams Univ Cairo 31:1241–1259

Mozafar A (1989) Boron effect on mineral nutrients of maize. Agron J 81:285–290

Mukhopadhyaya M, Ghosh PD, Mondal TK (2013) Effect of boron deficiency on photosynthesis and antioxidant responses of young tea plantlets. Russ J Plant Physiol 60:633–639

Nable RO (1989) Effect of boron toxicity upon the mineral nutrient composition of barley and wheat cultivars. Div Rep Div Soils, CSIRO No. 104: 1–10

Nable RO, Bañuelos GS, Paull JG (1997) Boron toxicity. Plant Soil 193:181–198

Nadia MA, Badran M, Abd El-Hamide AF (2006) Response of peanut to foliar spray with boron and/or rhizobium inoculum. J Appl Sci Res 2:1330–1337

Nakagawa Y, Hanaoka H, Kobayashi M, Miyoshi K, Miwa K, Fujiwara T (2007) Cell-type specificity of the expression of *OsBOR*1, a rice efflux boron transporter gene, is regulated in response to boron availability for efficient boron uptake and xylem loading. Plant Cell 19:2624–2635

Noppakoonwong RN, Rerkasem B, Bell RW, Dell B, Loner Gan JF (1997) Prognosis and diagnosis of boron deficiency in blackgram (*Vigna mungo* L. Hepper) in the field by using plant analysis. In: Bell RW, Rerksaem B (eds) Proceedings of boron in soil and plants. Kluwer Academic Publishers, Dordrecht, pp 89–93

North KA, Ehlting B, Kopriva A, Rennenberg H, Kopriva S (2009) Natural variation in *Arabidopsis* adaptation to growth at low nitrogen conditions. Plant Physiol Biochem 47:912–918

Oertli JJ, Richardson WF (1970) The mechanism of boron immobility in plants. Plant Physiol 23:108–116

Pal B, Jadaun SPS, Raghav CS (1989) Effect of phosphorous and boron application on dry matter yield and nutrient contents in berseem. J Indian Soc Soil Sci 37:579–581

Pandey N, Archana (2012) Effect of boron on seed germination and biochemical changes in linseed at seeling stage. Indian J Agric Biochem 25:167–170

Pandey DK, Pandey N (2008) Screening of wheat genotypes for their susceptibility to boron deficiency. Res Environ Life Sci 1:37–42

Parks RQ, Lyon CB, Hood SL (1944) Some effect of boron supply on the chemical composition of tomato leaflets. Plant Physiol 19:404–419

Patel MS, Golakia BA (1986) Effect of calcium carbonate and boron application on yield and nutrient uptake by groundnut. J Indian Soc Soil Sci 34:815–820

Pollard AS, Parr AJ, Loughman BC (1977) Boron in relation to membrane function in higher plants. J Exp Bot 28:831–834

Power PP, Woods WG (1997) The chemistry of boron and its speciation in plants. Plant Soil 193:1–13

Ruiz JM, Baghour M, Bretones G, Belakbir A, Romero L (1998) Nitrogen metabolism in tobacco plants (*Nicotiana tabacum* L.): role of boron as a possible regulatory factor. Int J Plant Sci 159:121–126

Ryden P, Sugimoto-Shirasu K, Smith AC, Findlay K, Reiter WD, McCann MC (2003) Tensile properties of arabidopsis cell walls depend on both a xyloglucan cross-linked microfibrillar network and rhamnogalacturonan II-borate complexes. Plant Physiol 132:1033–1040

Sathya S, Mahendran PP, Arulmozhiselven K (2013) Influence of soil and foliar application of borax on fractions of boron under tomato cultivation in boron deficient soil of typic Haplustalf. Afr J Agric Res 8:2567–2571
Sharma PN, Ramchandra T (1990) Water relations and photosynthesis in mustard plants subjected to boron deficiency. Indian J Plant Physiol 33:150–154
Shelp BJ, Shattuck VI (1987) Boron nutrition and mobility, and its relation to hollow stem and the elemental composition of greenhouse grown cauliflower. J Plant Nutr 10:143–162
Simojoki P (1972) Boron deficiency, pollen sterility and ergot disease of barley. Ann Agric Fenn 11:333–341
Singh AL (1994) Micronutrient nutrition and crop productivity in groundnut. In: Singh K, Purohit SS (eds) Plant productivity under environmental stress. Agro Botanical Publishers, Bikaner, pp 67–72
Singh V, Singh SP (1983) Effect of applied boron on the chemical composition of lentil plants. J Indian Soc Soil Sci 31:169–170
Singh JP, Dahia DJ, Narwal RP (1990) Boron uptake and toxicity in wheat in relation to zinc supply. Fert Res 24:105–110
Stangoulis JCR, Brown PH, Bellaloui N, Reid RJ, Graham RD (2001) The efficiency of boron utilization in canola. Aust J Plant Physiol 28:1109–1114
Steinburg RA, Specht AW, Roller EM (1955) Effect of micronutrient deficiency on mineral composition, nitrogen fraction, ascorbic acid and burn of tobacco grown flowering in water culture. Plant Physiol 30:123–129
Stewart WM, Dibb DW, Johnston AE, Smyth TJ (2005) The contribution of commercial fertilizer nutrients to food production. Agron J 97:1–6
Takano J, Yamagami M, Noguchi K, Hayashi H, Fujiwara T (2001) Preferential translocation of boron to young leaves in *Arabidopsis thaliana* regulated by the *BOR1* gene. Soil Sci Plant Nutr 47:345–357
Takano J, Noguchi K, Yasumori M, Kobayashi M, Gajdos Z, Miwa K, Hayashi H, Yoneyama T, Fujiwara T (2002) *Arabidopsis* boron transporter for xylem loading. Nature 420:337–339
Tariq M (1997) Effect of boron supply on the availability of nutrients in soil and uptake by radish (*Raphanus sativas* L.) Ph.D. thesis, University of Reading, England
Tariq M, Mott CJB (2006) Effect of applied boron on the accumulation of cations and their ratios to boron in radish (*Raphanus sativus* L.). Soil Environ 25:40–47
Tariq M, Mott CJB (2007) Effect of boron on behavior of nutrients in soil-plant systems – a review. Asian J Plant Sci 6:195–202
Thurzo S, Szabo Z, Nyeki J, Silva AP, Nagy PT, Goncalves B (2010) Effect of boron and calcium sprays on photosynthetic pigments, total phenols and flavonoid content of sweet cherry (*Prunus avium* L.). Acta Hortic 868:457–461
Tolgyesi G, Kozma A (1974) Investigation on factors affecting boron uptake by grasses. Agrokemia es Talajatan 23:83–98
Touchton JT, Boswell FC (1975) Effects of boron on soyabean yield, chemical composition and related characteristics. Agron J 67:417–420
Vaughan AKF (1977) The relation between the concentration of boron in the reproductive and vegetative organ of maize plants and their development. Rhod J Agric Res 15:163–170
Wallace A, Romney EM, Alexander GV, Kinnear J (1977) Phytotoxic and some interactions of the essential metals iron, manganese, molybdenum, zinc, copper and boron. Commun Soil Sci Plant Anal 8:741–750
Wang ZY, Tang YL, Zhang FS, Wang H (1999) Effect of boron and low temperature on membrane integrity of cucumber leaves. J Plant Nutr 22:543–550
Warington K (1923) The effect of boric acid and borax on the broad bean and certain other plants. Ann Bot 37:629–672
Wimmer MA, Eichert T (2013) Review: mechanism for boron deficiency mediated changes in plant water relations. Plant Sci 203–204:25–32

Wimmer MA, Lochnit G, Bassil E, Mühling KH, Goldbach HE (2009) Membrane associated, boron-interacting proteins isolated by boronate affinity chromatography. Plant Cell Physiol 50:1292–1304

Woodbridge CG, Venegas A, Carndall PC (1971) The boron content of developing pear, apple and cherry flower buds. J Am Soc Hortic Sci 96:613–615

Woods WG (1994) An introduction to boron: history, sources, uses, and chemistry. Environ Health Perspect 102:5–11

Yadav OP, Manchanda HR (1979) Boron tolerance studies in gram and wheat grown on a sierozem sandy soil. J Indian Soc Soil Sci 27:174–180

Yang YH, Gu HJ (2004) Effects of boron on aluminum toxicity on seedlings of two soybean cultivars. Water Air Soil Pollut 154:239–248

Yang L, Zhang Q, Dou J, Li L, Guo L, Shi L, Xu F (2013) Characteristics of root boron nutrition confer high boron efficiency in *Brassica napus* cultivars. Plant Soil 371:95–104

Zhao D, Oosterhuis DM (2002) Cotton carbon exchange, non-structural carbohydrates, and boron distribution in tissues during development of boron deficiency. Field Crop Res 78:75–87

Chapter 7
Role of Autophagy in Plant Nutrient Deficiency

Milagros Collados Rodríguez, Katarzyna Zientara-Rytter, and Agnieszka Sirko

Abstract One of the environmental stresses frequently encountered by plants is nutrient deficiency. Therefore, reuse of valuable cellular nutrients is an important trait in nutrient use efficiency (NUE). High NUE is a desired trait in plants at all developmental steps to reach maximum potentials with minimum inputs. Two highly conserved evolutionary mechanisms are responsible for protein turnover at the cellular level, the ubiquitin-proteasome system (UPS) and the autophagy pathway. Generally, UPS recycles short-lived regulatory proteins while autophagy recycles long-lived proteins, protein aggregates or organelles. The proteins, which are destined for degradation, are marked by a special polypeptide tag, ubiquitin. The features of this tag, as well as activity of ubiquitinating and deubiquitinating enzymes, are determinants that allocate the protein into one or the other degradation systems. Apart from the common subset of over 30 proteins required for the "core autophagy", there exist selective autophagy cargo receptors. These proteins perform the quality control function by recognizing ubiquitinated cargoes (ready for degradation) and linking them to the autophagy machinery. Adequate knowledge of the processes of selective autophagy will be beneficial for agricultural production and the environment by delivering the methods and means for obtaining crops with improved NUE, higher yield and better stress tolerance.

Keywords NUE (nutrient use efficiency) • Autophagy • UPS (ubiquitin proteasome system) • Nutrient deficiency • Protein turnover • Protein homeostasis • PCD (programmed cell death)

M. Collados Rodríguez • K. Zientara-Rytter • A. Sirko (✉)
Institute of Biochemistry and Biophysics, Polish Academy of Sciences, ul, Pawinskiego 5A, 02-106 Warsaw, Poland
e-mail: asirko@ibb.waw.pl

M.J. Hawkesford et al. (eds.), *Nutrient Use Efficiency in Plants*,
Plant Ecophysiology 10, DOI 10.1007/978-3-319-10635-9_7

Introduction

Response to Nutrient Deficiency in Plants

Plants convert CO_2, light and water into biomass, however they also require essential nutrients to complete their lifecycle. The sessile nature of plants makes them vulnerable to the environmental fluctuations that cause nutrient limitation (among other stresses). In plants, response to nutrient deficit is complex and depends on the kind of limiting nutrient (macro- or micronutrient) and the extent of the shortage. Nutrient deficits result in phenotypic adaptation, such as root length and branching, shoot biomass reduction and faster transfer from vegetative stage of growth into the generative one (Amtmann and Armengaud 2009; Gruber et al. 2013). Plant responses to nutrient deficit may also be monitored at various molecular levels, such as gene expression, posttranslational modification of various proteins, qualitative and quantitative changes of proteins and metabolites. Researchers from many laboratories have revealed that transcriptome, metabolome and proteome profiles are reprogrammed in plants which encounter nutrient deficit (for example, Hoefgen and Nikiforova 2008; Howarth et al. 2008; Liang et al. 2013). Limitation of the particular nutrient (and/or an increased internal requirement for that nutrient) is sensed by plants and this information is transferred to the appropriate effectors, which modify the pathway responsible for assimilation and metabolism of the nutrient. The sensing and regulatory mechanisms are not completely characterised in most nutrients. In addition, the need to coordinate assimilation and metabolism of various nutrients makes this regulation even more complicated. The growing body of evidence suggests that protein degradation processes not only adjust plant metabolism to long lasting starvation periods, but they are also important in early responses to short-term nutrient deficit.

Ubiquitin and Protein Turnover Processes

Ubiquitin (Ub) is a 76 amino acid polypeptide that folds up into a compact globular structure and is heat-stable. It mediates selective proteolysis after enzymatic conjugation to the target proteins that can be either monoubiquitinated, when a single Ub unit is attached through its C-terminal glycine (G-76) to one lysine (K) residue of the substrate, or multi-monoubiquitinated, when single Ub units are attached through G-76 to several K residues of the substrate or polyubiquitinated when multiple Ub units are attached to the single K residue (Schreiber and Peter 2013; Vierstra 2012). Ub is evolutionary conserved and contains seven highly conserved lysines that can be used for poly(Ub) chain formation in vivo. Some of these K residues are traditionally associated with particular degradation systems, for example, K-48-linked (poly)Ub appears to be linked rather to proteasomal degradation while K-63–linked (poly)Ub appears to be more linked to autophagic degradation. In addition, the role of unconventional ubiquitination in targeting proteins for degradation indicates the complexity of the process (Xu et al. 2009).

Ubiquitins are encoded by a small-to-medium-sized multigene family comprising two gene classes, monomeric and polymeric. Monomeric ubiquitin genes consist of 228 nucleotides (76 codons) with an additional C-terminal sequence that encodes a ribosomal protein. By contrast, polymeric genes known as polyubiquitins are composed of tandem repeats of a 228-bp gene with no spacer sequence between them (Nei et al. 2000). In Arabdopsis, the Ub gene family consists of at least 14 members (Callis et al. 1995).

Three enzymes (E1-E3), working in a hierarchical cascade, attach a Ub tag to the target proteins (Sadanandom et al. 2012). E1 (ubiquitin-activating enzyme) hydrolyses ATP to adenylate the C-terminal carboxyl of Ub and form a thiolester bond with E1 cysteinyl sulphydyl residue (E1-Ub intermediate). After completion of this activation, E2 (ubiquitin-conjugating enzyme) accepts Ub on its cysteinyl sulphydryl group and a thiolester bond is formed again (E2-Ub intermediate). When conjugation is completed, E2 can either bind to E3 (ubiquitin ligase) to transfer Ub to the protein substrate or transfer Ub to E3, which subsequently transfers Ub to the target. Usually a Lys residue of the substrate accepts the C-terminal Gly-76 of Ub. The E3 Ub ligases recognise and bind specific degradation signals in substrate proteins and thus confer specificity to the ubiquitination-mediated degradation.

This ubiquitination cycle can be repeated multiple times. To add more complexity, the protein family of deubiquitinating enzymes form an additional regulatory step. A comprehensive repository of ubiquitinating and deubiquitinating enzymes from 50 distinct genomes belonging to four of the five major phylogenetic supergroups of eukaryotes was recently completed (Hutchins et al. 2013) and it is publicly available at the Database of Ubiquitinating and Deubiquitinating Enzymes [http://www.DUDE-db.org].

Proteins targeted for recycling by linkage of the Ub tag are degraded in two evolutionary conserved degradation systems. The ubiquitin-proteasome system (UPS) takes over the short-lived regulatory proteins (Dantuma et al. 2000; Vierstra 2009), while the autophagy pathway sequesters within double membrane structures the longer-lived and bigger proteinaceous material, such as protein aggregates (Yang and Klionsky 2010). Autophagy and the UPS are critical in the maintenance of cellular homeostasis, thus their activities need to be carefully orchestrated (Korolchuk et al. 2010). The limited degradative capacity of UPS is complemented with autophagy and crosstalk mechanisms between both systems exist (Schreiber and Peter 2013). Moreover, both recycling mechanisms are in crosstalk with senescence, programmed cell death (PCD) and prematurate aging (Madeo et al. 2010). The literature shows that when some autophagy related genes (ATG) are impaired, accelerated senescence overtakes the plant destiny (Hanaoka et al. 2002; Thompson and Vierstra 2005) presumably by impairment of autophagic recycling that contributes to plant energy availability (Izumi et al. 2013).

An overview of the ubiquitination cascade and UPS is shown in Fig. 7.1. The 26S proteasome structures in mammals, yeast and plants indicate a similar overall design (Sadanandom et al. 2012). However, UPS-dependent regulation of signaling and metabolic pathways appears to be more complex and prevalent in plants than in yeast and animals.

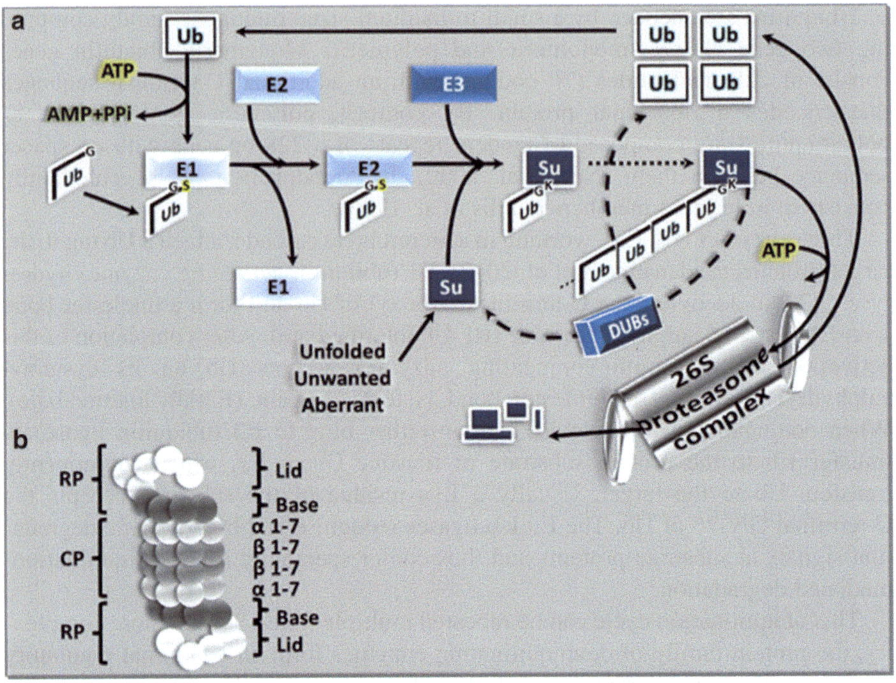

Fig. 7.1 Ubiquitin/26S proteasome system (*UPS*) overview (**a**) and schematic structure of 26S proteasome (**b**). (**a**) Ubiquitin (*Ub*) is in an ATP-dependent manner activated by E1 enzyme, transferred to an E2 enzyme and delivered to the target protein due to E3 attachment. Rounds of ubiquitin conjugation cascades polyubiquitinates the substrate to be recognised by 26S proteasome complexes, which hydrolyse peptide bonds and releases oligopeptides (consisting of 2–10 residues) to the cytoplasm; the Ub moieties are reused. *Su* substrate protein, *G* Gly-76 C-terminal residue of Ub, *K* Lys residue of the target protein, *S* thiolester bond, *DUBs* deubiquitinating enzymes. (**b**) One or two 19S regulatory complexes (*RP*) composed of base and lid subcomplexes cap the 20S core complex (*CP*) formed by two external and two internal rings of seven α-subunits and seven β-subunits respectively

The proteasome exists in multiple forms, and contains two major assemblies, the 28-subunit core particle (CP, also known as the 20S core) and the regulatory subunit (RP, also known as the 19S complex and PA700) consisting of 19 subunits in yeast, and probably the same number in other eukaryotes (Finley 2009). The CP is composed of four rings: two external rings made of seven α-subunits (α 1-7) flanking two rings of β-subunits (β 1-7 (A-G in plants). The CP functions as a non-specific ATP and Ub-independent protease with the capacity to cleave most peptide bonds due to peptidylglutamyl, trypsin-like and chymotrypsin-like activities. One or two RP complexes, each consisting of two parts base and lid, can be attached to the CP. The base consists of a ring of six related AAA+ATPase subunits (RPT1-6) and three non-ATPase subunits (RPN1, 2 and 10) and is attached directly to the CP. The lid contains non-ATPase subunits (RPN3, 5–9 an 11–12). RPN11 is a metaloprotease, which allows recycling of the substrate recognition tag, ubiquitin. The RP is responsible for recognition of K48 linked poly(Ub) chains, removal of

Ub moieties, unfolding and pore gating the substrate. The proteasome has the ability to bind different lids or lid-like structures depending on the cellular needs. A growing body of evidence indicates that monoubiquitinated proteins can be also targeted to proteasomes indicating that monoubiquitination is sufficient for degradation of small proteins (shorter than 150 aa) containing the unstructured region (Shabek et al. 2012). For more information about 26S proteasomes in plants, see reviews and references within (Sullivan et al. 2003; Vierstra 2003, 2009; Sadanandom et al. 2012).

What Is Autophagy?

Autophagy is a an important cellular process, and one of the two major degradation pathways along with UPS, which are involved in maintaining homeostasis during normal development and response to environmental stresses. Different kinds of autophagy have been reported, including microautophagy, macroautophagy and chaperone-mediated autophagy. Here (as in majority of other publications) the term "autophagy" refers to the best-characterised type, macroautophagy. The general scheme of the process is common in all eukaryotic organisms (Fig. 7.2). The process starts by the formation of a cup-like structure that engulfs selected cytoplasmic content (cargo) and progresses into a double-membrane autophagosome. The outer membrane of the autophagosome then fuses with the tonoplast (in yeast or plant cells) or lysosomal membrane (animal cells), while the inner membrane, together with the cargo composes the autophagic body (or autolysosome in animal cells). When such an organelle is present in the vacuolar lumen, its contents are attacked by vacuolar proteases and then released through vacuolar permeases to the cytosol for reuse (Klionsky 2007).

Over 30 evolutionary conserved autophagy-related (ATG) proteins play a role in the "core" autophagy pathway in yeast in mammals. Due to the high conservation of autophagy among eukaryotes, mechanistic explanations for plants are often taken from yeast or mammalian models (Thompson and Vierstra 2005; Diaz-Troya et al. 2008; Nakatogawa et al. 2009; Klionsky et al. 2011). However, some ATG proteins (e.g. ATG29 and ATG31) are not found in plants. The ATG11 protein was also considered nonexistent in plants (Suzuki et al. 2007) but a current TAIR search shows its potential existence and hybrid similarity to ATG17; both proteins have scaffolding properties in yeast.

Autophagy has an important role in regulating PCD. Both, excessive autophagy as well as inefficient or defective autophagy may lead to cell death either by self-eating (excessive autophagy) or by accumulation of damaged proteins and organelles (defective autophagy) (Guiboileau et al. 2010). For clarification of the nomenclature relating to different types of PCD in animals, the reader is referred to the published guidance (Kroemer et al. 2009) and for classification of plant PCD and their merging with categories of PCD recognised in animals, to the review by van Doorn (2011).

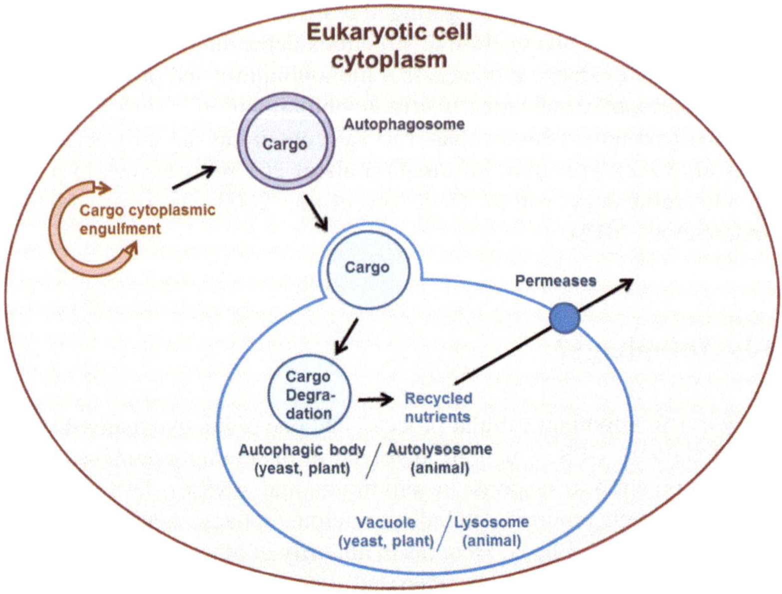

Fig. 7.2 Macroautophagy steps. The *cup shaped* double membrane (phagophore or isolation membrane) forms and expands trapping portions of the cell material. When the phagophore seals and produces the (double membrane) autophagosome, this encloses the cargo, which is transported along with the inner membrane of autophagosome into the vacuole (yeast or plants) or into the lysosome (animals), where it is recycled

TOR Signaling Pathway and Autophagosome Biogenesis

In yeasts, animals and plants, induction of autophagy is possible after inactivating the Target of Rapamycin (TOR) Ser/Thr kinase belonging to the family of phosphatidynositol kinase-related kinases (Noda and Ohsumi 1998; Pattingre et al. 2008; Liu and Bassham 2010). The TOR's name results from sensitivity to macrocyclic bacterial lactone rapamycin. TOR is involved in switching cellular responses depending on the nutritional status (John et al. 2011). During nutrient-rich conditions an active TOR (the TORC1 complex) consists of TOR, LST8-1, LST8-2 and RAPTOR (Regulatory Associated Protein of TOR) (John et al. 2011). TORC1 together with other kinases promote anabolic processes and cell growth while keeping autophagy switched off. *Arabidopsis* contains two RAPTOR proteins, RAPTOR1A and RAPTOR1B, whose function is to recruit TOR substrate proteins. The complex maintains the ATG1 (Autophagy Related Gene1) and ATG13 kinases, the latter in complex with ATG17 and ATG101 accessory proteins, dissociated through hypophosphorylation of ATG1 (one phosphorylated site) and hyperphosporylation of ATG13 (three phosphorylated sites). In nutrient deficient

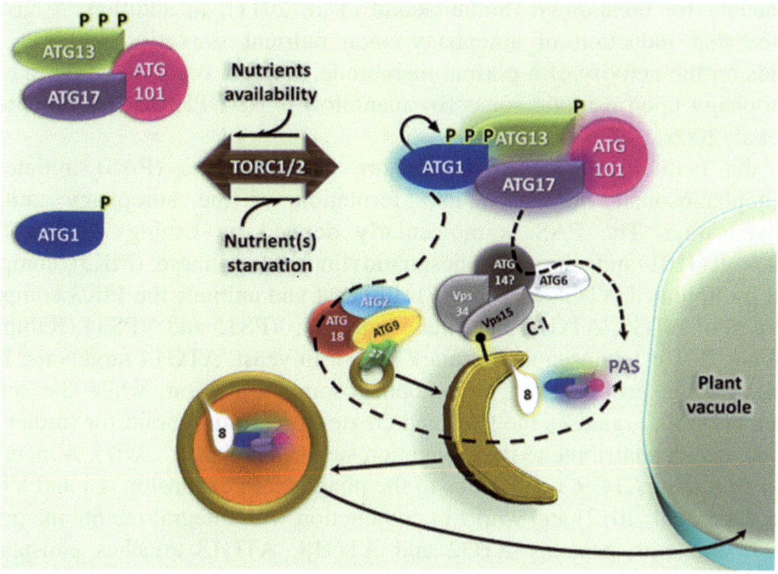

Fig. 7.3 Initiation of autophagy and phagophore formation. When nutrients are available, TOR is active in the complex TORC1 (see text for details) and keeps ATG1 dissociated from ATG13 (and associated proteins) trough hypo- and hyper-phosphorylation, respectively. During nutrient starvation the TORC2 complex is formed and the phosphorylation status reverses, allowing the complex ATG1-ATG13-ATG17-ATG101 to promote the autophagosome nucleation to the PAS by C-I and ATG8-PE docking to the phagopore inner surface. The C-I is composed of ATG6, hypothetically ATG14 and their kinase subunits VPS15 and VPS34. The Atg17 scaffold protein is a recruiting point for further ATG proteins as ATG9 (in complex with ATG2-ATG18-ATG27), which mediates membrane delivery contributing to the phagophore elongation. Once the autophagosome closes it can fuse to the plant vacuole for cargo degradation. *Dashed arrows* passing through proteins indicate the proteins, which are governed by the protein present in the *tail of the arrow*. *TOR* Target of rapamycin, *PAS* phagophore assembly site, *C-I* phosphatidylinositol 3-kinase complex I, *Vps* vacuolar protein sorting

conditions, an inactive TOR (the TORC2 complex) includes RICTOR (rapamycin-insensitive companion of TOR) instead of RAPTORs allowing it to swap the ATG1 and ATG13 phosphorylation status to the active one (autophosphorylation of ATG1 and dephosphorylation of ATG13). This, in turn, promotes the nucleation of the autophagosome by involvement of a downstream hypothetical (not yet identified) phosphorylation target. *Arabidopsis* contains four ATG1 proteins (ATG1a-c and a truncated ATG1t, present only in plants) and two ATG13 proteins (ATG13a and ATG13b). The phosphorylation-dependent regulatory mechanism has been confirmed for ATG1a and ATG13a (Suttangkakul et al. 2011; Li and Vierstra 2012b). The activated ATG1-ATG13 kinase complex relocates to the phagophore assembly site (PAS) and promotes ATG9-mediated membrane delivery to the expanding phagophore (Fig. 7.3). The ATG1-ATG13 kinase complex also docks to ATG8–PE bound to the inner surface of the phagophore, and is captured and delivered to the vacuole for breakdown (Suttangkakul et al. 2011). In addition, it has been reported that induction of autophagy upon nutrient starvation and salt stress depends on the activity of a plasma membrane NADPH oxidase, while induction of autophagy upon osmotic stress (by mannitol) is NADPH oxidase-independent (Liu et al. 2009).

At the cellular level, the phagophore assembly site (PAS) initiates the pre-autophagosomal nucleation and formation of the autophagy unit, the autophagosome. The PAS is molecularly defined by having ATG1-ATG13-ATG17-ATG101 and class III phosphatidylinositol 3-kinase (PIK3) complexes bound by lipidated ATG8 (ATG8-PE). In yeast and animals the PIK3 complex is composed of ATG6, ATG14 and kinase subunits, VPS15 and VPS34 (Rabinowitz and White 2010; Johansen and Lamark 2011). In yeast, ATG14 targets the kinase complex to the probable site of autophagosome formation, while the scaffold protein, ATG17, organises the PAS and creates a recruiting point for further ATG proteins, which contribute to the elongation step (Suzuki et al. 2007). A membrane delivery protein, ATG9, contributes to the phagophore expansion (Li and Vierstra 2012a; Orsi et al. 2012) and works in conjunction with integral membrane protein, ATG27, and with proteins ATG2 and ATG18. ATG18 attaches phosphatidylinositol 3-phosphate (PI3P) and phosphatidylinositol 3,5-bisphosphate (PI(3,5)P2) and together with ATG2 is involved in retrograde transport of ATG9 (Klionsky et al. 2011). ATG2 and ATG18 are functionally related, independently localise to the PAS by means of ATG1, and can interact and form a complex but do not recruit other proteins to the PAS; ATG1 is drawn by ATG13 to the PAS. The plant orthologues of ATG14, ATG17, ATG22 and ATG27 remain unidentified (see also Table 7.1).

The steps described above, namely autophagy initiation and nucleation (phagophore assembly) are followed by a step called vesicle elongation, where two pathways utilising UPS-like E1-E3 enzymatic cascades are involved in activation of two ubiquitin-like proteins, ATG8 and ATG12. In the first cascade the Cys-protease, ATG4, exposes the C-terminal Gly of ATG8. It allows ATG7 (E1) to bind to the C-terminus of ATG8 that, in turn, permits ATG3 (E2) to lipidate the C-terminus of ATG8 with phosphatidylethanolamine (PE). The PE tag enables ATG8 anchoring into the autophagosomal membrane. In the second cascade, ATG7 (E1) also participates by activating the ubiquitin like protein ATG12, enabling its conjugation to ATG10 (E2) which in turn, conjugates ATG12 to ATG5 (Noda et al. 2013). The ATG12-ATG5 platform oligomerizes with ATG16 and forms the E3-like ATG12-ATG5-ATG16 ligase complex. This complex further promotes ATG8 lipidation and phagophore growth (Fig. 7.4; Maiuri et al. 2007; Johansen and Lamark 2011). On the other hand, ATG7 activates also the C-terminal residue of ATG3 (E2) enabling its conjugation to ATG12. In addition, ATG4 protease acts as deubiquitinating (DUB) enzyme and recycles ATG8 through PE deconjugation from the outer autophagosomal membrane in a reactive oxygen species (ROS) regulated way (Rabinowitz and White 2010; Johansen and Lamark 2011).

Table 7.1 Components of the core and selective autophagy in *Saccharomyces cerevisiae*, *Homo sapiens* and *Arabidopsis thaliana* and the examples of loss of function phenotypes of autophagy mutants in plants

Saccharomyces cerevisiae (number of residues)	*Homo sapiens* (number of residues)	*Arabidopsis thaliana* (number of residues, gene)	Loss of function phenotype in plants	Selected references related to plant phenotypes
Initiation (**ATG1** kinase complex)[a]				
TOR1 (2470); **TOR2** (2474)	**FRAP1/mTOR** (2549)	**TOR** (2841, At1g50030)	Cell, organ, seed production and resistance to osmotic stress decrease. Post-germinative arrest in growth and development. Early senescence	Leiber et al. (2010), Johansen and Lamark (2011), and Deprost et al. (2007)
ATG1 (897)	**ULK1** (1050), **ULK2** (1036)	**ATG1a** (626, At3g61960); **ATG1b** (711, At3g53930); **ATG1c** (733, At2g37840); **ATGt** (408, At1g49180)	Accelerated senescence, both fixed C and N starvation sensitivity	Suttangkakul et al. (2011)
ATG13 (738)	**HARBI1/ mATG13** (349)	**ATG13a** (618, At3g49590); **ATG13b** (625, At3g18770)	Accelerated senescence, both fixed C and N starvation sensitivity	Suttangkakul et al. (2011) and Kim et al. (2012)
ATG17 (417)	**FIP200/RBCC1** (1594)	**ATG17** (three potential isoforms: 1022, At2g37420; 376, At1g59580 [weak interaction with ATG9]; 1516, At4g28710)) –see next column	No phenotype data. Exact yeast homologue still unclear[b]	–
ATG29 (213)	Not found	Not found	–	–
ATG31 (196)	Not found	Not found	–	–
Not found	**ATG101** (218)	**ATG101** (215, At5g66930) –see next column	No phenotype data. Exact yeast homologue still unclear[b]	–
KOG1/RAPTOR (1557)	**RAPTOR** (1335)	**RAPTOR1/1b** (1344, At3g08850); **RAPTOR2/1a** (1336, At5g01770)	Post-embryonic meristem-driven growth affected in double mutants. Single mutant phenotypes differ	Anderson et al. (2005) and Deprost et al. (2005)

(continued)

Table 7.1 (continued)

Saccharomyces cerevisiae (number of residues)	*Homo sapiens* (number of residues)	*Arabidopsis thaliana* (number of residues, gene)	Loss of function phenotype in plants	Selected references related to plant phenotypes
LST8 (303)	**mLST8** (260)	**LST8-1** (305, At3g18140); **LST8-2** (313; At2g22040)	Altered apical dominance, meristem cell proliferation and leaf development (bushy)	Moreau et al. (2012)
			Accumulation of amino acid and sensitivity to osmotic stress	
ATG9 complex[c]				
ATG2 (1592)	**mATG2a** (1938); **mAtg2b** (2078)	**ATG2** (1839; At3g19190)	Accelerated senescence. Unrestricted PCD	Yoshimoto et al. (2009), Wang et al. (2011a, b), Kim et al. (2012), and Liu and Bassham (2012)
ATG9 (997)	**mATG9a** (839); **mATG9b** (924)	**ATG9** (866; At2g31260)	Unrestricted PCD	Hofius et al. (2009), Hanaoka et al. (2002), and Kim et al. (2012)
ATG18 (500)	**WIPI1** (previously WIPI49) (446)	**ATG18a** (425, At3g62770); **ATG18b** (312, At4g30510); **ATG18c** (393, At2g40810); **ATG18d** (391, At3g56440); **ATG18e** (374, At5g05150); **ATG18f** (763, At5g54730); **ATG18g** (959, At1g03380); **ATG18h** (927, At1g54710)	Accelerated senescence, both fixed C and N starvation sensitivity; enhanced sensitivity to osmotic, oxidative, drought and salt stresses. Unrestricted PCD	Xiong et al. (2005), (2007), Liu et al. (2009), Wang et al. (2011b), and Liu and Bassham (2012)
ATG21 (496)	**WIPI2/Atg21 a-f** (454- 310)	Not found	–	–
ATG23 (453)	Not found	Not found	–	–

(continued)

Table 7.1 (continued)

Saccharomyces cerevisiae (number of residues)	*Homo sapiens* (number of residues)	*Arabidopsis thaliana* (number of residues, gene)	Loss of function phenotype in plants	Selected references related to plant phenotypes
Nucleation (class III phosphatidylinositol 3-phosphate kinase complex)[d]				
ATG6/VPS30 (557)	**BECN1** (450)	**ATG6** (517, At3g61710)	Male sterility	Harrison-Lowe and Olsen (2008) and Kim et al. (2012)
ATG14 (344)	**ATG14L/ BARKOR** (492)	**ATG14** (506, At4g01270)	No phenotype data. Exact homologue still unclear	–
VPS15 (1454)	**PIK3R4/VPS15** **BCL2** (1358)	**VPS15** (1494, At4g29380)	Defective pollen germination	Wang et al. (2012)
VPS34 (224)	**PIK3c3/VPS34** (887)	**VPS34a** (814, At1g60490)	Inhibited growth and development	Welters et al. (1994)
ATG20 (640)	Not found	**ATG20** (402; At5g06140)	No phenotype data	
ATG27 (271)	**ATG27**	**ATG27** (276; At2g40316)	Not studied in plants	–
Expansion and autophagosome formation (**ATG8** and **ATG12** conjugation systems)[e]				
ATG3 (310)	**ATG3** (314)	**ATG3** (313, At5g61500)	No phenotype data	–
Atg4 (507)	**ATG4a-d/ autophagin1-4** (398; 392; 458; 474 respectively)	**ATG4a** (467, At2g44140); **ATG4b** (477, At3g59950)	Accelerated senescence, oxidative stress, fixed C and N starvation sensitivity	Yoshimoto et al. (2004), Scherz-Shouval et al. (2007), Chung et al. (2010), Kim et al. (2012), and Tsai et al. (2012)
ATG5 (294)	**ATG5** (275)	**ATG5** (337, At5g17290)	Accelerated senescence, both fixed C and N starvation sensitivity; enhanced sensitivity to heat, oxidative, drought and salt stresses. Delayed differentiation of tracheary elements. Unrestricted PCD	Thompson et al. (2005), Phillips et al. (2008), Yoshimoto et al. (2009), Chung et al. (2010), Kwon et al. (2010), Suttangkakul et al. (2011), Kim et al. (2012), Liu and Bassham (2012),

(continued)

Table 7.1 (continued)

Saccharomyces cerevisiae (number of residues)	*Homo sapiens* (number of residues)	*Arabidopsis thaliana* (number of residues, gene)	Loss of function phenotype in plants	Selected references related to plant phenotypes
				Tsai et al. (2013), Minina et al. (2013), and Zhou et al. (2013)
ATG7 (630)	**ATG7** (703; 676; 623)	**ATG7** (697, At5g45900)	Accelerated senescence, both fixed C and N starvation sensitivity; enhanced sensitivity to heat, oxidative, drought and salt stresses. Unrestricted PCD	Doelling et al. (2002), Thompson et al. (2005), Phillips et al. (2008), Chung et al. (2010), Lenz et al. (2011), Suttangkakul et al. (2011), Kim et al. (2012), Li and Vierstra (2012b), Liu and Bassham (2012), Minina et al. (2013), Zhou et al. (2013), and Hofius et al. (2009)
			Enhanced ROS production	
ATG8 (117)	**MAPLC3A** (125)	**ATG8a** (137, At4g21980); **ATG8b** (122, At4g04620); **ATG8c** (133, At1g62040); **ATG8d** (120, At2g05630); **ATG8e** (122, At2g45170); **ATG8f** (121, At4g16520); **ATG8g** (121, At3g60640); **ATG8h** (120, At3g06420); **ATG8i** (116, At3g15580)	Accelerated senescence, both fixed C and N starvation sensitivity	Thompson and Vierstra (2005) and Liu and Bassham (2012)
	MAPLC3B (125)		No phenotype data. All family mutant necessary	
	MAPLC3C (147)			
	GABARAP (117)			
	GABARAP-L1/GEC-1/ATG8L (117)			
	GABAPAP-L2/GATE-16/GEF2 (117)			
	GABARAP-L3 (117)			

(continued)

Table 7.1 (continued)

Saccharomyces cerevisiae (number of residues)	*Homo sapiens* (number of residues)	*Arabidopsis thaliana* (number of residues, gene)	Loss of function phenotype in plants	Selected references related to plant phenotypes
ATG10 (167)	**ATG10** (220)	**ATG10** (226, At3g07525)	Accelerated senescence, both fixed C and N starvation sensitivity. Unrestricted PCD	Tsai et al. (2012), Phillips et al. (2008), Chung et al. (2010), Kim et al. (2012), and Liu and Bassham (2012)
ATG12 (186)	**ATG12** (140; 74)	**ATG12a** (96, At1g54210); **ATG12b** (994, At3g13970)	Accelerated senescence, both fixed C and N starvation sensitivity	Thompson et al. (2005), Chung et al. (2010), and Kim et al. (2012)
ATG16 (150)	**ATG16 1 a-f** (607-284); **ATG16 2 a-f** (619-131)	**ATG16** (509, At5g50230)	No phenotype data	–
Autophagy receptors and cargo ligands[f]				
ATG11 (1178)	**ATG11** (1591)	**ATG11** (1148, At4g30790)	No phenotype data Exact yeast homologue still unclear[b]	
ATG19 (415)	Not found	Not found	–	
ATG26 (1198)	Not found	Not found	–	
ATG30 (384)	Not found	Not found	–	
ATG32 (529)	**NIX/BNIP3L/ ATG32** (219)	**ATG32** (953; At4g29060)	No phenotype data Exact homologue still unclear	
ATG34 (412)	Not found	**ATG34** (614; AT1G20950)	No phenotype data Exact homologue still unclear	
Not found	**p62/SQSTM** (440)	**NBR1** (704; At4g24690)	Increased sensitivity to some abiotic stresses	Zhou et al. (2013)
Not found	**NBR1** (966)			
Not found (only NDP52 gives some low hits)	**NDP52** (446) **optineurin** (577) **TBK** (729) **UBQL4** (601)	Not found	–	

(continued)

Table 7.1 (continued)

Saccharomyces cerevisiae (number of residues)	*Homo sapiens* (number of residues)	*Arabidopsis thaliana* (number of residues, gene)	Loss of function phenotype in plants	Selected references related to plant phenotypes
Bph1p (2167)	**ALFY** (3601)	At4g02660; At1g03060	Not studied in plants. No phenotype data. Exact homologue still unclear	

Some of the proteins cited in this table are represented by more than one alternative spliced isoform; only the longest one is indicated in brackets. For the function description of most of the proteins shown here see Klionsky et al. (2011)

[a]***FRAP1*** [FKBP12-rapamycin complex-associated protein (FK506-binding protein 12-rapamycin complex-associated protein 1)], ***mTOR*** (mammalian target of rapamycin), ***ATG*** (autophagy-related protein), ***ULK1*** (Unc-51-like kinase 1), ***ATG1s*** ATG1t is represented longer than in Suttangkakul et al. (2011) because manual removing of the introns in this locus reveals cDNA predicted to encode 408 aa protein (instead of 267 aa). After ELM (EukarioticLinear Motif) domain search a conflict with Li and Vierstra 2012 appears because the regulatory domain exists although the protein is shorten in the middle (mostly disordered region), ***C*** (carbon), ***N*** (nitrogen), ***LST8*** (Lethal with SEC13 protein 8), ***RAPTOR*** (regulatory-associated protein of TOR), ***FIP200*** (FAK family kinase-interacting protein of 200 kDa), ***RBCC1*** (RB1-inducible coiled-coil protein 1)

[b]Suttangkakul et al. (2011) comment to have found possible ATG17 and ATG101 orthologues but do not state exactly which one. Similarly we failed to identify the ATG29 and ATG31 plant orthologues but we propose homologues of AtATG101, AtATG11, AtATG17, and ATG32 (At4g29060 has mitochondrial and chloroplastic location based on BLAST, MOTIFSCAN, TMPRED and ELM search/predictions). Souval et al. (2007) showed oxidative stress sensitivity in atg4 mutants (a and b) of mammal cells

[c]***ATG9*** in plant ATG9 no transmembrane region is predicted by ELM (weakly by TMPRED), hypothetically plant ATG9 may attach to the membrane by means of ATG27, ***WIPI1*** (WD repeat domain, phosphoinositide interacting 1)

[d]***BECN1*** (BECLIN 1), **BARKOR** (Beclin 1-associated autophagy-related key regulator), ***PIK3R4*** (phosphoinositide 3-kinase regulatory subunit 4), ***VPS15*** (vacuolar protein sorting-associated protein 15), ***ATG27*** according to Yen et al. (2007) the N terminus is in the vesicular lumen while that C terminus is in the cytosolic part in yeast; TM and ELM predicts that and plant Nt also ends into the membrane and has a Ct TM. ***PIK3c3*** (phosphatidylinositol 3-kinase catalytic/class subunit type 3)

[e]***LC3*** (light chain 3), ***GABARAP*** (gamma-aminobutyrate receptor associated protein), ***GATE16*** (golgi-associated ATPase enhancer of 16 kDa), ***GEF2*** (ganglioside expression factor 2)

[f]***BNIP3L*** (BCL2/NIP3-like), ***SQSTM*** (Sequestrosome), ***NBR1*** [next to BRCA1 (breast cancer) gene 1], ***TBK*** (tank binding kinase), ***UBQL4*** (Ubiquilin4), ***Bph1p*** (beige protein homologue 1), ***ALFY*** (autophagy linked FYVE protein)

How Many Autophagy Forms Exist in Plants?

The process of autophagy in plants was initially considered as non-selective bulk degradation of cellular contents, however now it is clear that the process is selective (Floyd et al. 2012; Li and Vierstra 2012a). For example, the following specific forms of autophagy were reported each involved in degradation of a specific target,

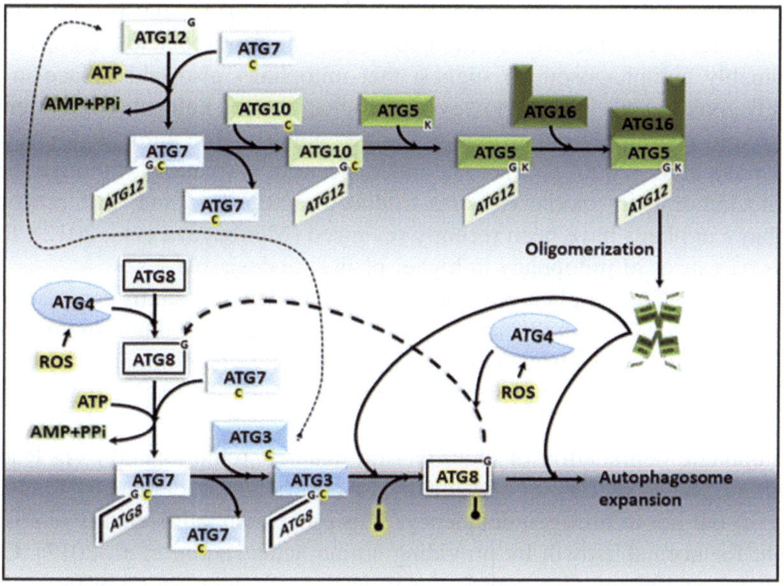

Fig. 7.4 Recruitments of ATG8 and ATG12 proteins to the autophagosome. The ubiquitin-like proteins, ATG12 and ATG8, are processed in a UPS-like E1-E3 manner through two separate conjugation cascades with some overlapping elements. In the first, ATG12 covalently linked to ATG5 (by subsequent action of E1-like ATG7 and E2-like ATG10) promotes formation of ATG12-ATG5-ATG16 complex needed for creation of pre-autophagosomal structure. In the second system, cleavage of ATG8 by ATG4 enables attachment of phosphatidylethanolamine (*PE*) to the C-terminus of ATG8 by subsequent action of E1-like ATG7 and E2-like ATG3. Activated ATG8 (ATG8-PE) associates with the expanding autophagosome membranes and remains through autophagosome maturation until fusion with the vacuole. This property is commonly exploited in many labs for reliable autophagy monitoring (Figure based on Suzuki et al. (2007) and Li and Vierstra (2012a))

such as mitochondria (mitophagy), peroxisomes (pexophagy), ribosomes (ribophagy), ER (reticulophagy or ERphagy) lipid droplets (lipophagy), pathogens (xenophagy), protein aggregates (aggrephagy) and choroplast material (chlorophagy). In addition, in yeast a mechanism also exists called CVT (Cytoplasm to Vacuole Targeting; (Bassham et al. 2006), which is only hypothesised in plants (Nakatogawa et al. 2009; Li and Vierstra 2012a), leaving open the question of how functional vacuolar proteins reach the plant vacuole.

Work on mitophagy in yeast indicated that the ATG32 mitochondrial outer membrane protein is the main factor involved in selective degradation in fungal mitochondria. Following the induction of mitophagy, Atg32 binds Atg11, an adaptor protein for selective types of autophagy, then it is recruited to mitochondria and imported into the vacuole along with mitochondria. Therefore, in yeast Atg32 confers selectivity for mitochondrial sequestration as a cargo and is necessary for recruitment of this organelle by the autophagy machinery for mitophagy

(Kanki et al. 2009). In plants, much less work on mitophagy has been performed. Electron microscopy pictures of mitochondria being fused to acidic compartments (presumably autophagosomes) suggest that mitophagy also takes place in plant cells (Toyooka et al. 2006). However, autophagosomes colocalise preferentially with protein aggregates rather than with mitochondria. The autophagy postulated by Toyooka et al. (2006) seems to be independent of starvation. Although the role of mitochondria in oxidative stress-induced autophagy and the process of mitophagy in plants have been recently reviewed (Minibayeva et al. 2012), further studies on causes of mitophagy in higher plants are desirable.

Literature related to plant pexophagy (selective degradation of peroxisomes) is limited. It has been shown that the process of pexophagy contributes to virulence of fungal pathogen, *Colletotrichum orbiculare*, since the *atg26* mutant of this pathogen, specifically affected in peroxisomes failed to penetrate the host cells (Asakura et al. 2009).

Autophagic degradation of chloroplastic material (chlorophagy) via Rubisco-containing bodies (RCBs) is specifically linked to carbon deficiency (caused by darkness) but not to nitrogen deficiency. This process supports energy availability at night for normal growth by providing amino acids (Izumi et al. 2013). On the other hand, autophagy also contributes to leaf starch degradation at night (Wang et al. 2013). This process takes place in vacuoles and is independent of the classic chloroplast pathway of starch degradation. No proof for degradation of entire chloroplasts is available. On the other hand, vacuolar localisation of the chloroplasts in protoplasts from the mesophyll leaves of the senescing wheat leaves seem to support such possibility (Wittenbach et al. 1982). Plastid degradation has mostly been studied in the conditions of starvation or in leaves undergoing senescence. It has been suggested that remobilisation of Rubisco in RCBs represents the first step of chloroplasts degradation i.e. chloroplast downsizing due to RCB formation and subsequent deposition into the vacuole (Wada et al. 2009). Rubisco is the most abundant plant protein and catalyses carboxylation and oxidation competing reactions of photosynthesis. It is a substantial nitrogen source stored in chloroplasts that the plant may use in case of N deficiency. Plants rely on chloroplasts to survive; therefore they undergo senescence only when the day becomes short and the respiratory period surpasses the photosynthetic period. During stress and senescence a compromise is made between maintaining functional chloroplasts and reusing the nutrients accumulated in the form of Rubisco (Wada et al. 2009). The extraplastidic degradation processes, namely chlorophagy and RCB rely on autophagy, demonstrated as some autophagy-deficient mutants (*atg5*) do not show RBCs (Ishida et al. 2008). Moreover, it was recently shown that degradation of stromal proteins takes place within chloroplasts largely via the autophagy-independent process. However, autophagy is required for the complete degradation of these proteins (Lee et al. 2013).

The existence in plants of the ribophagy-like mechanism, targeting ribosomes for recycling under normal growth conditions and similar to ribophagy observed in yeast under nutrients deficiency, has been suggested (Hillwig et al. 2011). The authors suggest that this mechanism is not only active during nutritional stress, but also has a housekeeping role. In addition, autophagy was demonstrated to be responsible for delivering endoplasmic reticulum (ER) to the vacuole during ER stress (Liu et al. 2012).

Some differences in autophagy seem to depend on the plant species (Bassham et al. 2006; Toyooka et al. 2006). These observations point to the possible existence of lysosome/endosome-like organelles in plants such as tobacco (which also fuse with the central vacuole) but not in most species such as barley or *Arabidopsis*. Toyooka et al. (2006) observed two transport pathways in BY-2 tobacco cells: direct delivery of autophagosomes to the central vacuole (in the absence of E-64 autophagy inhibitor) versus engulfment of autophagosomes by small vacuoles after E-64 treatment. This second pathway may be a consequence of incomplete autophagy inhibition, which could be solved with 3MA (3-methyl-adenine) but further research on this possibility is required. It is possible that the time of observation of E-64 treated plants has to be varied (to extract the same conclusions) depending on the generation time of the plant. In other words, 24 h of E-64 treatment for *Arabidopsis* (a plant with a 3 month generation time) may have proportionally stronger effect than such treatment for tobacco (a plants with 6 month generation time). Another reason for discrepancies may be the comparison of protoplast or BY-2 cells obtained results with those derived from plant organs.

Genes for Plant Autophagy and Typical Phenotypes of *atg* Mutants

Nomenclature of autophagy related genes (ATG, some were previously called *APG*) originates from yeast heterologous comparisons (Bassham et al. 2006), occasionally leading to confusion between unique yeast and metazoan proteins with plant ones. A common subset of over 30 proteins is required for all forms of autophagy (Table 7.1). The significant subset of these proteins represent the components of the so called "core autophagy" machinery required for both selective and non-selective autophagy (Lynch-Day and Klionsky 2010). These proteins belong to five large complexes responsible for the subsequent steps of autophagosome formation, namely induction, nucleation and extension. In addition to this "core" subset more proteins are required for certain pathways. Some proteins are present only in one group of organisms and it is difficult to identify their orthologues because of low sequence conservation. These proteins are involved in, so called, selective autophagy and perform a quality control function by distinguishing between cargoes ready for degradation and their functional counterparts (see review: Schreiber and Peter 2013). Mutations in ATG proteins impair autophagy in different ways. The effects depend to which complex the impaired protein belongs. For example, *atg5* or *atg7* mutations cause the accumulation of downstream pathway proteins and speed up the plant life span (Minina et al. 2013). Some typical phenotypes of *Arabidopsis* autophagy mutants are shown in Table 7.1.

Role of Autophagy in Plant Response to Nutrient Deficiency

Plants deficient in autophagy (*atg* mutants) do not show evident phenotypic differences from wild type during normal non-stress conditions (see Table 7.1). However, some *atg* mutants show accelerated life cycle and early senescence when the light intensity is low and during short day photoperiods (Thompson et al. 2005). This suggests that a basal or constitutive autophagy is important for normal plant growth (Minina et al. 2013). The senescence features, such as leaf yellowing in a consequence of Rubisco breakdown and chlorophyll dismantling, are more evident in *atg* mutants than in the wild type grown under nitrogen (N) or carbon (C) deficiency (Doelling et al. 2002; Hanaoka et al. 2002; Thompson et al. 2005; Phillips et al. 2008). Apparently, autophagy is important for recycling and delivering nutrients to the reproductive organs during senescence and as such it can be considered as a nutrient redistribution process. In addition, autophagy plays an important role in N management at the whole-plant level through the control of remobilisation, under both limiting and ample nitrate conditions (Guiboileau et al. 2012). For example, heterologous expression of soybean ATG8c in *Arabidopsis* led to better performance of the transgenic lines under both starvation and normal conditions (Xia et al. 2012). Although the protein and soluble sugar concentrations were similar in the wild type and transgenic line, the fresh weight of the transgenic lines was significantly larger in both, N sufficient and N deficient conditions. The transgenic plants survived the period of carbon-limiting conditions induced by extended darkness much better than the wild type and promptly recovered and resumed growth when moved back to normal conditions. Comparison of the growth parameters in soil under a long-day photoperiod indicated that the transgenic plants grew faster than the wild type plants, reached a larger size before bolting, entered the reproductive stage slightly earlier, and produced more flowers, siliques and seeds. The authors concluded that over-expression of soybean ATG8 in *Arabidopsis* promoted the vegetative growth and facilitated the transition into reproductive stage.

Similarly, *Arabidopsis* plants overproducing ATG8 protein fused on its N terminus to green fluorescent protein (GFP) showed enhanced growth and leaf size, accelerated flowering under nutrient-limiting and short-day growth, however they were also slightly more sensitive to mild salt and osmotic stress (Slavikova et al. 2008). This study also suggested that ATG8 participates in cytokinin-mediated root-shoot communication possibly by sequestration of proteins involved in cytokinin transport or signaling.

It has been shown that ATG18a protein from *Arabidopsis* is required for the formation of autophagosomes under nitrogen deficiency and sucrose starvation (Xiong et al. 2005). Interestingly, the mutants with silenced expression of

ATG18a are more sensitive to high salt and drought than the wild-type plants, demonstrating a role for autophagy in the response to these stresses (Liu et al. 2009). The same authors also reported that induction of autophagy upon nutrient starvation and salt stress depends on the activity of plasma membrane NADPH oxidase and that induction of autophagy upon osmotic stress (caused by mannitol) is NADPH oxidase-independent.

Nutrient deficit up-regulates autophagy. For example, induced transcription of some *ATG* genes (e.g. *ATG4*, *ATG8a-i*, *ATG3*, *ATG7*) was observed upon C starvation in *Arabidopsis* (Rose et al. 2006), and of *ATG8* upon N and sulfur starvation in tobacco (Zientara-Rytter et al. 2011). Additionally, some *atg* mutants (e.g. *atg7*, *atg5*, *atg10*) are hypersensitive to nutrient deficit (Li and Vierstra 2012a).

Autophagy participation in plant responses to other nutrient stresses such as phosphorus or iron starvation has been discussed only in the context of participating Ub-ligases and the UPS (Lyzenga and Stone 2012; Rojas-Triana et al. 2013). However, no attempts to test the role of autophagy in this process were undertaken.

Interestingly, it has been recently reported that sulfide, which is generated from cysteine by cysteine desulfhydrase in the cytosol of *Arabidopsis* plants, acts as a negative regulator of autophagy and that this effect is independent from the sulfur status of the plant (Alvarez et al. 2012a, b). Such a conclusion came from the observation that plants defective in the activity of cysteine desulfhydrase (*des1* mutants) show premature leave senescence and increased expression of senescence associated genes. In addition, externally added hydrogen sulfide was able to alleviate some of the observed phenotypes. The mechanism of action of sulfide is completely unknown. However, the authors suggested that some enzymes necessary for autophagy process might be possible targets of such regulation (e.g. through reversible S-thiolation), for example E1 and E2 enzymes involved in ubiquitination or ATG4 cysteine protease. It is worthwhile mentioning that hydrogen sulfide is already recognised as an important signaling molecule in mammalian systems (Lowicka and Beltowski 2007; Szabo 2007; Gadalla and Snyder 2010).

In conclusion, autophagy is an important process involved in plant response to nutritional stresses. Emerging data suggest that the controlled enhancement of autophagy can be exploited for improvement of crop performance in normal and nutrient-limiting conditions.

Selective Autophagy

The process of selective autophagy was first described in yeast (see above, CVT). Paradoxically, this process is a part of biosynthetic pathway and it is involved in a selective delivery of two lysosomal enzymes to the vacuole described neither for animals nor plants (Lynch-Day and Klionsky 2010). As mentioned before, a number of organelles, storage structures, pathogens or protein aggregates are known to be specifically targeted by autophagy (Weidberg et al. 2011). Each specific cargo was used as a base of providing the names for different autophagy types: mitophagy, pexophagy, ribophagy, xenophagy, aggrephagy etc. (see also above). In most cases the molecular machinery involved in specificity of cargo recognition is not characterised. However, it is evident that the various cargos in animal cells are specifically recognised and targeted for lysosomal degradation by the action of specific cargo-recognising receptors, called selective autophagy receptors (Lynch-Day and Klionsky 2010; Johansen and Lamark 2011; Behrends and Fulda 2012; Isakson et al. 2013). The main feature of these proteins is their ability to interact directly with both the ATG8 proteins through the LIR (LC3-interating region) motif and the cargo. The selective cargo receptors are degraded along with the cargo in the autophagosomes (Komatsu and Ichimura 2010). Recognition of the cargo designed for degradation from the one needed in the cell requires Ub as a specific ligand. The ubiquitinated cargo is specifically recognised by cargo receptors, such as p62, NBR1, NDP52, possess the Ub-binding domain (Johansen and Lamark 2011). Yet, some cargoes can be targeted for lysosomal degradation in a Ub-independent mechanism and some selective autophagy receptors, which have a Ub-binding domain, directly recognise cargo in a Ub-independent mechanism (Komatsu et al. 2010). In addition to cargo receptors able to bind to both the cargo and ATG8 proteins, another category of proteins, autophagy adaptors (defined as scaffold proteins capable of binding both the cargo-receptor complex and the core components of autophagic machinery) have been characterised. The best described examples of such proteins are the mammalian ALFY (autophagy-linked FYVE protein) and ATG11 from yeast (Lynch-Day and Klionsky 2010; Isakson et al. 2013).

Figure 7.5 illustrates selective autophagy types reported for plants. Some have already been discussed above. There is no indication that in every case a specific autophagy receptor exists. For example, although importance of lipophagy in animals is well documented (Christian et al. 2013), few data for plants are available and the presence of this process in plants can only be inferred by accumulation of lipid peroxides in *atg* mutants (Xiong et al. 2007). The involvement of mammalian p62 and NBR1 proteins in degradation of protein aggregates is well known and therefore it seems reasonable that the plant NBR1-like proteins work as selective receptors for protein aggregates and possibly also for single Ub-tagged proteins. The involvement of autophagy in response to biotic stresses has been well studied in plants but the authors of this review are not aware of any report about direct targeting of viral or bacterial pathogens by selective cargo receptors. Xenophagy involving specific cargo receptors is reported for animal pathogens only (Mostowy et al. 2011). An interesting example are anthocyanins, plant pigments of various functions maintained in the vacuole: a significant change in anthocyanin profiles (overall reduction) was observed in *atg* mutants (Pourcel et al. 2010). Although no direct evidence exists for the specific authophagy-dependent vacuolar transport of anthocyanins, it is tempting to speculate that it might be an example of autophagy involvement in a plant anabolic process (compare to CVT in yeast as an example of such a process).

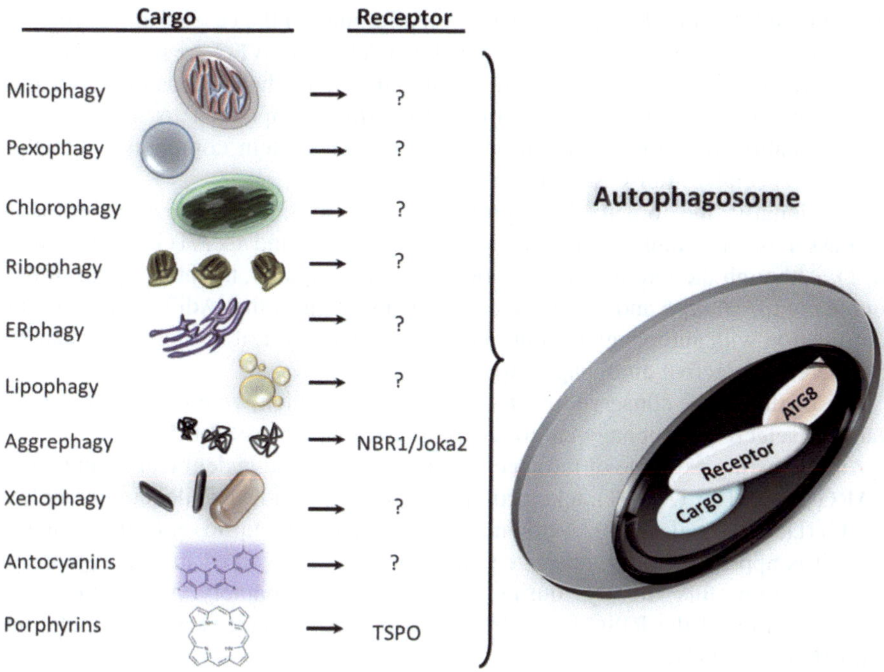

Fig. 7.5 Examples of selective autophagy in plants

The last example of selective autophagy shown in Fig. 7.5 deals with haem and porphyrins degradation (Vanhee et al. 2011). The authors indicated that the membrane-located tryptophan-rich sensory protein (TPSO), which is induced by various stresses, is in fact involved in haeme scavenging and targeting for degradation via autophagy. TSPO has the LIR motif and co-localizes with ATG8 in plant cells. In addition, haem-binding and functional LIR motif are necessary for autophagy-dependent degradation of TSPO in vitro. It is probable that the TSPO protein is also involved in binding and degradation of porphyrins, because over-expression of TSPO in plants alleviated porphyrin-induced cytotoxity in plant cells. Although the study still leaves many unanswered questions, such as how the plastid – synthesised haem reaches the cytosol, it revealed the novel autophagy-based regulatory mechanisms involved in switching off the response to stress when no longer needed. The increased requirement for ROS scavengers (appearing as a result of stress) leads to the increased requirements for porphyrin cofactors and up-regulated tetrapyrrole biosynthesis. These compounds are cytotoxic and their level needs to be carefully regulated. TSPO binds porphyrins and is along with this cargo, degraded via autophagy.

The plant-specific ATI1 and ATI2 proteins (autophagy interacting proteins 1 and 2), identified in young seedlings upon carbon starvation were not included in Fig. 7.5, however they both should be considered a part of selective autophagy and both bind ATG8 (Honig et al. 2012a, b). Although in the case of these proteins the cargo is unknown, and it was speculated that ATI1 and ATI2 are involved either in removal of germination inhibiting compounds and their deposition into the vacuole or in transport of germination-promoting compounds to the vacuole. They probably transport membrane proteins or non-protein compounds into the vacuole, especially after C starvation.

Additionally, the tomato AGC protein kinase Adi3, known to function as a suppressor of programmed cell death, was shown to bind to ATG8 (Devarenne 2011). Although the interaction was not further investigated, the phenotypes of the double *atg* (*atg3*, *atg6* and *atg7*) and *adi3* mutants suggest that Adi3 might work in coordination with autophagy to control cell death. However, it is unclear if it can be considered a selective autophagy cargo receptor.

The evolutionary conserved NBR1-like proteins, which were described also in plants as selective autophagy cargo receptors, will be discussed below.

An interesting example of the specific target of autophagic degradation in plants is ARGONAUTE1 (AGO1), a key component of RNA-induced silencing complex (RISC) (Derrien et al. 2012). Autophagy is involved in the turnover of this protein. Thus, it is appealing to hypothesize that autophagy is involved in degradation of proteins from this family during cellular stress, namely when the fast reprogramming of the RISC is needed following rapid changes of the population of miRNA and siRNA.

Knowledge of the selective autophagy cargo receptors in plants is very limited and extensive work is needed to fill the multiple gaps in this field.

NBR1-Like Proteins in Plants

The plant NBR1-like selective cargo receptors, AtNBR1 and Joka2 were identified independently by two research groups in *Arabidopsis* (Svenning et al. 2011) and tobacco (Zientara-Rytter et al. 2011), respectively. Plant NBR1-like proteins seem to be good autophagy markers (Svenning et al. 2011; Zientara-Rytter et al. 2011; Minina et al. 2013). Moreover, it has been shown that AtNBR1, similarly to the mammalian p62 and NBR1, is degraded by the autophagy pathway (Minina et al. 2013). With respect to the phenotype, the enhanced sensitivity of *nbr1* mutants to heat, oxidative, drought and salt stress was reported (Zhou et al. 2013).

NBR1-like proteins identified in plants are functional hybrids of the mammalian NBR1 (next to breast cancer 1 gene 1) and p62/SQSTM1 (Sequestosome 1) and have a modular composition (Fig. 7.6). They have four main domains: PB1 (Phox and Bem1), ZZ (ZZ-type Zinc finger domain), NBR1/FW and double UBA (ubiquitin-associated domain), UBA1 and UBA2. The roles of individual domains and the protein partners specifically binding to these domains are characterised in mammalian cargo receptors (Kirkin et al. 2009; Mardakheh et al. 2010; Salminen et al. 2012).

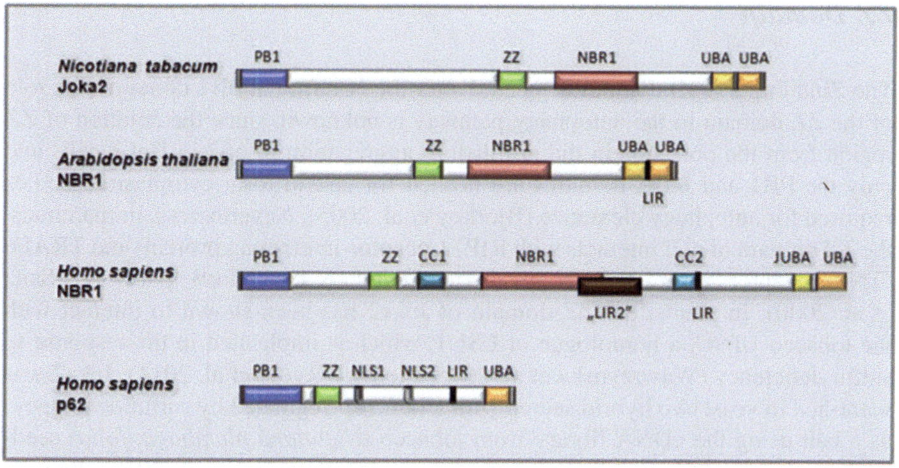

Fig. 7.6 The proteins are drawn to scale. The PB1, ZZ, NBR1, UBA1 and UBA2 domains are marked. The experimentally verified nuclear localization signals (*NLS*) and mapped LIR motifs are shown

PB1 Domain

The PB1 domain is a well-known interaction module, highly conserved among animals, plants, fungi and amoebae (Sumimoto et al. 2007) that can interact with various proteins by creating salt bridges between basic and acidic residues located on the PB1 domain to modify their functions. The PB1 domains are present in nearly 200 proteins participating in diverse biological processes in all eukaryotes (Letunic et al. 2002). The PB1 domains could be classified into three groups based on structure. Type-A is represented by PB1 domains possessing acidic OPCA motif. Type-B includes PB1 domains with lysine residue/s on the first beta strand carrying basic charge. The third group contains PB1 domains with both acidic and basic clusters (Type-A+B).

The PB1 domain of human NBR1 has been classified as Type-A (Muller et al. 2006). No homodimers can be formed by this domain, while the PB1 domain of human p62 is classified as Type-A+B, having both acidic and basic clusters, and is involved in the formation of p62-p62 homodimers and in p62-NBR1 heterodimers. Similarly to p62, AtNBR1 (NBR1 from *Arabidopsis*) and Joka2 (a tobacco homologue) have been shown to form homodimers through the PB1 domain (Svenning et al. 2011; Zientara-Rytter et al. 2011).

Oligomerisation of the selective autophagy cargo receptors is crucial for degradation process since mutants of p62 and AtNBR1, with substituted amino acids that are particularly important for polymerisation, lost the ability to aggregate and are not degraded by autophagy. Consequently, the human NBR1, possessing only an OPCA motif is not able to self-interact through PB1 and an additional CC (coiled-coil) region is required for self-oligomerisation.

ZZ Domain

The Zinc finger (ZZ) domain is located after the N-terminal PB1 domain. The role of the ZZ domain in the autophagy pathway is unknown, since the deletion of ZZ region from the p62 protein did not disturb aggregation of p62 in HeLa cells and only the PB1 and UBA domains are needed for p62 to form cytoplasmic bodies required for autophagy clearance (Bjorkoy et al. 2005). Nevertheless, in mammals, the ZZ domain of p62 interacts with RIP1 (receptor-interacting protein) and TRAF6 (TNF receptor associated factor 6) to modulate the NFκB pathway (Sanz et al. 2000). In plants, the ZZ domain of Joka2 has been shown to interact with the tobacco UP9C, a homologue of LSU1, which is implicated in the response to sulfur deficiency (Wawrzynska et al. 2005; Zientara-Rytter et al. 2011). Joka2 was identified in yeast two hybrid search with UP9C (up-regulated by sulfur deficiency) as a bait using the cDNA library from tobacco (*Nicotiana plumbaginifolia*) seedlings grown in nutrient sufficient conditions (Lewandowska et al. 2010). Interactions between proteins from both families have been confirmed *in planta* using *N. tobaccum* and *Arabidopsis* orthologues by bimolecular fluorescence complementation (BiFC) (K. Zientara-Rytter, A. Sirko – unpublished). Due to the unknown molecular function of UP9/LSU the role of such binding has yet to be discovered.

NBR1/FW Domain

The NBR1/FW domain, (consisting of two sets of three β-strands separated by an unstructural linker) is located between the ZZ and UBA domains and contains four highly conserved tryptophan (W) residues. The NBR1/FW domain is present in all NBR1-like proteins throughout the eukaryotic kingdom (Svenning et al. 2011). The function of the NBR1/FW domain is unknown. However, it has been recently shown that the mammalian selective autophagy receptor, NBR1, binds to the light chain of the microtubule-associated protein MAP1B through the NBR1/FW domain (Marchbank et al. 2012).

UBA Domains

The ubiquitin-binding domain (UBA) is short, containing only about 45 residues. It is a commonly present in many proteins, which are involved in degradation pathways requiring Ub tagged to the substrate as a signal for its degradation. In the selective autophagy receptors the UBA domain is located at the C-terminus as a single domain as in p62 or it could be duplicated as in AtNBR1, Joka2 and other plant homologues. The UBA domain present in selective autophagy cargo receptors prefers (poly)Ub tails bound by lysine in position 63 *(K-63)* and has a moderate affinity to the Ub bound to the lysine residue in position 48 (Johansen and Lamark 2011). The UBA domains serve to ensure transport of cargo to the appropriate process (autophagy or UPS) (Raasi et al. 2005). Deletion of UBA domains completely inhibit the function of selective autophagy receptors since they are unable to recognise and bind substrates for autophagy clearance (Seibenhener et al. 2004).

LIR Motif

The LIR (LC3-interacting region) motif is a short (eight amino acids) sequence responsible for directly binding of the selective autophagy receptors to ATG8 proteins attached to autophagosomal membranes. Interaction between the LIR motif of the receptor and the ATG8 proteins is crucial for targeting cargo for degradation. In a typical LIR motif ($X_{-1}X_{-2}X_{-3}W/F/YX_1X_2L/I/VX_3$) the hydrophobic amino acids (W, F or Y and L, I or V) are essential for binding and occupy the W-site and L-site created on the surface of ATG8 proteins. Amino acids in positions X_{-1}, X_{-2}, X_{-3}, X_1, X_2, X_3 are not crucial for interaction but they can enhance the strength of binding, particularly if acidic residues (D or E) are located at positions X_{-1}, X_{-2} and X_{-3} or X_1 (Noda et al. 2008, 2010; Yamaguchi et al. 2010; Johansen and Lamark 2011). For the AtNBR1 protein, an established LIR motif (VSEWDPIL) was identified in a strongly conserved region between the UBA1 and UBA2 domains.

A high conservation of the interaction between autophagy receptors and ATG8 proteins has been revealed. In HeLa cells, AtNBR1 is sequestered into autophagosomes when human GABARAP subfamily members are also overexpressed, demonstrating high evolutionary conservation of the selective autophagy pathway and high structural similarity of ATG8 proteins among different kingdoms (Svenning et al. 2011).

Nuclear Localisation Signals

Selective autophagy cargo receptors can be involved in several different pathways in addition to autophagy. Human p62, for example, also transports substrates designated for degradation to the proteasome and continuously shuttles between cytoplasm and nucleus. Proteins larger than 40 kDa, such as p62, are actively transported between these two cellular compartments and need nuclear import and export signals. Both importins and exportins transfer cargo protein into the nucleus or cytoplasm in cooperation with the nuclear pore complex. Nuclear localization signals (NLS) and nuclear export signal (NES) are short well-described sequences recognised by nuclear-cytosolic transport receptors. Two functional basic monopartite NLS signals (NLS1, NLS2) and one NES motif are present in p62 (Pankiv et al. 2010), however the role of p62 in the nucleus is unclear. It is speculated that, since in the nucleus in contrast to the cytoplasm, only proteasomal degradation system takes place, p62 may act as a polyubiquitin targeting factor carrying substrates to a nuclear pool of proteasomes. It could be also involved in the transport of other proteins or it could act as a sensor of nuclear and cytosolic proteotoxic stress. The role of NLS and NES motifs and the significance of nuclear localization have not yet been investigated in NBR1-like proteins.

Conclusions and Future Perspectives

Within this chapter multiple reasons to continue studies on autophagy in plants have been highlighted. Autophagy plays a protective role during nutrient starvation (and other environmental stresses) by enhancing the recycling of unwanted cellular materials. The process is evolutionarily conserved. In animals, induction of autophagy results in extension of the life span (Madeo et al. 2010; Rubinsztein et al. 2011), while defective autophagy has been related to cancer, various neurodegenerative and immune-related diseases (Todde et al. 2009). Various genetic, pharmacological or nutritional approaches designed to improve autophagy in human may be a future strategy of choice to avoid or delay aging-associated pathologies. Such potential therapies still need further extensive investigation. Considering the strong conservation of the process and the participating proteins, studies on autophagy in plants can give clues to this area of medical research by providing inexpensive testing models for novel drugs and treatments. Additionally, in plants, autophagy is involved in regulation of the lifespan (Minina et al. 2013).

Sufficient knowledge of these mechanisms may allow for controlled genetic or molecular manipulations of plant metabolism. All modifications leading to moderate up-regulation of autophagy may have a positive impact on plant biomass and crop yield. Studies on regulation of the specificity of autophagy and its role in maintaining proteostasis (protein homeostasis), as well as overall homeostasis, have become very active areas of research. Nevertheless many important questions remain unanswered. The problems still to be elucidated include the regulation of the specificity of the autophagy process and its role in maintaining proteostasis (protein homeostasis), as well as overall homeostasis in plants and other living organisms, including humans.

Acknowledgements M.C.-R. was supported by the Marie Curie Initial Training Network BIONUT (project No. 264296). K. Z-R is supported by National Science Centre, Poland (project No. 201/05/N/NZ1/00699). Research in A.S. lab is also supported by the Polish Ministry of Science and Higher Education (project No. W16/7.PR/2011).

References

Alvarez C, Garcia I, Romero LC, Gotor C (2012a) Mitochondrial sulfide detoxification requires a functional isoform *O*-acetylserine(thiol)lyase C in *Arabidopsis thaliana*. Mol Plant 5:1217–1226

Alvarez C, Garcia I, Moreno I, Perez-Perez ME, Crespo JL, Romero LC, Gotor C (2012b) Cysteine-generated sulfide in the cytosol negatively regulates autophagy and modulates the transcriptional profile in *Arabidopsis*. Plant Cell 24:4621–4634

Amtmann A, Armengaud P (2009) Effects of N, PK and S on metabolism: new knowledge gained from multi-level analysis. Curr Opin Plant Biol 12:275–283

Anderson GH, Veit B, Hanson MR (2005) The *Arabidopsis* AtRaptor genes are essential for post-embryonic plant growth. BMC Biol 3:12

Asakura M, Ninomiya S, Sugimoto M, Oku M, Yamashita S, Okuno T, Sakai Y, Takano Y (2009) Atg26-mediated pexophagy is required for host invasion by the plant pathogenic fungus *Colletotrichum orbiculare*. Plant Cell 21:1291–1304

Bassham DC, Laporte M, Marty F, Moriyasu Y, Ohsumi Y, Olsen LJ, Yoshimoto K (2006) Autophagy in development and stress responses of plants. Autophagy 2:2–11

Behrends C, Fulda S (2012) Receptor proteins in selective autophagy. Int J Cell Biol 2012: 673290

Bjorkoy G, Lamark T, Brech A, Outzen H, Perander M, Overvatn A, Stenmark H, Johansen T (2005) p62/SQSTM1 forms protein aggregates degraded by autophagy and has a protective effect on huntingtin-induced cell death. J Cell Biol 171:603–614

Callis J, Carpenter T, Sun CW, Vierstra RD (1995) Structure and evolution of genes encoding polyubiquitin and ubiquitin-like proteins in *Arabidopsis thaliana* ecotype Columbia. Genetics 139:921–939

Christian P, Sacco J, Adeli K (2013) Autophagy: emerging roles in lipid homeostasis and metabolic control. Biochim Biophys Acta 1831:819–824

Chung T, Phillips AR, Vierstra RD (2010) ATG8 lipidation and ATG8-mediated autophagy in *Arabidopsis* require ATG12 expressed from the differentially controlled ATG12A and ATG12B loci. Plant J 62:483–493

Dantuma NP, Lindsten K, Glas R, Jellne M, Masucci MG (2000) Short-lived green fluorescent proteins for quantifying ubiquitin/proteasome-dependent proteolysis in living cells. Nat Biotechnol 18:538–543

Deprost D, Truong HN, Robaglia C, Meyer C (2005) An *Arabidopsis* homolog of RAPTOR/KOG1 is essential for early embryo development. Biochem Biophys Res Commun 326:844–850

Deprost D, Yao L, Sormani R, Moreau M, Leterreux G, Nicolai M, Bedu M, Robaglia C, Meyer C (2007) The *Arabidopsis* TOR kinase links plant growth, yield, stress resistance and mRNA translation. EMBO Rep 8:864–870

Derrien B, Baumberger N, Schepetilnikov M, Viotti C, De Cillia J, Ziegler-Graff V, Isono E, Schumacher K, Genschik P (2012) Degradation of the antiviral component ARGONAUTE1 by the autophagy pathway. Proc Natl Acad Sci U S A 109:15942–15946

Devarenne TP (2011) The plant cell death suppressor Adi3 interacts with the autophagic protein Atg8h. Biochem Biophys Res Commun 412:699–703

Diaz-Troya S, Perez-Perez ME, Florencio FJ, Crespo JL (2008) The role of TOR in autophagy regulation from yeast to plants and mammals. Autophagy 4:851–865

Doelling JH, Walker JM, Friedman EM, Thompson AR, Vierstra RD (2002) The APG8/12-activating enzyme APG7 is required for proper nutrient recycling and senescence in *Arabidopsis thaliana*. J Biol Chem 277:33105–33114

Finley D (2009) Recognition and processing of ubiquitin-protein conjugates by the proteasome. Annu Rev Biochem 78:477–513

Floyd BE, Morriss SC, Macintosh GC, Bassham DC (2012) What to eat: evidence for selective autophagy in plants. J Integr Plant Biol 54:907–920

Gadalla MM, Snyder SH (2010) Hydrogen sulfide as a gasotransmitter. J Neurochem 113:14–26

Gruber BD, Giehl RF, Friedel S, von Wiren N (2013) Plasticity of the *Arabidopsis* root system under nutrient deficiencies. Plant Physiol 163:161–179

Guiboileau A, Sormani R, Meyer C, Masclaux-Daubresse C (2010) Senescence and death of plant organs: nutrient recycling and developmental regulation. C R Biol 333:382–391

Guiboileau A, Yoshimoto K, Soulay F, Bataille MP, Avice JC, Masclaux-Daubresse C (2012) Autophagy machinery controls nitrogen remobilization at the whole-plant level under both limiting and ample nitrate conditions in *Arabidopsis*. New Phytol 194:732–740

Hanaoka H, Noda T, Shirano Y, Kato T, Hayashi H, Shibata D, Tabata S, Ohsumi Y (2002) Leaf senescence and starvation-induced chlorosis are accelerated by the disruption of an *Arabidopsis* autophagy gene. Plant Physiol 129:1181–1193

Harrison-Lowe NJ, Olsen LJ (2008) Autophagy protein 6 (ATG6) is required for pollen germination in *Arabidopsis thaliana*. Autophagy 4:339–348

Hillwig MS, Contento AL, Meyer A, Ebany D, Bassham DC, Macintosh GC (2011) RNS2, a conserved member of the RNase T2 family, is necessary for ribosomal RNA decay in plants. Proc Natl Acad Sci U S A 108:1093–1098

Hoefgen R, Nikiforova VJ (2008) Metabolomics integrated with transcriptomics: assessing systems response to sulfur-deficiency stress. Physiol Plant 132:190–198

Hofius D, Schultz-Larsen T, Joensen J, Tsitsigiannis DI, Petersen NH, Mattsson O, Jorgensen LB, Jones JD, Mundy J, Petersen M (2009) Autophagic components contribute to hypersensitive cell death in *Arabidopsis*. Cell 137:773–783

Honig A, Avin-Wittenberg T, Galili G (2012a) Selective autophagy in the aid of plant germination and response to nutrient starvation. Autophagy 8:838–839

Honig A, Avin-Wittenberg T, Ufaz S, Galili G (2012b) A new type of compartment, defined by plant-specific Atg8-interacting proteins, is induced upon exposure of *Arabidopsis* plants to carbon starvation. Plant Cell 24:288–303

Howarth JR, Parmar S, Jones J, Shepherd CE, Corol DI, Galster AM, Hawkins ND, Miller SJ, Baker JM, Verrier PJ, Ward JL, Beale MH, Barraclough PB, Hawkesford MJ (2008) Co-ordinated expression of amino acid metabolism in response to N and S deficiency during wheat grain filling. J Exp Bot 59:3675–3689

Hutchins AP, Liu S, Diez D, Miranda-Saavedra D (2013) The repertoires of ubiquitinating and deubiquitinating enzymes in eukaryotic genomes. Mol Biol Evol 30:1172–1187

Isakson P, Holland P, Simonsen A (2013) The role of ALFY in selective autophagy. Cell Death Differ 20:12–20

Ishida H, Yoshimoto K, Izumi M, Reisen D, Yano Y, Makino A, Ohsumi Y, Hanson MR, Mae T (2008) Mobilization of rubisco and stroma-localized fluorescent proteins of chloroplasts to the vacuole by an ATG gene-dependent autophagic process. Plant Physiol 148:142–155

Izumi M, Hidema J, Makino A, Ishida H (2013) Autophagy contributes to nighttime energy availability for growth in *Arabidopsis*. Plant Physiol 161:1682–1693

Johansen T, Lamark T (2011) Selective autophagy mediated by autophagic adapter proteins. Autophagy 7:279–296

John F, Roffler S, Wicker T, Ringli C (2011) Plant TOR signaling components. Plant Signal Behav 6:1700–1705

Kanki T, Wang K, Cao Y, Baba M, Klionsky DJ (2009) Atg32 is a mitochondrial protein that confers selectivity during mitophagy. Dev Cell 17:98–109

Kim SH, Kwon C, Lee JH, Chung T (2012) Genes for plant autophagy: functions and interactions. Mol Cells 34:413–423

Kirkin V, Lamark T, Sou YS, Bjorkoy G, Nunn JL, Bruun JA, Shvets E, McEwan DG, Clausen TH, Wild P, Bilusic I, Theurillat JP, Overvatn A, Ishii T, Elazar Z, Komatsu M, Dikic I, Johansen T (2009) A role for NBR1 in autophagosomal degradation of ubiquitinated substrates. Mol Cell 33:505–516

Klionsky DJ. (2007) Autophagy: from phenomenology to molecular understanding in less than a decade. Nat Rev Mol Cell Biol 8:931–937

Klionsky DJ, Baehrecke EH, Brumell JH, Chu CT, Codogno P, Cuervo AM, Debnath J, Deretic V, Elazar Z, Eskelinen EL, Finkbeiner S, Fueyo-Margareto J, Gewirtz D, Jaattela M, Kroemer G, Levine B, Melia TJ, Mizushima N, Rubinsztein DC, Simonsen A, Thorburn A, Thumm M, Tooze SA (2011) A comprehensive glossary of autophagy-related molecules and processes (2nd edition). Autophagy 7:1273–1294

Komatsu M, Ichimura Y (2010) Physiological significance of selective degradation of p62 by autophagy. FEBS Lett 584:1374–1378

Komatsu M, Kurokawa H, Waguri S, Taguchi K, Kobayashi A, Ichimura Y, Sou YS, Ueno I, Sakamoto A, Tong KI, Kim M, Nishito Y, Iemura S, Natsume T, Ueno T, Kominami E, Motohashi H, Tanaka K, Yamamoto M (2010) The selective autophagy substrate p62 activates the stress responsive transcription factor Nrf2 through inactivation of Keap1. Nat Cell Biol 12:213–223

Korolchuk VI, Menzies FM, Rubinsztein DC (2010) Mechanisms of cross-talk between the ubiquitin-proteasome and autophagy-lysosome systems. FEBS Lett 584:1393–1398

Kroemer G, Galluzzi L, Vandenabeele P, Abrams J, Alnemri ES, Baehrecke EH, Blagosklonny MV, El-Deiry WS, Golstein P, Green DR, Hengartner M, Knight RA, Kumar S, Lipton SA, Malorni W, Nunez G, Peter ME, Tschopp J, Yuan J, Piacentini M, Zhivotovsky B, Melino G, Nomenclature Committee on Cell, D (2009) Classification of cell death: recommendations of the Nomenclature Committee on Cell Death 2009. Cell Death Differ 16:3–11

Kwon SI, Cho HJ, Jung JH, Yoshimoto K, Shirasu K, Park OK (2010) The Rab GTPase RabG3b functions in autophagy and contributes to tracheary element differentiation in *Arabidopsis*. Plant J 64:151–164

Lee TA, Vande Wetering SW, Brusslan JA (2013) Stromal protein degradation is incomplete in *Arabidopsis thaliana* autophagy mutants undergoing natural senescence. BMC Res Notes 6:17

Leiber RM, John F, Verhertbruggen Y, Diet A, Knox JP, Ringli C (2010) The TOR pathway modulates the structure of cell walls in *Arabidopsis*. Plant Cell 22:1898–1908

Lenz HD, Vierstra RD, Nurnberger T, Gust AA (2011) ATG7 contributes to plant basal immunity towards fungal infection. Plant Signal Behav 6:1040–1042

Letunic I, Goodstadt L, Dickens NJ, Doerks T, Schultz J, Mott R, Ciccarelli F, Copley RR, Ponting CP, Bork P (2002) Recent improvements to the SMART domain-based sequence annotation resource. Nucleic Acids Res 30:242–244

Lewandowska M, Wawrzynska A, Moniuszko G, Lukomska J, Zientara K, Piecho M, Hodurek P, Zhukov I, Liszewska F, Nikiforova V, Sirko A (2010) A contribution to identification of novel regulators of plant response to sulfur deficiency: characteristics of a tobacco gene *UP9C*, its protein product and the effects of *UP9C* silencing. Mol Plant 3:347–360

Li F, Vierstra RD (2012a) Autophagy: a multifaceted intracellular system for bulk and selective recycling. Trends Plant Sci 17:526–537

Li F, Vierstra RD (2012b) Regulator and substrate: dual roles for the ATG1-ATG13 kinase complex during autophagic recycling in *Arabidopsis*. Autophagy 8:982–984

Liang C, Tian J, Liao H (2013) Proteomics dissection of plant responses to mineral nutrient deficiency. Proteomics 13:624–636

Liu Y, Bassham DC (2010) TOR is a negative regulator of autophagy in *Arabidopsis* thaliana. PLoS One 5:e11883

Liu Y, Bassham DC (2012) Autophagy: pathways for self-eating in plant cells. Annu Rev Plant Biol 63:215–237

Liu Y, Xiong Y, Bassham DC (2009) Autophagy is required for tolerance of drought and salt stress in plants. Autophagy 5:954–963

Liu Y, Burgos JS, Deng Y, Srivastava R, Howell SH, Bassham DC (2012) Degradation of the endoplasmic reticulum by autophagy during endoplasmic reticulum stress in *Arabidopsis*. Plant Cell 24:4635–4651

Lowicka E, Beltowski J (2007) Hydrogen sulfide (H2S) – the third gas of interest for pharmacologists. Pharmacol Rep 59:4–24

Lynch-Day MA, Klionsky DJ (2010) The Cvt pathway as a model for selective autophagy. FEBS Lett 584:1359–1366

Lyzenga WJ, Stone SL (2012) Abiotic stress tolerance mediated by protein ubiquitination. J Exp Bot 63:599–616

Madeo F, Tavernarakis N, Kroemer G (2010) Can autophagy promote longevity? Nat Cell Biol 12:842–846

Maiuri MC, Zalckvar E, Kimchi A, Kroemer G (2007) Self-eating and self-killing: crosstalk between autophagy and apoptosis. Nat Rev Mol Cell Biol 8:741–752

Marchbank K, Waters S, Roberts RG, Solomon E, Whitehouse CA (2012) MAP1B interaction with the FW domain of the autophagic receptor Nbr1 facilitates its association to the microtubule network. Int J Cell Biol 2012:208014

Mardakheh FK, Auciello G, Dafforn TR, Rappoport JZ, Heath JK (2010) Nbr1 is a novel inhibitor of ligand-mediated receptor tyrosine kinase degradation. Mol Cell Biol 30:5672–5685

Minibayeva F, Dmitrieva S, Ponomareva A, Ryabovol V (2012) Oxidative stress-induced autophagy in plants: the role of mitochondria. Plant Physiol Biochem 59:11–19

Minina EA, Sanchez-Vera V, Moschou PN, Suarez MF, Sundberg E, Weih M, Bozhkov PV (2013) Autophagy mediates caloric restriction-induced lifespan extension in *Arabidopsis*. Aging Cell 12:327–329

Moreau M, Azzopardi M, Clement G, Dobrenel T, Marchive C, Renne C, Martin-Magniette ML, Taconnat L, Renou JP, Robaglia C, Meyer C (2012) Mutations in the *Arabidopsis* homolog of LST8/GbetaL, a partner of the target of Rapamycin kinase, impair plant growth, flowering, and metabolic adaptation to long days. Plant Cell 24:463–481

Mostowy S, Sancho-Shimizu V, Hamon MA, Simeone R, Brosch R, Johansen T, Cossart P (2011) p62 and NDP52 proteins target intracytosolic Shigella and Listeria to different autophagy pathways. J Biol Chem 286:26987–26995

Muller S, Kursula I, Zou P, Wilmanns M (2006) Crystal structure of the PB1 domain of NBR1. FEBS Lett 580:341–344

Nakatogawa H, Suzuki K, Kamada Y, Ohsumi Y (2009) Dynamics and diversity in autophagy mechanisms: lessons from yeast. Nat Rev Mol Cell Biol 10:458–467

Nei M, Rogozin IB, Piontkivska H (2000) Purifying selection and birth-and-death evolution in the ubiquitin gene family. Proc Natl Acad Sci U S A 97:10866–10871

Noda T, Ohsumi Y (1998) Tor, a phosphatidylinositol kinase homologue, controls autophagy in yeast. J Biol Chem 273:3963–3966

Noda NN, Kumeta H, Nakatogawa H, Satoo K, Adachi W, Ishii J, Fujioka Y, Ohsumi Y, Inagaki F (2008) Structural basis of target recognition by Atg8/LC3 during selective autophagy. Genes Cells 13:1211–1218

Noda NN, Ohsumi Y, Inagaki F (2010) Atg8-family interacting motif crucial for selective autophagy. FEBS Lett 584:1379–1385

Noda NN, Fujioka Y, Hanada T, Ohsumi Y, Inagaki F (2013) Structure of the Atg12-Atg5 conjugate reveals a platform for stimulating Atg8-PE conjugation. EMBO Rep 14:206–211

Orsi A, Razi M, Dooley HC, Robinson D, Weston AE, Collinson LM, Tooze SA (2012) Dynamic and transient interactions of Atg9 with autophagosomes, but not membrane integration, are required for autophagy. Mol Biol Cell 23:1860–1873

Pankiv S, Lamark T, Bruun JA, Overvatn A, Bjorkoy G, Johansen T (2010) Nucleocytoplasmic shuttling of p62/SQSTM1 and its role in recruitment of nuclear polyubiquitinated proteins to promyelocytic leukemia bodies. J Biol Chem 285:5941–5953

Pattingre S, Espert L, Biard-Piechaczyk M, Codogno P (2008) Regulation of macroautophagy by mTOR and Beclin 1 complexes. Biochimie 90:313–323

Phillips AR, Suttangkakul A, Vierstra RD (2008) The ATG12-conjugating enzyme ATG10 is essential for autophagic vesicle formation in *Arabidopsis* thaliana. Genetics 178:1339–1353

Pourcel L, Irani NG, Lu Y, Riedl K, Schwartz S, Grotewold E (2010) The formation of anthocyanic vacuolar inclusions in *Arabidopsis* thaliana and implications for the sequestration of anthocyanin pigments. Mol Plant 3:78–90

Raasi S, Varadan R, Fushman D, Pickart CM (2005) Diverse polyubiquitin interaction properties of ubiquitin-associated domains. Nat Struct Mol Biol 12:708–714

Rabinowitz JD, White E (2010) Autophagy and metabolism. Science 330:1344–1348

Rojas-Triana M, Bustos R, Espinosa-Ruiz A, Prat S, Paz-Ares J, Rubio V (2013) Roles of ubiquitination in the control of phosphate starvation responses in plants(f). J Integr Plant Biol 55:40–53

Rose TL, Bonneau L, Der C, Marty-Mazars D, Marty F (2006) Starvation-induced expression of autophagy-related genes in *Arabidopsis*. Biol Cell 98:53–67

Rubinsztein DC, Marino G, Kroemer G (2011) Autophagy and aging. Cell 146:682–695

Sadanandom A, Bailey M, Ewan R, Lee J, Nelis S (2012) The ubiquitin-proteasome system: central modifier of plant signalling. New Phytol 196:13–28

Salminen A, Kaarniranta K, Haapasalo A, Hiltunen M, Soininen H, Alafuzoff I (2012) Emerging role of p62/sequestosome-1 in the pathogenesis of Alzheimer's disease. Prog Neurobiol 96:87–95

Sanz L, Diaz-Meco MT, Nakano H, Moscat J (2000) The atypical PKC-interacting protein p62 channels NF-kappaB activation by the IL-1-TRAF6 pathway. EMBO J 19:1576–1586

Scherz-Shouval R, Shvets E, Fass E, Shorer H, Gil L, Elazar Z (2007) Reactive oxygen species are essential for autophagy and specifically regulate the activity of Atg4. EMBO J 26:1749–1760

Schreiber A, Peter M (2013) Substrate recognition in selective autophagy and the ubiquitin-proteasome system. Biochim Biophys Acta. doi:10.1016/j.bbamcr.2013.03.019 [Epub ahead of print]

Seibenhener ML, Babu JR, Geetha T, Wong HC, Krishna NR, Wooten MW (2004) Sequestosome 1/p62 is a polyubiquitin chain binding protein involved in ubiquitin proteasome degradation. Mol Cell Biol 24:8055–8068

Shabek N, Herman-Bachinsky Y, Buchsbaum S, Lewinson O, Haj-Yahya M, Hejjaoui M, Lashuel HA, Sommer T, Brik A, Ciechanover A (2012) The size of the proteasomal substrate determines whether its degradation will be mediated by mono- or polyubiquitylation. Mol Cell 48:87–97

Slavikova S, Ufaz S, Avin-Wittenberg T, Levanony H, Galili G (2008) An autophagy-associated Atg8 protein is involved in the responses of *Arabidopsis* seedlings to hormonal controls and abiotic stresses. J Exp Bot 59:4029–4043

Sullivan JA, Shirasu K, Deng XW (2003) The diverse roles of ubiquitin and the 26S proteasome in the life of plants. Nat Rev Genet 4:948–958

Sumimoto H, Kamakura S, Ito T (2007) Structure and function of the PB1 domain, a protein interaction module conserved in animals, fungi, amoebas, and plants. Sci STKE 2007:re6

Suttangkakul A, Li F, Chung T, Vierstra RD (2011) The ATG1/ATG13 protein kinase complex is both a regulator and a target of autophagic recycling in *Arabidopsis*. Plant Cell 23:3761–3779

Suzuki K, Kubota Y, Sekito T, Ohsumi Y (2007) Hierarchy of Atg proteins in pre-autophagosomal structure organization. Genes Cells 12:209–218

Svenning S, Lamark T, Krause K, Johansen T (2011) Plant NBR1 is a selective autophagy substrate and a functional hybrid of the mammalian autophagic adapters NBR1 and p62/SQSTM1. Autophagy 7:993–1010

Szabo C (2007) Hydrogen sulphide and its therapeutic potential. Nat Rev Drug Discov 6:917–935

Thompson AR, Vierstra RD (2005) Autophagic recycling: lessons from yeast help define the process in plants. Curr Opin Plant Biol 8:165–173

Thompson AR, Doelling JH, Suttangkakul A, Vierstra RD (2005) Autophagic nutrient recycling in *Arabidopsis* directed by the ATG8 and ATG12 conjugation pathways. Plant Physiol 138:2097–2110

Todde V, Veenhuis M, van der Klei IJ (2009) Autophagy: principles and significance in health and disease. Biochim Biophys Acta 1792:3–13

Toyooka K, Moriyasu Y, Goto Y, Takeuchi M, Fukuda H, Matsuoka K (2006) Protein aggregates are transported to vacuoles by a macroautophagic mechanism in nutrient-starved plant cells. Autophagy 2:96–106

Tsai YC, Koo Y, Delk NA, Gehl B, Braam J (2012) Calmodulin-related CML24 interacts with ATG4b and affects autophagy progression in *Arabidopsis*. Plant J. doi:10.1111/tpj.12043 [Epub ahead of print]

Tsai IT, Chen YH, Chen YH, Wang YH (2013) Amikacin-induced fin reduction is mediated by autophagy. J Toxicol Pathol 26:79–82

van Doorn WG (2011) Classes of programmed cell death in plants, compared to those in animals. J Exp Bot 62:4749–4761

Vanhee C, Zapotoczny G, Masquelier D, Ghislain M, Batoko H (2011) The *Arabidopsis* multistress regulator TSPO is a heme binding membrane protein and a potential scavenger of porphyrins via an autophagy-dependent degradation mechanism. Plant Cell 23:785–805

Vierstra RD (2003) The ubiquitin/26S proteasome pathway, the complex last chapter in the life of many plant proteins. Trends Plant Sci 8:135–142

Vierstra RD (2009) The ubiquitin-26S proteasome system at the nexus of plant biology. Nat Rev Mol Cell Biol 10:385–397

Vierstra RD (2012) The expanding universe of ubiquitin and ubiquitin-like modifiers. Plant Physiol 160:2–14

Wada S, Ishida H, Izumi M, Yoshimoto K, Ohsumi Y, Mae T, Makino A (2009) Autophagy plays a role in chloroplast degradation during senescence in individually darkened leaves. Plant Physiol 149:885–893

Wang Y, Wu Y, Tang D (2011a) The autophagy gene, ATG18a, plays a negative role in powdery mildew resistance and mildew-induced cell death in *Arabidopsis*. Plant Signal Behav 6:1408–1410

Wang Y, Nishimura MT, Zhao T, Tang D (2011b) ATG2, an autophagy-related protein, negatively affects powdery mildew resistance and mildew-induced cell death in *Arabidopsis*. Plant J 68:74–87

Wang WY, Zhang L, Xing S, Ma Z, Liu J, Gu H, Qin G, Qu LJ (2012) *Arabidopsis* AtVPS15 plays essential roles in pollen germination possibly by interacting with AtVPS34. J Genet Genomics 39:81–92

Wang Y, Yu B, Zhao J, Guo J, Li Y, Han S, Huang L, Du Y, Hong Y, Tang D, Liu Y (2013) Autophagy contributes to leaf starch degradation. Plant Cell 25:1383–1399

Wawrzynska A, Lewandowska M, Hawkesford MJ, Sirko A (2005) Using a suppression subtractive library-based approach to identify tobacco genes regulated in response to short-term sulphur deficit. J Exp Bot 56:1575–1590

Weidberg H, Shvets E, Elazar Z (2011) Biogenesis and cargo selectivity of autophagosomes. Annu Rev Biochem 80:125–156

Welters P, Takegawa K, Emr SD, Chrispeels MJ (1994) AtVPS34, a phosphatidylinositol 3-kinase of *Arabidopsis* thaliana, is an essential protein with homology to a calcium-dependent lipid binding domain. Proc Natl Acad Sci U S A 91:11398–11402

Wittenbach VA, Lin W, Hebert RR (1982) Vacuolar localization of proteases and degradation of chloroplasts in mesophyll protoplasts from senescing primary wheat leaves. Plant Physiol 69:98–102

Xia T, Xiao D, Liu D, Chai W, Gong Q, Wang NN (2012) Heterologous expression of ATG8c from soybean confers tolerance to nitrogen deficiency and increases yield in *Arabidopsis*. PLoS One 7:e37217

Xiong Y, Contento AL, Bassham DC (2005) AtATG18a is required for the formation of autophagosomes during nutrient stress and senescence in *Arabidopsis thaliana*. Plant J 42:535–546

Xiong Y, Contento AL, Bassham DC (2007) Disruption of autophagy results in constitutive oxidative stress in *Arabidopsis*. Autophagy 3:257–258

Xu P, Duong DM, Seyfried NT, Cheng D, Xie Y, Robert J, Rush J, Hochstrasser M, Finley D, Peng J (2009) Quantitative proteomics reveals the function of unconventional ubiquitin chains in proteasomal degradation. Cell 137:133–145

Yamaguchi M, Noda NN, Nakatogawa H, Kumeta H, Ohsumi Y, Inagaki F (2010) Autophagy-related protein (Atg) 8-family interacting motif in Atg3 mediates the Atg3-Atg8 interaction and is crucial for the cytoplasm-to-vacuole targeting pathway. J Biol Chem 285:29599–29607

Yang Z, Klionsky DJ (2010) Eaten alive: a history of macroautophagy. Nat Cell Biol 12:814–822

Yen WL, Legakis JE, Nair U, Klionsky DJ (2007) Atg27 is required for autophagy-dependent cycling of Atg9. Mol Biol Cell 18: 581–593

Yoshimoto K, Hanaoka H, Sato S, Kato T, Tabata S, Noda T, Ohsumi Y (2004) Processing of ATG8s, ubiquitin-like proteins, and their deconjugation by ATG4s are essential for plant autophagy. Plant Cell 16:2967–2983

Yoshimoto K, Jikumaru Y, Kamiya Y, Kusano M, Consonni C, Panstruga R, Ohsumi Y, Shirasu K (2009) Autophagy negatively regulates cell death by controlling NPR1-dependent salicylic acid signaling during senescence and the innate immune response in *Arabidopsis*. Plant Cell 21:2914–2927

Zhou J, Wang J, Cheng Y, Chi YJ, Fan B, Yu JQ, Chen Z (2013) NBR1-mediated selective autophagy targets insoluble ubiquitinated protein aggregates in plant stress responses. PLoS Genet 9:e1003196

Zientara-Rytter K, Lukomska J, Moniuszko G, Gwozdecki R, Surowiecki P, Lewandowska M, Liszewska F, Wawrzynska A, Sirko A (2011) Identification and functional analysis of Joka2, a tobacco member of the family of selective autophagy cargo receptors. Autophagy 7:1145–1158

Chapter 8
Mineral Nutrient Depletion Affects Plant Development and Crop Yield

Sarah J. Whitcomb, Elmien Heyneke, Fayezeh Aarabi, Mutsumi Watanabe, and Rainer Hoefgen

Abstract Optimal plant development depends on the availability of light, water, favourable temperatures and mineral nutrients. Insufficient availability of plant mineral nutrients leads to growth impairments and yield depressions. In natural environments as well as in agricultural systems, mineral nutrient availability is changing in space and time over the growth season of a plant. Therefore, plants have developed adaptation strategies to cope with nutrient deficiencies. Fully understanding these mechanisms at the molecular level is a necessity for breeding nutrient use efficient crops. Plant systems biology approaches contribute to this endeavour as agriculture and plant breeding face-increasing challenges to achieve sustainable and effective agricultural production.

Keywords Nutrient deficiency • Development • Senescence • Omics • Systems biology • Nitrogen • Phosphorus • Sulfur

Introduction – The Importance of Mineral Nutrient Use Efficiency for a Sustainable Agriculture

Sustainable crop production will be one of the most critical challenges in the next 50 years due to predicted worldwide population growth and increasing meat consumption, especially in the developing world (Dyson 1999; Long and Ort 2010; Tester and Langridge 2010; Mueller et al. 2012). Conventional breeding and improved agronomical practices are currently producing only incremental and insufficient increases in crop yield potential, and factors such as global climate change will increasingly compromise the ability to reach this theoretical yield potential (Jaggard et al. 2010). With respect to climate change, increasing night temperatures and altered water availability patterns are particularly problematic (Semenov and Halford 2009; Long and Ort 2010; Dolferus et al. 2011;

S.J. Whitcomb • E. Heyneke • F. Aarabi • M. Watanabe • R. Hoefgen (✉)
Max Planck Institute of Molecular Plant Physiology, Science Park Potsdam-Golm, 14424 Potsdam, Germany
e-mail: hoefgen@mpimp-golm.mpg.de

M.J. Hawkesford et al. (eds.), *Nutrient Use Efficiency in Plants*,
Plant Ecophysiology 10, DOI 10.1007/978-3-319-10635-9_8

Li et al. 2011). Additionally, the necessary shift from a fossil fuel to a biofuel based economy and from petrochemicals to biochemicals increases the pressure on crop productivity and will demand development of crop cultivars with a wider range of agronomic features. The biofuel and biochemical industries require regenerative plant commodities with high vegetative biomass or high seed content of specific compounds such as particular modified starches, while food and feed production rely mostly upon annual crops bred for high seed yield and complex quality traits.

For food and feed production, annual, hapaxanthic crop plants dominate agriculture, top among them the cereals wheat, rice and maize (Dyson 1999). As they complete their life cycle from seed to seed in one vegetative period, crop productivity and quality is tightly coupled to senescence. During the growth season, plants follow a genetically controlled developmental programme leading to organ and plant senescence, including regulated allocation of nutrients from vegetative tissues to the seeds, and finally to plant death (Himelblau and Amasino 2001; Lim et al. 2007; Wu et al. 2012; Watanabe et al. 2013). Importantly, developmental senescence programmes are responsive to environmental conditions such as light, temperature, water and nutrient availability (Watanabe et al. 2012). Adverse environmental conditions outside of the plant's ecologically optimal niche usually result in stress for the plant and trigger developmental acceleration, premature senescence, and eventually result in reduced yields. This chapter discusses plant responses to mineral nutrient deficiencies and whether the ability of crop plants to utilise such nutrients can be improved to meet the global need for more nutrient use efficient cultivars.

Agroecosystems are open systems where nutrient ions are mobilised from the soil into crop biomass and removed with the harvest. Crop productivity depends on mineral nutrient ions in the root environment, the availability of which is determined by soil parameters such as water content and pH as well as microbial degradation of biomass such as straw, humus or manure, generally termed mineralisation (von Liebig 1840; Marschner 2012; Buchanan et al. 2007; Haensch and Mendel 2009; Amtmann and Blatt 2009; Amtmann and Armengaud 2009). Mineral fertilisers are applied to compensate for nutrient losses in conventional agriculture. Usually the provision of macronutrients (N, P, S, Mg, K, Ca) is sufficient, while micronutrients or trace elements (Fe, Cu, Mn, Ni, Zn, Cl, B, Mo) are supplemented depending upon the needs of specific soil types or crops (Watanabe et al. 2012). Nitrogen and phosphate have the largest effect on production costs due to the quantities applied and their cost. Optimal crop production requires a balanced supply of nutrients, as a deficiency of one nutrient cannot be compensated for by others according to Liebig's law of the minimum (von Liebig 1840). Furthermore, both deficiency and over-accumulation can lead to negative effects on plants due to the interaction and competition between minerals at the level of uptake and assimilation (Hoefgen and Hesse 2008). For example iron uptake in barley and tomato has been shown to be dependent on sulfate availability. Under sulfate depletion iron transporters and iron reductase are downregulated, preventing uptake even in the presence of iron. Further, phytosiderphore biosynthesis and internal iron transport are impaired as nicotianamine synthase is essentially switched off (Astolfi et al. 2010; Zuchi et al. 2009; Cassin et al. 2009; Klatte et al. 2009). Especially among the micronutrients, including REDOX-active transition metals such as copper or zinc, the concentration range between deficiency and toxicity is often narrow, both having negative effects on plant growth (He et al. 2005).

Plants are able to respond flexibly to varied environmental conditions. In addition to light, water and temperature, mineral nutrient availability determines plant growth and propagation, or, in an agricultural context, crop yield. The ecological niche of a plant is determined by the specific range of conditions a certain plant species can adapt to. Crop cultivars are usually the result of adapted breeding for agricultural conditions and, as a result of this selection process, they are disposed to narrow genome variability (Tanksley and McCouch 1997). There are different avenues to re-introduce increased genetic variability, but all approaches have to take into account the need to retain the very high yield potential of modern crop cultivars, otherwise the necessary goal of sustainably and affordably providing more food for a growing population on less land will not be met.

Mineral nutrient deficiencies typically result in yield depression and quality impairment in crop plants. Here we compare what is known about the effects of mineral depletion on plant development and metabolic composition to natural developmental senescence of plants. The model plant *Arabidopsis thaliana* provides all the necessary tools to perform such studies at a systems biology level. The conclusions from such investigations will be helpful for developing concepts for breeding of nutrient use efficient crops and will provide directives for more sustainable agricultural practice.

Senescence – A Developmentally Controlled Process Critical for Seed Development

Leaf and whole plant senescence is a genetically programmed and highly regulated process termed developmental senescence (Wu et al. 2012; Thomas 2013; Watanabe et al. 2013). This is essentially a self-destructive programme, which directs breakdown of cellular structures and components in source tissues for export to sink tissues such as young leaves, roots, flowers, tubers or seeds.

The developmental cycle of an annual plant (Fig. 8.1) starts with germination and a growth phase leading to leaf expansion, during which young leaves act as a sink for carbohydrates and root-derived mineral ions. This is followed by a phase of maturity, in which fully expanded leaves are photosynthetically and biochemically most active and provide metabolites to developing leaves or other sink tissues such as roots and developing seeds. Older leaves successively senesce, but after anthesis and fertilisation, developing seeds act as the major sink and the plant is primed for general senescence. Upon the switch from vegetative to generative growth, no new leaves develop and existing leaves enter senescence to provide nutrients to the seeds. During this phase the plant has to find a compromise between degradation of macromolecules to remobilise nutrients and leaf functional integrity for photosynthesis and carbohydrate production. For example in rice, wheat and barley, up to

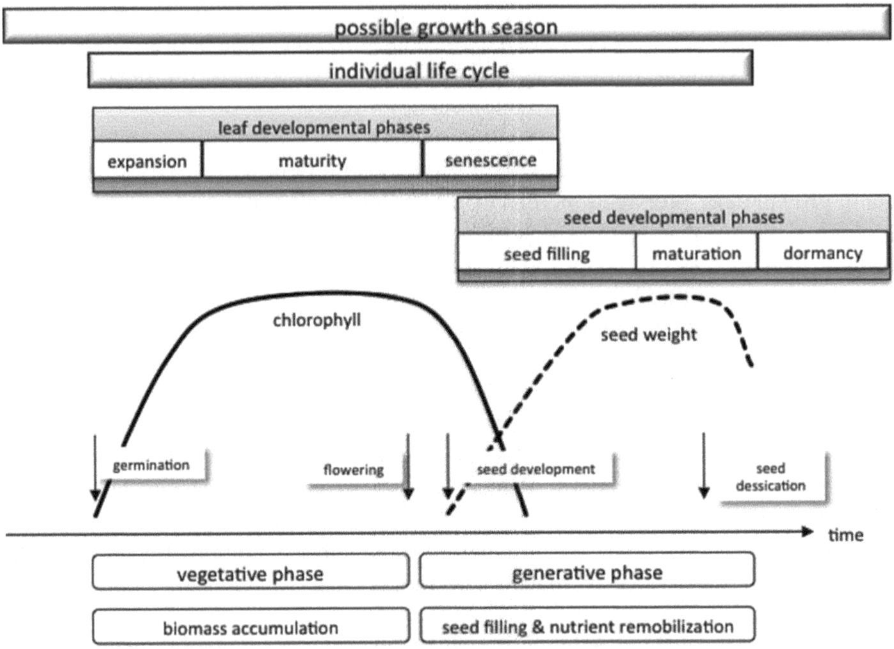

Fig. 8.1 Schematic life cycle of an annual plant. Light and temperature determine the potential growth season. The life cycle of an annual plant starts with germination. Seedling leaf development initially depends on resources from the seed, and young leaves act as a sink organs until they develop sufficient photosynthetic activity to function as source tissue by exporting sugars and nutrients to other parts of the plant. Here the chlorophyll content of leaves is graphed as a proxy for photosynthetic capacity in leaves. Individual leaves eventually enter a senescence phase resulting essentially in a programmed self-destruction. During this phase leaf constituents such as proteins, lipids and mineral nutrients are degraded and exported to sinks. Each subsequent leaf growing from the shoot apical meristem follows this developmental progression. During the vegetative phase, the plant accumulates biomass, establishes a canopy and grows. Changes in light or other environmental conditions trigger flower formation from the shoot apical meristem, constituting a switch from a vegetative to generative growth phase and eventually results in seed production and seed maturation. During the seed developmental phase, leaves undergo senescence and remobilize necessary nutrients to the seeds. Mature seeds then enter a dormancy phase until germination in the next growth season

90 % of nitrogen in the seed is remobilised from vegetative tissues (Hirel et al. 2007; Masclaux-Daubresse et al. 2010; Masclaux-Daubresse and Chardon 2011). As a consequence, in these crops grain yield and quality depend primarily on pre-anthesis uptake of nitrogen. In contrast, 35–55 % of seed nitrogen in maize is taken up from the soil during grain filling (Hirel et al. 2007; Gregersen et al. 2008).

Starch accumulation is the major factor determining seed yield, but other grain parameters are equally important for farmers and consumers. For example, nutritional quality is primarily a function of seed protein content and amino acid balance. Therefore the supply and remobilisation of nitrogen, and to a lesser extent sulfur for the sulfur-containing amino acids cysteine and methionine, represent critical control points for crop quality (Hoefgen and Galili 2002; Hoefgen et al. 2004). Chloroplasts contain approximately 80 % of total leaf nitrogen, mostly in the form of Rubisco, and so not surprisingly, the major nitrogen resources for export to the developing seeds are derived from chloroplast degradation in senescing leaves (Feller et al. 2008; Kato und Sakamoto 2010). Leaf senescence is further characterised by the formation of senescence-associated vacuoles, involvement of the autophagy machinery (see chapter of Collados-Rodriguez et al. in this book), and activation of proteolytic and other degradative enzymes (Ishida et al. 2008; Wada et al. 2009; Izumi et al. 2010; Hörtensteiner 2012; Yamada et al. 2001; Martinez et al. 2008; Watanabe et al. 2010). As a result of these degradation processes, cellular integrity is compromised and reactive oxygen species (ROS) accumulate, which may regulate senescence associated gene (SAG) expression and additionally induce stress mitigation responses to maintain cellular function as long as possible (Navabpour et al. 2003; Guo and Crawford 2005).

For breeders focused on increasing crop yields, prolongation of leaf integrity has long been a trait of interest. Some cultivars with extremely late onset of senescence, so called "stay-green" cultivars, do indeed show higher starch filling and yields (Richards 2000). However, as robust senescence is required for maximal remobilisation of nutrients, seed quality might be impaired in late-senescing varieties (Bogard et al. 2011; Gong et al. 2005; Hirel et al. 2007; Masclaux-Daubresse and Chardon 2011; Gregersen et al. 2008). Further, the effects of late onset senescence have been found to vary depending on the species, genotype, and environmental conditions. In wet and cool conditions wheat yield was negatively correlated with the stay-green characteristic (Naruoka et al. 2012), while on the other hand, the stay-green trait is often correlated with increased drought tolerance or accentuated by low nitrogen supply (Yan et al. 2004; Robson et al. 2004; Jordan et al. 2012). Thus, the relation of gross physiological characteristics such as stay-green to final biomass and seed yield is still not well understood. We are still far from being able to provide general rules, particularly as the stay-green phenotype is caused by different mutations, some of which do not affect senescence parameters only chlorophyll content (for summary see: Gregersen et al. 2013).

The case of stay-green cultivars highlights the need to consistently describe plant development and senescence processes at the systems level. This kind of integrative knowledge would greatly facilitate breeding schemes as it may identify those processes most relevant for crop breeding.

Transcriptomic and Metabolomic Consequences of Plant Maturation and Senescence

Arabidopsis thaliana provides a tractable and sufficiently valid model for systems level analysis of senescence processes. Several detailed transcriptomics and metabolomic studies (Guo et al. 2004; Buchanon-Wollaston et al. 2005; Breeze et al. 2011; Guo and Gan 2012; Masclaux-Daubresse et al. 2005; Watanabe et al. 2013), and numerous targeted investigations have greatly expanded our understanding of the shift from anabolic to catabolic metabolism and senescence-associated gene (SAG and SDG) expression changes.

Arabidopsis thaliana is a long-day plant with a basal rosette and a life cycle of about 2 months belonging to the brassicaceae family (Fig. 8.2a). Older leaves successively senesce during development in a highly ordered physiological process with new appearing leaves acting as sinks. Upon inflorescence formation, a set of upper rosette leaves as well as small leaves at the flower stem are generated. As seeds ripen in siliques, the leaves further senesce and finally die off. The senescence process at the level of genes or metabolites is the same in all leaves and appears to follow a time shift according to the sequence of emergence (Breeze et al. 2011; Watanabe et al. 2013). Even within a single leaf, a trajectory of senescence is established along the basipetal axis with the (older) tip senescing earlier than the base.

The characteristic shift from anabolic to catabolic processes is readily apparent when analysing the metabolic composition of senescing plants or tissues over a time course (Fig. 8.2c; Watanabe et al. 2013). Over the course of senescence, chlorophyll contents and chloroplast lipids such as mono- and di galactosyldiglycerides (MGDGs and DGDGs) decrease, which point to the long-described senescence-associated chloroplast degradation and decrease in photosynthetic capacity. Despite this fact, the major sugars remain constant (sucrose) or increase (glucose and fructose) in senescing leaves, and storage lipids such as triacylglycerides (TAGs) increase. Protein contents are reduced due to proteolysis. Free branched-chain amino acids, aromatic amino acids, and stress-related amino acids such as proline, beta-alanine, and gamma-aminobutyric acid (GABA) accumulate. Conversely, the major N-containing amino acids glutamine, arginine and their precursors glutamate and aspartate decrease, probably due to export to sinks.

Furthermore, when looking at nutrient ions, nitrate and phosphate are efficiently exported from the senescing leaf. In addition to inorganic phosphate, nucleic acids are a likely source of phosphate for export, as indicated by the induction of RNases (Himelblau and Amasino 2001; Morcuende et al. 2007; Lers et al. 2006, Watanabe et al. 2013) and the concomitant induction of phosphate transporters (Chapin and Jones 2009). The regulatory response network of *Arabidopsis* to phosphate

Fig. 8.2 (continued) are presented as fold-change from stage 1 in log2 scale. No change to stage 1 is indicated by the *dashed line*, decrease by negative and increase by positive values. Typically chlorophyll contents are reduced as well as chloroplast associated lipids as mono- and di galactosyldiglycerides (MGDS, DGDS) and storage lipids as triacylglycerides (TAGs) accumulate. Protein degradation is accompanied by typical accumulation patterns of amino acids, most importantly the phloem-transported amino acids (asp, glu, gln, and arg) are reduced, while branched chain amino acids rather accumulate. Sulfur-rich glucosinolates are reduced and reduced sulfur probably mobilized to the seeds and the nutrient ions phosphate and nitrate are mobilised to the seeds, while sulfate is not mobilised. Sugars and anthocyanins as well as stress related amino acids accumulate, these latter processes being linked to senescence induced reactive oxygen species (ROS) accumulation. The slight increase of most metabolites at the end of the senescence phase is due to a decrease of fresh weight due to water loss (wilting)

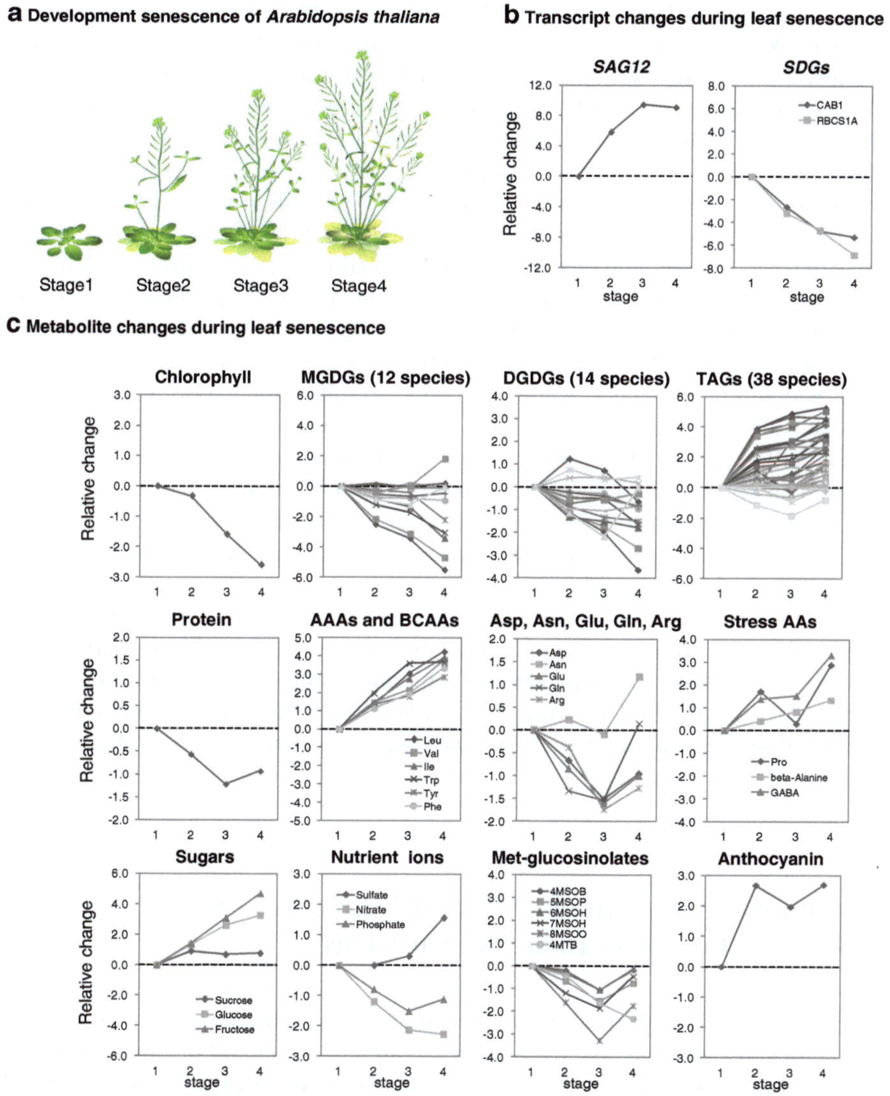

Fig. 8.2 Developmental senescence of *Arabidopsis thaliana*. (**a**) The senescence process in *Arabidopsis* is schematically depicted to illustrate the changes accompanied with progressing senescence during the life cycle for a whole plant. A commonly visible phenotype is the development of leaf chlorosis due to degradation of chlorophyll in the leaves. The Fig. is modified from Watanabe et al. 2013 (www.plantphysiol.org.; Copyright American Society of Plant Biologists). (**b**) Genes associated with senescence belong to various functional classes. Typical senescence associated genes (SAGs, induced upon senescence) and genes down regulated during senescence (SDGs) are used as markers for senescence in plants. Expression changes of SAG12 and two *SDGs*, *CAB* and *RBCS1A*, are presented here for an individual life cycle of *Arabidopsis thaliana* as fold change to stage 1 in log2 scale. No change to stage 1 is indicated by the *dashed line*, decrease by negative and increase by positive values. (**c**) Metabolite changes during leaf senescence are depicted. Metabolite contents in rosette leaves during developmental senescence

starvation is already highly resolved (Scheible et al. 2004; Morcuende et al. 2007; Doerner 2008). In contrast, sulfate is often poorly mobilised from the vacuole, probably because soil sulfate uptake is sufficient for seed filling and vacuolar sulfate export transporters, which are known to be responsive to sulfate starvation, are not always induced (Davidian and Kopriva 2010). When different regimes of sulfate were applied to soil grown wheat, persistent sulfate deficiency led to yield reductions and poor protein quality of the seeds. Application of sulfate, even at the seed filling stage, was able to revert the phenotype with respect to improved starch deposition and seed protein quality, especially of sulfur-rich seed proteins though it could not compensate for previous losses in biomass accumulation (Zörb et al. 2012; Steinfurth et al. 2012). Interestingly, the sulfur-rich defense compounds of brassicaceae, the glucosinolates, are actively degraded in leaves during senescence and might provide a source of reduced sulfur for export in contrast to vacuolar sulfate (Yoshimoto and Saito 2012; Watanabe et al. 2013). However, various sulfated compounds are not mobilised. Therefore sulfate and sulfurous organic compounds are a substantial part of the leaf litter and contribute to the fact that in humus rich soils the majority of nutrient sulfate is released by bacteria from organic sources (Kertesz 1999; Kertesz and Mirleau 2004; Schmalenberger et al. 2009).

Senescence processes are clearly extremely complex, and the functions of individual genes, not to mention the regulatory networks, are still not fully understood. However, transcriptomic and metabolomic studies on developmental senescence in *Arabidopsis thaliana* (Fig. 8.2) are building a framework at the molecular level for gluing together more detailed and targeted observations from *Arabidopsis* and other species including crops.

Inadequate Availability of Mineral Nutrients Leads to Nutrient Depletion Induced Senescence

Crops in an agricultural environment are exposed to varying and often non-optimal environmental conditions, which usually negatively affect carbohydrate synthesis and nutrient uptake. Moreover, stresses typically occur in combinations: drought stress is usually accompanied by high temperatures and high light (Long and Ort 2010; Krasensky and Jonak 2012), while unavailability of water in the root zone hinders nutrient ion uptake and stomatal closure inhibits transpiration and *in planta* transport of nutrients. Deficiencies of mineral nutrients leads to nutrient depletion induced senescence (NuDIS) in plants (Watanabe et al. 2010). NuDIS is accompanied by a mixture of transcriptomic and metabolomic responses depending on which nutrient ion or combination of ions is depleted. These responses are to some extent specific for particular nutrients, *e.g.* the induction of high affinity transport systems (Hawkesford 2000, 2003; Gojon et al. 2009; Davidian and Kopriva 2010; Stitt et al. 2002; Xu et al. 2012; Bouguyon et al. 2012). Interestingly, overlapping molecular responses occur under various nutrient depletions

(Fig. 8.5a). In particular, later responses all channel into a general senescence programme comparable to developmental senescence (Fig. 8.5b; Watanabe et al. 2010, 2012, 2013).

Nutrient ion availability in the field depends on physical parameters such as soil type and soil chemistry, weather conditions, leaching and run-off and on agricultural parameters such as the amount of minerals removed from the field with the harvest and fertilisation to replenish resources (Hawkesford 2012). Fertilisation is usually employed to counteract deficits, though this becomes increasingly costly due to rising energy costs, *e.g.* for nitrogen production, and the fact that some minerals, such as phosphate, might become limiting in the future which would likely increase prices (Gamuyao et al. 2012). Furthermore, fertilisation carries the risk of polluting the environment, and it is therefore necessary to improve application procedures and to breed mineral nutrient-efficient crop cultivars (McAllister et al. 2012). In order to strategically support breeding efforts, it is necessary to understand the processes and identify the genes affecting nutrient use efficiency. However, the situation in a field can be very complex as nutrients can be available in varying amounts and combinations. Deficiency might exist from germination on poor soils or might become established during growth due to active depletion of soil minerals by the plants or by environmental conditions such as heavy rainfalls. Sufficient minerals might even be present in a given soil but not accessible to the plant due to unfavourable soil pH, structure or drought. The effect of distinct mineral nutrient stresses on plants is well documented from agricultural practice and from long-term, systematic field experiments, such as the Broadbalk field trial at Rothamsted Research, UK (Lu et al. 2005; Shinmachi et al. 2010). Plants react to mineral nutrient depletion with gross phenotypic responses (Fig. 8.3a), such as reduction of biomass and seed yield, alteration of leaf pigment contents and distribution, and alteration of root morphology (Lopez-Bucio et al. 2006; Gruber et al. 2013, Nikiforova et al. 2004; Walch-Liu et al. 2006; Marschner 2012). It is suggested that the plant response to nutrient deprivation can be divided into two major phases (Fig. 8.3b).

During the initial rescue phase, the plant tries to counteract mineral deficiencies by inducing reversible adaptation mechanisms including increased uptake in the root zone, mobilisation of internal resources, and by preventing biosynthetic investments dependent on the depleted nutrient. High-affinity transporters get induced in roots, which allow the plant to accumulate ions against steep concentration gradients present between soil with very low ion concentrations and root tissue. Lateral root development is decreased in favour of exploratory growth of the main roots (Gruber et al. 2013; Hubberten et al. 2009, 2012b; Drew and Saker 1975; Walch-Liu et al. 2006), assimilatory enzymes are activated and the respective genes are induced (Hoefgen and Hesse 2008; Davidian and Kopriv 2010). Additionally, previously deposited internal stores of limiting elements such as those found in the vacuole are utilised by inducing vacuolar transporters in the tonoplast. In the face of continued deficiency, these primary responses are supported by strategies promoting degradation processes or by delaying anabolic processes. Even mild nitrogen starvation quickly affects polysome loading and

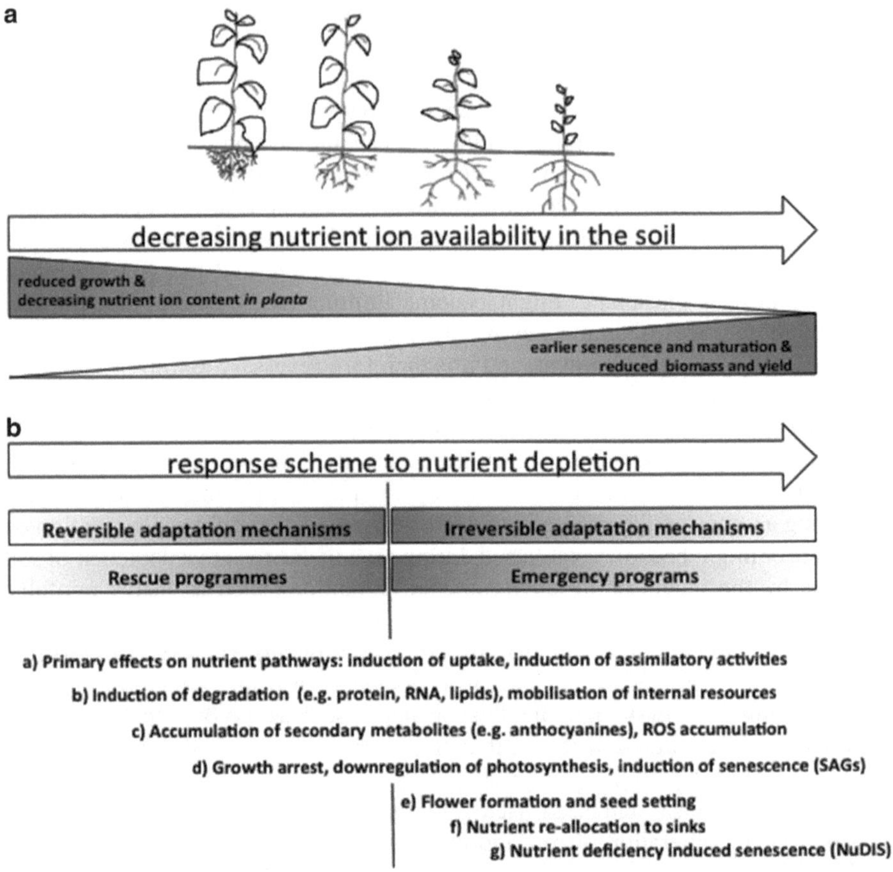

Fig. 8.3 Generalised response scheme of annual plants to mineral nutrient availability. Plants have evolved complex and flexible response mechanisms to cope with variable mineral nutrient availability, from optimal supply to complete depletion of one or several nutrients. (**a**) With decreasing nutrient availability, the relative growth rate, and hence biomass and eventually yield is reduced. As the plant at the same time "invests" in exploratory root growth, a relative increase of the shoot to root ratio can be observed. Severe starvation leads to alterations in the plant life cycle including earlier senescence. (**b**) When nutrient depletion occurs during the growth phase, the plant seeks to maintain homeostasis by activating rescue programmes to adapt its metabolism (points a–d). If nutrient starvation persists, a point of no return is reached when metabolism is irreversibly shifted to emergency programmes including senescence and early flower formation and seed production while cannibalizing internal resources (points *e.g.*). The "goal" is to produce as many seeds as possible even under continuing mineral nutrient starvation

hence reduces protein synthesis rates (Tschoep et al. 2009) and in the case of sulfur starvation in *Arabidopsis*, degradation of glucosinolates (sulfur-rich secondary metabolites) increases and their synthesis rates decrease (Yoshimoto and Saito 2012). Additionally, any macronutrient starvation immediately negatively affects photosynthesis (Fig. 8.6), thereby decreasing carbon supply, essentially stalling growth (Wulff-Zottele et al. 2010). Mineral nutrient deficiencies also result in ROS accumulation, which triggers stress and senescence responses such as the accumulation of anthocyanins. During this initial response phase, resupply or acquisition of nutrients can restore normal cellular functions and rescue the plant, though yield reductions are likely as biomass accumulation was reduced during the time of nutrient deprivation.

Continued starvation of one or more mineral nutrients leads to an emergency response phase. The plant irreversibly switches its developmental programme to maturation and senescence (NuDIS; Watanabe et al. 2010). This transition marks a point of no return. Resupply of nutrients in this phase does not reverse the maturation programme leading to flower formation, seed ripening and terminal leaf senescence. The plant apparently invests all remaining resources in producing as many seeds as possible. The programme actually resembles that of developmental senescence (Fig. 8.1), which also irreversibly propagates from the vegetative to the generative phase when the decision for flower formation has been made. However, mineral nutrient starvation shifts the timing for senescence to an earlier time point, thus shortening the individual life span.

The effect of NuDIS on yield can be dramatic in crop plants. This has been demonstrated for example for potassium deficiency (Armengaud et al. 2004; Amtmann and Armengaud 2009), phosphate deficiency (Batten and Wardlaw 1987) or nitrogen depletion (Parrot et al. 2010). For example, potassium exerts multiple functions in metabolism and is the second most abundant mineral in plants after nitrogen. Deficiency leads to multiple negative effects on osmoregulation, photosynthesis, protein synthesis, enzyme activities and results in light-dependent chlorosis and necrosis of leaf tissues, all of which impair growth and yield. Wheat grown on insufficient phosphate senesce far more rapidly and have drastically reduced seed yield (Batten and Wardlaw 1987). The flag leaf, which provides 51–89 % of grain phosphorus (Batten et al. 1986; Fig. 8.4), starts to rapidly senesce about 20 days post- anthesis (DPA). This early and steep decay in chlorophyll content is associated with slower grain filling, probably due to inadequate carbon supply from reduced photosynthetic capacity. About 40 DPA, flag leaves of phosphate-depleted plants are essentially devoid of chlorophyll and unable to perform photosynthesis for carbon supply to the grain. Grain filling plateaus at about 60 % of the potential grain dry weight result in a yield penalty of about 40 %. Under sufficiently high phosphate supply, chlorophyll decay also starts about 20 DPA, although at a much slower rate and accelerates only at about 40 DPA when grains have already reached about 80 % of their final dry weight. Under phosphate-depleted conditions seed filling stops at about 35 DPA while under phosphate-sufficient conditions grain filling continues for an additional 20 days. Thus, early senescence in vegetative tissues truncates seed maturation and reduces yield (Batten and Wardlaw 1987; Lynch and White 1992; Lauer et al. 1989; Fig. 8.4).

Notably, under optimal phosphate supply conditions, approximately 30 % of wheat flag leaf chlorophyll is intact when grain filling is complete, and therefore a substantial amount of leaf nitrogen remains immobilised in the leaf (Fig. 8.4; Batten and Wardlaw 1987). Similarly, it has been shown in *Arabidopsis* that N and P are substantially but incompletely mobilised during leaf senescence, while significant

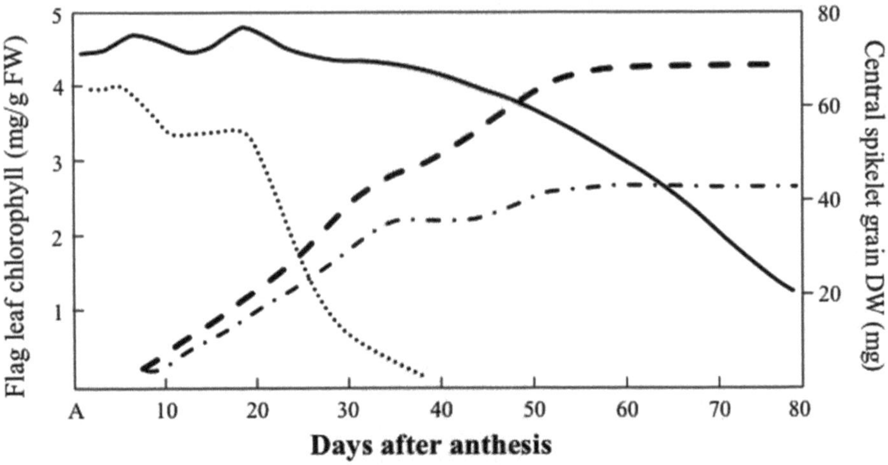

Fig. 8.4 Effect of phosphate starvation on plant senescence and yield (Adapted from Batten and Wardlaw 1987). Under ample phosphate supply (*solid line*) chlorophyll content, which is a proxy for photosynthetic activity, decreases slowly for the first 40 days post anthesis (DPA). Grain dry weight (*dashed line*) increases during this period and reaches a plateau at about 50 DPA. Only in the latest stages of grain filling does chlorophyll breakdown accelerate, even though complete degradation is not reached. Under phosphate deficient conditions (*dotted line*) chlorophyll content declines steeply starting at about 20 DPA and is completely degraded by 30–40 DPA, indicating that the flag leaf is fully senesced. Due to this early and enhanced senescence, the flag leaf is unable to provide photosynthates to the developing grain. Grain filling (*dash – dotted line*) is truncated by about 20 days, and final grain dry weight reaches only about 60 % of the weight under full phosphate supply

quantities of sulfate actually remain in the leaf, though organic sulfur compounds appear to be exported from the senescing leaf (Watanabe et al. 2013). Breeding efforts to improve nutrient use efficiency need to take into account the inevitable compromise between nutrient mobilisation and maintenance of leaf functionality, especially photosynthesis, as already discussed for the stay-green phenotype. Rather than gross changes, fine-tuning of senescence processes to the actual growth condition and for each crop are more likely to produce reliable increases in crop nutrient use efficiency.

Plant Systems Biology Approaches to Nutrient Depletion Responses

In order to understand and eventually to manipulate nutrient use efficiency in crops it is necessary to understand the interplay of nutrient availability, nutrient depletion induced senescence and nutrient use efficiency. As these are not simple causal relationships but rather dynamic networks of multiple factors which cross-influence each other, systems biology approaches will be necessary to unravel these relationships. Complicating this analysis, the biochemistry of a plant under nutrient deprivation typically shows pleiotropic effects, which are difficult to directly connect to the initial nutrient stress (Nikiforova et al. 2004, 2005). Metabolites related to a distinct nutrient are often involved in diverse biochemical processes different from the immediate assimilatory pathway. For example, sulfate starvation results in reduced *S*-adenosylmethionine (SAM) levels in plants. As SAM is one of the main methyl group donors for biochemical reactions, pleiotropic effects of sulfate starvation are inevitable (Nikiforova et al. 2005; Morcuende et al. 2007). Therefore, it is necessary to differentiate between early response processes, such as the induction of high affinity transporters or assimilatory genes, later processes triggered by already downstream effects (Fig. 8.3b) and general processes such as the NuDIS induced induction of SAGs (Figs. 8.2b, 8.3b, and 8.5b).

Arabidopsis thaliana, due to its amenability to systems biology approaches, has been widely used to investigate the effects of single nutrient starvations on the transcriptome (Wang et al. 2003; Amtmann and Armengaud 2009; Watanabe et al. 2010, 2013) and the metabolome (Nikiforova et al. 2005; Morcuende et al. 2007). Taken together, these studies reveal that responses to distinct nutrient stresses overlap to a certain extent (Figs. 8.5a and 8.6), even when focusing on the subfraction of the SAGs (Fig. 8.5b). Some trends are obvious and corroborate previous findings. During senescence protein and chlorophyll levels are reduced (Fig. 8.2c). Anthocyanin accumulation, a typical feature of stressed plants, is shared across N, P, and S starvation due to the induction of the senescence-associated transcription factor PAP1 (MYB75; Fig. 8.5a, b; Tohge et al. 2005; Watanabe et al. 2010, 2012). Also, one of the most responsive genes to sulfate starvation, low sulfur induced (LSU; Hubberten et al. 2012a), is induced under all three nutrient conditions compared here. LSU has been postulated to be involved in autophagy (Zientara-Rytter et al. 2011), a degradatory process, which mobilises nutrients from cellular macromolecules, even organelles (Fig. 8.3b) during senescence and nutrient deficiency rescue programmes (Wu et al. 2012). See also the chapter of Collados-Rodriguez et al. in this book.

The transcriptome responses to N, P, and S starvation overlap substantially (Fig. 8.5; Watanabe et al. 2012). Among the 1,240 genes induced at least twofold under phosphate starvation, 12 % are shared with sulfate and 43 % with nitrate starvation, and from 1,602 sulfate starvation induced genes, 39 % are shared with nitrate and 9 % with phosphate starvation. When assigning such expression profiles to functional categories (Fig. 8.6) using MAPMAN annotations (Thimm et al. 2004), complex patterns of up- and down-regulation within individual categories and even within gene families are observed which are not easy to interpret. It becomes obvious that certain categories are over-represented, *e.g.* genes of photosynthesis and carbohydrate biosynthesis are down-regulated and that this is a common trend for all nutrient starvations and also senescence (Fig. 8.6). Furthermore, the strength of the response mirrors the severity of symptoms (Figs. 8.5a and 8.6) as nitrogen starvation is usually stronger than phosphate, followed by sulfate starvation (Wulff-Zottele et al. 2010). The molecular basis and the physiological

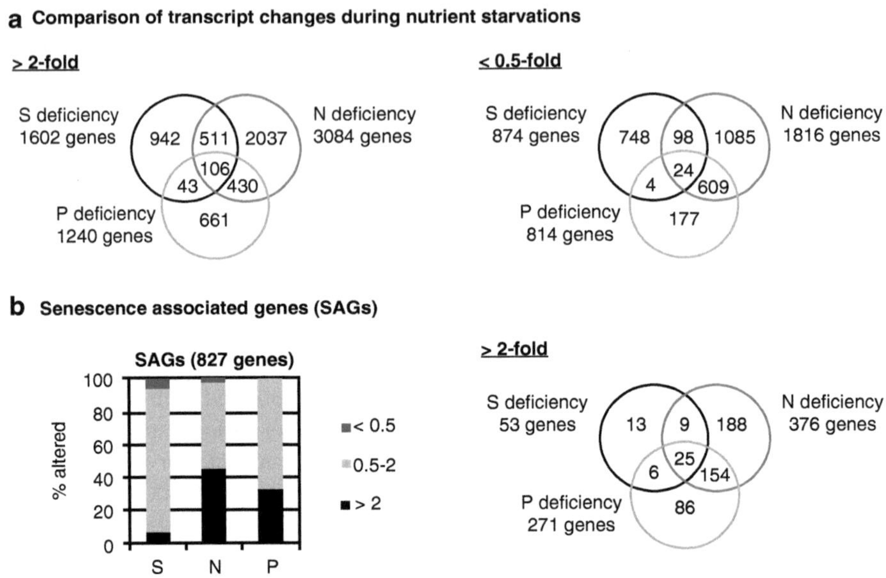

Fig. 8.5 Transcriptomic responses to sulfate, nitrate and phosphate (S, N and P) deficiencies. (**a**) Transcriptome data of *Arabidopsis* seedlings exposed to nitrate, phosphate and sulfate starvation were compared. The Venn diagrams show the overlap respectively, the specificity between the data sets in terms of genes induced or reduced by a factor of 2, respectively. The data sets were obtained from the following publications: S, low-sulfate for 7 days + sulfate-0 mM for 2 days (Bielecka, unpublished data); N, full nutrient for 7 days + nitrate-0 mM for 2 days (Scheible et al. 2004); P, low-phosphate for 7 days + phosphate-0 mM for 1 day (Morcuende et al. 2007). (**b**) Nutrient starvation induces senescence related responses (NuDIS) and the changes of the subfraction of senescence-associated genes (SAGs) are displayed. The percentage of SAGs (total 827 genes) with >2-fold and <0.5-fold changes is calculated (Buchanan-Wollaston et al. 2003). The Venn diagram shows the overlap in SAGs between data sets, sulfate, nitrate and phosphate deficiencies at a threshold of >2-fold. This depicts partial communalities between the responses recruiting shared response mechanisms

consequences of the observed responses to nutrient deficiency are far from understood. Under all three nutrient starvation conditions genes involved in the light reactions, the Calvin-Benson cycle and photorespiration are downregulated. This provides molecular support of previous findings on the effect of nutrient starvation on photosynthesis. Nutrients are direct constituents of the photosynthetic apparatus, such as N and S in protein and N and Mg in chlorophyll, which result in the fact that 24 % of leaf total N is located in the thylakoid membranes (Terashima and Evans 1988). Fe and S are part of the iron-sulfur clusters of the photosynthetic machinery. P is obviously an essential part of photosynthetic reactions due to the centrality of energy-rich, phosphate-containing organic compounds (ATP, sugar phosphates). In addition to direct involvement in photosynthesis and carbohydrate synthesis, nutrient depletion usually results in retarded growth because carbohydrate backbones cannot be utilised for the biosynthesis of amino acids and proteins, vitamins and cofactors or energy-rich compounds containing P such as ATP (Hawkesford 2000,

2012; Nikiforova et al. 2005; Davidian and Kopriva 2010). As a result, despite the negative effect of nutrient depletion on photosynthesis (Fig. 8.6) photosynthesis-derived carbohydrates might accumulate in the leaves. Reduced sink strength due to reduced growth or seed development leads to a reduction of the carbohydrate export from the leaf further contributing to carbohydrate accumulation (Rao et al. 1990; Stitt 1991). This lack of sink strength and carbohydrate accumulation eventually leads to oxidative stress under high light conditions, as excess energy cannot be fully dissipated as heat. Consequently, nutrient starvation leads to photo oxidation, which further damages the remaining photosynthetic apparatus (Cakmak and Marschner 1992; Groot et al. 2003) and induction of senescence (Rolland et al. 2006), resulting in feedback down-regulation of photosynthesis (Pieters et al. 2001). A similar effect is caused when phloem loading is impaired leading

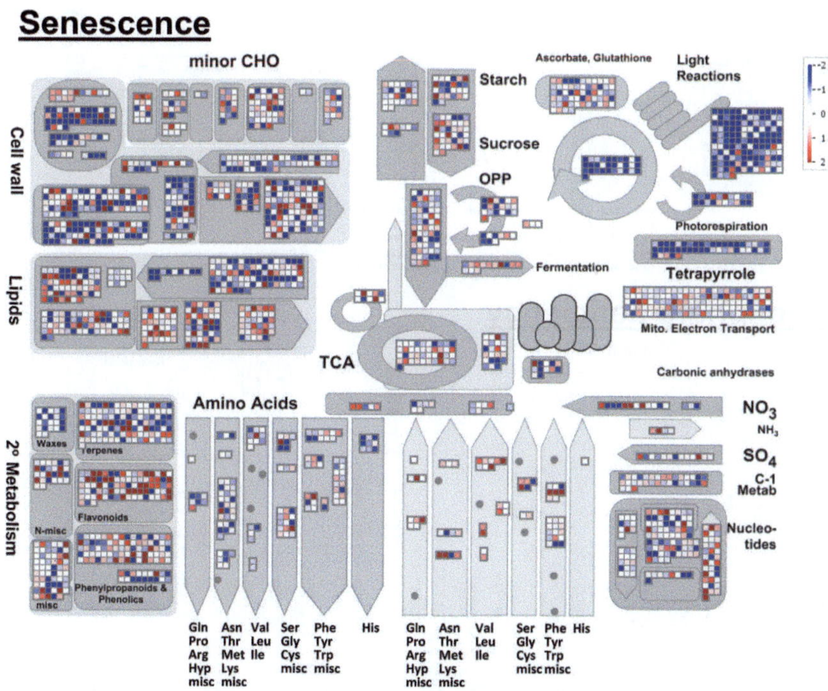

Fig. 8.6 A comparison of major metabolic pathway gene expression using MapMan. Data for senescence is from ATGE experiments. Ratio (senescent leaf/green leaf) is shown in log2 scale. Green leaf: L6 (ATGE_14A-C) and L8 (ATGE_15A-C). Senescent leaf: ATGE experiment senescent leaf (ATGE_25A-C; Buchanan-Wollaston et al. 2005). Data for sulfate, nitrate and phosphate deficiencies are from the same references as Fig. 8.5. Fold changes relative to the nutrient sufficient control are shown in log2 scale. *Red* and *Blue squares* indicate genes induced or repressed, respectively. First, different nutrient depletions show common response patterns, even with the senesce data, especially for light reactions, Calvin cycle, and photorespiration. Further, the strength of response to nutrient starvations affect plant metabolism in the order nitrogen, phosphate, and sulfate

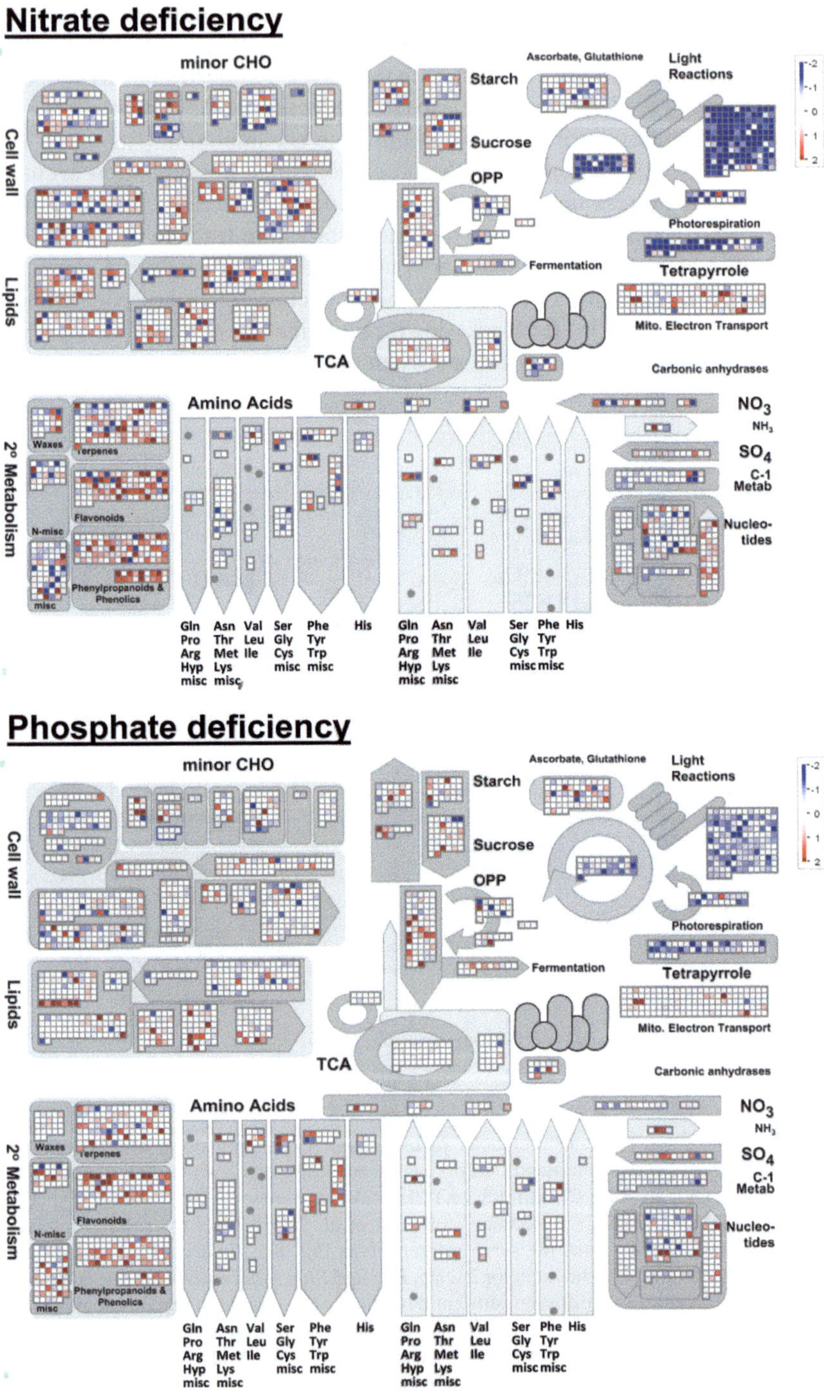

Fig. 8.6 (continued)

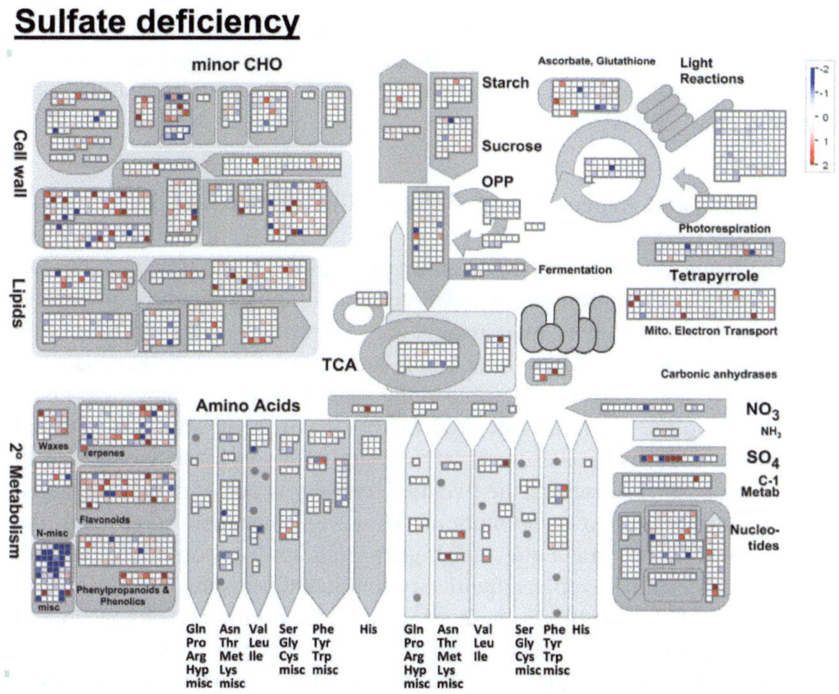

Fig. 8.6 (continued)

to photosynthate accumulation in the leaf accompanied by chlorosis and necrosis (von Schaewen et al. 1990). As shown in Figs. 8.5 and 8.6, established nutrient starvation overlaps with developmental senescence, hence the term NuDIS.

The plant response to nutrient stress is not only complex with respect to specific versus common responses, to the plant species and to specific nutrients but is additionally a dynamic process. Likewise, the ability of plants to positively respond to nutrient re-supply by fertilisation or if the plant manages to access nutrients is strictly dependent on the developmental status of the plant (Zörb et al. 2012; Steinfurth et al. 2012). Knowledge of the processes during alleviation of starvation responses is incomplete at the molecular level. To understand these processes and their molecular and regulatory bases it will be necessary to compare the responses of plant (model systems and crops) at various stages and severities of starvation or replenishment and also with combinations of mineral nutrient depletions, as exemplified in the chapter of Forieri et al. of this book. Furthermore, it will be helpful to correlate existing knowledge with improved profiling and phenotyping tools and to employ systems biology approaches such as transcriptomics, proteomics and metabolomics. It will also be necessary to devise novel bioinformatics and modelling tools. In order to disentangle these complex networks, it will probably be necessary to specify and subgroup the NuDIS-associated Genes (NAGs) and SAGs into functional categories, or response modules. One such effort has successfully been made in identifying the OAS-module as part of the sulfate starvation response (Hubberten et al. 2012a).

Conclusions and Expectations for Plant Mineral Nutrient Systems Research

The challenge facing twenty-first century agriculture is multifaceted. Growth of the world's population will necessitate producing significantly more food, while food production is increasingly in competition with non-food plant commodities for the world's finite arable land. Furthermore, food production is becoming increasingly vulnerable in the face of a rapidly changing climate. In addition to the climate-associated factors water, heat, and light, mineral nutrient availability is a critical determinant of agricultural yield. In fact, increased use of fertilisers in the twentieth century is one of the major factors contributing to the "green revolution". However, awareness is growing that short-term solutions in the quest for higher yield, such as copious application of fertiliser, need to be replaced with environmentally sustainable agricultural production. Modern cultivars were selected for their elite performance under "luxury" nutrient regimes. Therefore, development of new varieties that more efficiently convert soil nutrients into harvestable yield will be absolutely critical if global agriculture is to be able to feed the world sustainably.

One approach contributing to these goals is to understand the molecular bases of plant response programs to various nutrient conditions, especially to nutrient levels lower than typically used in intensive agriculture. Time resolved systems biology analyses on single and multiple nutrient stresses and nutrient replenishment will yield information about processes and candidate genes or alleles. Obviously, such analysis is currently still largely based on research employing model organisms, as much at the screening level as at the proof of concept level. To transfer this knowledge, field studies will be necessary to prove the validity of the concepts, particularly as different crop species follow different strategies (*i.e.* have different ecological niches). Systems biology-level tools will be needed for field experiments and breeding populations to ensure development of elite cultivars. This necessitates novel approaches at the analytical but also at the bioinformatics and modelling levels. Finally, systems level analysis needs to be extended beyond the individual plant into the ecophysiological level, as crop plants exist in a complex interaction web with geochemistry, climate, and the surrounding biome.

Additionally, renewable resources must provide the commodities for a bio-based chemistry. To this end, biodiversity can be exploited by crossing

(continued)

elite cultivars with wild members of the same species or even with more distant relatives. Biodiversity can also be generated through mutagenesis. Targeted engineering by transgenic approaches will allow breeders to exploit knowledge and gene pools, even across species borders. Agricultural practices such as mixed cropping, improved fertilisation regimes, and improved harvesting and storage procedures might also marginally contribute. With respect to plant mineral nutrients, the ultimate breeding goal is to improve uptake and internal use efficiency in novel crop cultivars in order to reduce losses, reduce costs and maintain high yields and high quality of the harvested material. All approaches have to take into account the paradigm of retaining the very high yield potential of modern crop cultivars or the goal of providing more food for a growing population on less land, environmentally sustainably and affordably, will not be met.

References

Amtmann A, Armengaud P (2009) Effects of N, P, K and S on metabolism: new knowledge gained from multi-level analysis. Curr Opin Plant Biol 12:275–283

Amtmann A, Blatt MR (2009) Regulation of macronutrient transport. New Phytol 181:35–52

Armengaud P, Breitling R, Amtmann A (2004) The potassium-dependent transcriptome of *Arabidopsis* reveals a prominent role of jasmonic acid in nutrient signaling. Plant Physiol 136:2556–2576

Astolfi S, Zuchi S, Hubberten H-M, Pinton R, Hoefgen R (2010) Supply of sulphur to S-deficient young barley seedlings restores their capability to cope with iron shortage. J Exp Bot 61:799–806

Batten GD, Wardlaw IF (1987) Senescence and grain development in wheat plants grown with contrasting phosphorus regimes. Austr J Plant Phys 14:253–265

Batten GD, Wardlaw IF, Aston MJ (1986) Growth and the distribution of phosphorus in wheat developed under various phosphorus and temperature regimes. Austr J Agric Res 37:459–469

Bogard M, Jourdan M, Allard V, Martre P, Perretant MR, Ravel C, Heumez E, Orford S, Snape J, Griffiths S, Gaju O, Foulkes J, Le Gouis J (2011) Anthesis date mainly explained correlations between post-anthesis leaf senescence, grain yield, and grain protein concentration in a winter wheat population segregating for flowering time QTLs. J Exp Bot 62:3621–3636

Bouguyon E, Gojon A, Nacry P (2012) Nitrate sensing and signaling in plants. Sem Cell Dev Biol 23:648–654

Breeze E, Harrison E, McHattie S, Hughes L, Hickman R, Hill C, Kiddle S, Kim YS, Penfold CA, Jenkins D, Zhang CJ, Morris K, Jenner C, Jackson S, Thomas B, Tabrett A, Legaie R, Moore JD, Wild DL, Ott S, Rand D, Beynon J, Denby K, Mead A, Buchanan-Wollaston V (2011) High-resolution temporal profiling of transcripts during *Arabidopsis* leaf senescence reveals a distinct chronology of processes and regulation. Plant Cell 23:873–894

Buchanan B, Gruissem W, Jones RL (2007) Biochemistry & molecular biology of plants. New York

Buchanan-Wollaston V, Earl S, Harrison E, Mathas E, Navabpour S, Page T, Pink D (2003) The molecular analysis of leaf senescence – a genomics approach. Plant Biotechnol J 1:3–22

Buchanan-Wollaston V, Page T, Harrison E, Breeze E, Lim PO, Nam HG, Lin JF, Wu SH, Swidzinski J, Ishizaki K, Leaver CJ (2005) Comparative transcriptome analysis reveals significant differences in gene expression and signalling pathways between developmental and dark/starvation-induced senescence in *Arabidopsis*. Plant J 42:567–585

Cakmak I, Marschner H (1992) Magnesium deficiency and high light intensity enhance activities of superoxide dismutase, ascorbate peroxidase, and glutathione reductase in bean leaves. Plant Physiol 98:1222–1227

Cassin G, Mari S, Curie C, Briat JF, Czernic P (2009) Increased sensitivity to iron deficiency in *Arabidopsis thaliana* overaccumulating nicotianamine. J Exp Bot 60:1249–1259

Chapin LJ, Jones ML (2009) Ethylene regulates phosphorus remobilization and expression of a phosphate transporter (PhPT1) during petunia corolla senescence. J Exp Bot 60:2179–2190

Davidian JC, Kopriva S (2010) Regulation of sulfate uptake and assimilation – the same or not the same? Mol Plant 3:314–325

De Groot CC, Van Den Boogaard R, Marcelis LF, Harbinson J, Lambers H (2003) Contrasting effects of N and P deprivation on the regulation of photosynthesis in tomato plants in relation to feedback limitation. J Exp Bot 54:1957–1967

Doerner P (2008) Phosphate starvation signaling: a threesome controls systemic P(i) homeostasis. Curr Opin Plant Biol 11:536–540

Dolferus R, Ji X, Richards RA (2011) Abiotic stress and control of grain number in cereals. Plant Sci 181:331–341

Drew MC, Saker LR (1975) Nutrient supply and the growth of the seminal root system of barley. II. Localized, compensatory increases in lateral root growth and rates of nitrate uptake when nitrate supply is restricted to only part of the root system. J Exp Bot 26:79–90

Dyson T (1999) World food trends and prospects to 2025. Proc Natl Acad Sci U S A 96:5929–5936

Feller U, Anders I, Mae T (2008) Rubiscolytics: fate of rubisco after its enzymatic function in a cell is terminated. J Exp Bot 59:1615–1624

Galili G, Hoefgen R (2002) Metabolic engineering of amino acids and storage proteins in plants. Metab Eng 4:3–11

Gamuyao R, Chin JH, Pariasca-Tanaka J, Pesaresi P, Catausan S, Dalid S, Slamet-Loedin I, Tecson-Mendoza EM, Wissuwa M, Heuer S (2012) The protein kinase Pstol1 from traditional rice confers tolerance of phosphorus deficiency. Nature 488:535–541

Gojon A, Nacry P, Davidian JC (2009) Root uptake regulation: a central process for NPS homeostasis in plants. Curr Opin Plant Biol 12:328–338

Gong YH, Zhang J, Gao JF, Lu JY, Wang JR (2005) Slow export of photoassimilate from stay-green leaves during late grain-filling stage in hybrid winter wheat (*Triticum aestivum* L.). J Agron Crop Sci 191:292–299

Gregersen PL, Holm PB, Krupinska K (2008) Leaf senescence and nutrient remobilisation in barley and wheat. Plant Biol 10:37–49

Gregersen PL, Culetic A, Boschian L, Krupinska K (2013) Plant senescence and crop productivity. Plant Mol Biol 82:603–622

Gruber BD, Giehl RF, Friedel S, von Wirén N (2013) Plasticity of the *Arabidopsis* root system under nutrient deficiencies. Plant Physiol 163:161–179

Guo FQ, Crawford NM (2005) *Arabidopsis* nitric oxide synthase1 is targeted to mitochondria and protects against oxidative damage and dark-induced senescence. Plant Cell 17:3436–3450

Guo Y, Gan S (2012) Convergence and divergence in gene expression profiles induced by leaf senescence and 27 senescence-promoting hormonal, pathological and environmental stress treatments. Plant Cell Environ 35:644–655

Guo Y, Cai Z, Gan S (2004) Transcriptome of *Arabidopsis* leaf senescence. Plant Cell Environ 27:521–549

Haensch R, Mendel RR (2009) Physiological functions of mineral micronutrients (Cu, Zn, Mn, Fe, Ni, Mo, B, Cl). Curr Opin Plant Biol 12:259–266

Hawkesford MJ (2000) Plant responses to sulphur deficiency and the genetic manipulation of sulphate transporters to improve S-utilization efficiency. J Exp Bot 51:131–138

Hawkesford MJ (2003) Transporter gene families in plants: the sulphate transporter gene family – redundancy or specialization? Physiol Plant 117:155–163

Hawkesford MJ (2012) Sulfate uptake and assimilation – whole plant regulation. In: De Kok LJ, Tausz M, Hawkesford MJ, Hoefgen R, McManus MT, Norton RM, Rennenberg H, Saito K, Schnug E, Tabe L (eds) Sulfur metabolism in plants: mechanisms and application to food security, and responses to climate change. Dordrecht Heidelberg London New York, pp 11–24

He ZL, Yang XE, Stoffella PJ (2005) Trace elements in agroecosystems and impacts on the environment. J Trace Elem Med Bio 19:125–140

Himelblau E, Amasino RM (2001) Nutrients mobilized from leaves of *Arabidopsis thaliana* during leaf senescence. J Plant Physiol 158:1317–1323

Hirel B, Le Gouis J, Ney B, Gallais A (2007) The challenge of improving nitrogen use efficiency in crop plants: towards a more central role for genetic variability and quantitative genetics within integrated approaches. J Exp Bot 58:2369–2387

Hoefgen R, Hesse H (2008) Sulfur and cysteine metabolism. In: Jez J (ed) Sulfur: a missing link between soils, crops and nutrition, vol 50. American Society of Agronomy, Madison, pp 83–104

Hoefgen R, Hesse H, Galili G (2004) Amino acid metabolism. In: Christou P, Klee H, (eds) Plants – quantitative and qualitative features. Handbook of plant biotechnology. New Jersey pp 577–608

Hörtensteiner S (2012) Update on the biochemistry of chlorophyll breakdown. Plant Mol Biol 82:505–517

Hubberten H-M, Hesse H, Hoefgen R et al (2009) Lateral root growth in sulfur enriched patches. In: Sirko A, De Kok LJ, Haneklaus S, Hawkesford MJ, Rennenberg H, Saito K, Schnug E, Stulen I (eds) Sulfur metabolism in plants: regulatory aspects, significance of sulfur in the food chain, agriculture and the environment. Backhuys Publishers/Margraf Publishers, Leiden/Weikersheim, pp 105–108

Hubberten H-M, Drozd A, Tran BV, Hesse H, Hoefgen R (2012a) Local and systemic regulation of sulfur homeostasis in roots of *Arabidopsis thaliana*. Plant J 72:625–635

Hubberten H-M, Klie S, Caldana C, Degenkolbe T, Willmitzer L, Hoefgen R (2012b) An additional role of *O*-acetylserine as a sulphur status independent regulator during plant growth. Plant J 70:666–677

Ishida H, Yoshimoto K, Izumi M, Reisen D, Yano Y, Makino A, Ohsumi Y, Hanson MR, Mae T (2008) Mobilization of rubisco and stroma-localized fluorescent proteins of chloroplasts to the vacuole by an ATG gene-dependent autophagic process. Plant Physiol 148:142–155

Izumi M, Wada S, Makino A, Ishida H (2010) The autophagic degradation of chloroplasts via rubisco-containing bodies is specifically linked to leaf carbon status but not nitrogen status in *Arabidopsis*. Plant Physiol 154:1196–1209

Jaggard KW, Qi A, Ober ES (2010) Possible changes to arable crop yields by 2050. Philos Trans R Soc London, Ser B 365:2835–2851

Jordan DR, Hunt CH, Cruickshank AW, Borrell AK, Henzell RG (2012) The relationship between the stay-green trait and grain yield in elite Sorghum hybrids grown in a range of environments. Crop Sci 52:1153–1161

Kato Y, Sakamoto W (2010) New insights into the types and function of proteases in plastids. Int Rev Cell Mol Biol 280:185–218

Kertesz MA (1999) Riding the sulfur cycle – metabolism of sulfonates and sulfate esters in Gram-negative bacteria. FEMS Microbiol Rev 24:135–175

Kertesz MA, Mirleau P (2004) The role of soil microbes in plant sulphur nutrition. J Exp Bot 55:1939–1945

Klatte M, Schuler M, Wirtz M, Fink-Straube C, Hell R, Bauer P (2009) The analysis of *Arabidopsis* nicotianamine synthase mutants reveals functions for nicotianamine in seed iron loading and iron deficiency responses. Plant Physiol 150:257–271

Krasensky J, Jonak C (2012) Drought, salt, and temperature stress-induced metabolic rearrangements and regulatory networks. J Exp Bot 63:1593–1608

Lauer MJ, Blevins DG, Sierzputowska-Gracz H (1989) ^{31}P-Nuclear magnetic resonance determination of phosphate compartmentation in leaves of reproductive soybeans (*Glycine max* L.) as affected by phosphate nutrition. Plant Physiol 89:1331–1336

Lers A, Sonego L, Green PJ, Burd S (2006) Suppression of LX ribonuclease in tomato results in a delay of leaf senescence and abscission. Plant Physiol 142:710–721
Li H, Chen Z, Hu M, Wang Z, Hua H, Yin C, Zeng H (2011) Different effects of night versus day high temperature on rice quality and accumulation profiling of rice grain proteins during grain filling. Plant Cell Rep 30:1641–1659
Lim PO, Kim HJ, Nam HG (2007) Leaf senescence. Annu Rev Plant Biol 58:115–136
Long SP, Ort DR (2010) More than taking the heat: crops and global change. Curr Opin Plant Biol 13:241–248
Lopez-Bucio J, Acevedo-Hernandez G, Ramirez-Chavez E, Molina-Torres J, Herrera-Estrella L (2006) Novel signals for plant development. Curr Opin Plant Biol 9:523–529
Lu C, Hawkesford MJ, Barraclough PB, Poulton PR, Wilson ID, Barker GL, Edwards KJ (2005) Markedly different gene expression in wheat grown with organic or inorganic fertilizer. Proc Royal Soc London – Series B: Biol Sci 272:1901–1908
Lynch J, White JW (1992) Shoot nitrogen dynamics in tropical common bean. Crop Sci 32:392–397
Marschner P (2012) In: Marschner P (ed) Marschner's mineral nutrition of higher plants, 3rd edn. Elsevier, Amsterdam
Martinez DE, Costa ML, Gomez FM, Otegui MS, Guiamet JJ (2008) 'Senescence-associated vacuoles' are involved in the degradation of chloroplast proteins in tobacco leaves. Plant J 56:196–206
Masclaux-Daubresse C, Chardon F (2011) Exploring nitrogen remobilization for seed filling using natural variation in *Arabidopsis thaliana*. J Exp Bot 62:2131–2142
Masclaux-Daubresse C, Carrayol E, Valadier MH (2005) The two nitrogen mobilisation- and senescence-associated GS1 and GDH genes are controlled by C and N metabolites. Planta 221:580–588
Masclaux-Daubresse C, Daniel-Vedele F, Dechorgnat J, Chardon F, Gaufichon L, Suzuki A (2010) Nitrogen uptake, assimilation and remobilization in plants: challenges for sustainable and productive agriculture. Ann Bot 105:1141–1157
McAllister CH, Beatty PH, Good AG (2012) Engineering nitrogen use efficient crop plants: the current status. Plant Biotechnol J 10:1011–1025
Morcuende R, Bari R, Gibon Y, Zheng W, Pant BD, Bläsing O, Usadel B, Czechowski T, Udvardi MK, Stitt M, Scheible W-R (2007) Genome-wide reprogramming of metabolism and regulatory networks of *Arabidopsis* in response to phosphorus. Plant Cell Environ 30:85–112
Mueller ND, Gerber JS, Johnston M, Ray DK, Ramankutty N, Foley JA (2012) Closing yield gaps through nutrient and water management. Nature 490:254–257
Naruoka Y, Sherman JD, Lanning SP, Blake NK, Martin JM, Talbert LE (2012) Genetic analysis of green leaf duration in spring wheat. Crop Sci 52:99–109
Navabpour S, Morris K, Allen R, Harrison E, A-H-Mackerness S, Buchanan-Wollaston V (2003) Expression of senescence-enhanced genes in response to oxidative stress. J Exp Bot 54:2285–2292
Nikiforova VJ, Gakière B, Kempa S, Adamik M, Willmitzer L, Hesse H, Hoefgen R (2004) Towards dissecting nutrient metabolism in plants: a systems biology case study on sulphur metabolism. J Exp Bot 55:1861–1870
Nikiforova VJ, Kopka J, Tolstikov V, Fiehn O, Hopkins L, Hawkesford MJ, Hesse H, Hoefgen R (2005) Systems re-balancing of metabolism in response to sulfur deprivation, as revealed by metabolome analysis of *Arabidopsis* plants. Plant Physiol 138:304–318
Parrott DL, Martin JM, Fischer AM (2010) Analysis of barley (*Hordeum vulgare*) leaf senescence and protease gene expression: a family C1A cysteine protease is specifically induced under conditions characterized by high carbohydrate, but low to moderate nitrogen levels. New Phytol 187:313–331
Pieters AJ, Paul MJ, Lawlor DW (2001) Low sink demand limits photosynthesis under P (i) deficiency. J Exp Bot 52:1083–1091

Rao M, Fredeen AL, Terry N (1990) Leaf phosphate status, photosynthesis, and carbon partitioning in sugar beet. III. Diurnal changes in carbon partitioning and carbon export. Plant Physiol 92:29–36

Richards RA (2000) Selectable traits to increase crop photosynthesis and yield of grain crops. J Exp Bot 51:447–458

Robson PRH, Donnison IS, Wang K, Frame B, Pegg SE, Thomas A, Thomas H (2004) Leaf senescence is delayed in maize expressing the Agrobacterium IPT gene under the control of a novel maize senescence-enhanced promoter. Plant Biotechnol J 2:101–112

Rolland F, Baena-Gonzalez E, Sheen J (2006) Sugar sensing and signaling in plants: conserved and novel mechanisms. Annu Rev Plant Biol 57:675–709

Scheible WR, Morcuende R, Czechowski T, Fritz C, Osuna D, Palacios-Rojas N, Schindelasch D, Thimm O, Udvardi MK, Stitt M (2004) Genome-wide reprogramming of primary and secondary metabolism, protein synthesis, cellular growth processes, and the regulatory infrastructure of *Arabidopsis* in response to nitrogen. Plant Physiol 136:2483–2499

Schmalenberger A, Hodge S, Hawkesford MJ, Kertesz MA (2009) Sulfonate desulfurization in Rhodococcus from wheat rhizosphere communities. FEMS Microbial Ecol 67:140–150

Semenov MZ, Halford NG (2009) Identifying target traits and molecular mechanisms for wheat breeding under a changing climate. J Exp Bot 60:2791–2804

Shinmachi F, Buchner P, Stroud JL, Parmar S, Zhao FJ, McGrath SP, Hawkesford MJ (2010) Influence of sulfur deficiency on the expression of specific sulfate transporters and the distribution of sulfur, selenium, and molybdenum in wheat. Plant Physiol 153:327–336

Steinfurth D, Zörb C, Braukmann F, Mühling KH (2012) Time-dependent distribution of sulphur, sulphate and glutathione in wheat tissues and grain as affected by three sulphur fertilization levels and late S fertilization. Plant Physiol 169:72–77

Stitt M (1991) Rising CO2 levels and their potential significance for carbon flow in photosynthetic cells. Plant Cell Environ 14:741–762

Stitt M, Müller C, Matt P, Gibon Y, Carillo P, Morcuende R, Scheible W-R, Krapp A (2002) Steps towards an integrated view of nitrogen metabolism. J Exp Bot 53:959–970

Tanksley SD, McCouch SR (1997) Seed banks and molecular maps: unlocking genetic potential from the wild. Science 277:1063–1066

Terashima I, Evans JR (1988) Effects of light and nitrogen nutrition on the organization of the photosynthetic apparatus in spinach. Plant Cell Phys 29:143–155

Tester M, Langridge P (2010) Breeding technologies to increase crop production in a changing world. Science 327:818–822

Thimm O, Blaesing O, Gibon Y, Nagel A, Meyer S, Krüger P, Selbig J, Müller LA, Rhee SY, Stitt M (2004) MAPMAN: a user-driven tool to display genomics data sets onto diagrams of metabolic pathways and other biological processes. Plant J 37:914–939

Thomas H (2013) Senescence, ageing and death of the whole plant. New Phytol 197:696–711

Tohge T, Nishiyama Y, Hirai MY, Yano M, Nakajima J, Awazuhara M, Inoue E, Takahashi H, Goodenowe DB, Kitayama M, Noji M, Yamazaki M, Saito K (2005) Functional genomics by integrated analysis of metabolome and transcriptome of *Arabidopsis* plants over-expressing an MYB transcription factor. Plant J 42:218–235

Tschoep H, Gibon Y, Carillo P, Armengaud P, Szecowka M, Nunes-Nesi A, Fernie AR, Koehl K, Stitt M (2009) Adjustment of growth and central metabolism to a mild but sustained nitrogen-limitation in *Arabidopsis*. Plant Cell Environ 32:300–318

Von Liebig J (1840) Die organische Chemie in ihrer Anwendung auf Agrikultur und Physiologie. Vieweg, Braunschweig

von Schaewen A, Stitt M, Schmidt R, Sonnewald U, Willmitzer L (1990) Expression of a yeast-derived invertase in the cell wall of tobacco and *Arabidopsis* plants leads to accumulation of carbohydrate and inhibition of photosynthesis and strongly influences growth and phenotype of transgenic tobacco plants. EMBO J 9:3033–3044

Wada S, Ishida H, Izumi M, Yoshimoto K, Ohsumi Y, Mae T, Makino A (2009) Autophagy plays a role in chloroplast degradation during senescence in individually darkened leaves. Plant Physiol 149:885–893

Walch-Liu P, Ivanov I, Filleur S, Gan Y, Remans T, Forde BG (2006) Nitrogen regulation of root branching. Ann Bot 97:875–881

Wang R, Okamoto M, Xing X, Crawford NM (2003) Microarray analysis of the nitrate response in *Arabidopsis* roots and shoots reveals over 1,000 rapidly responding genes and new linkages to glucose, trehalose-6-phosphate, iron, and sulfate metabolism. Plant Physiol 132:556–567

Watanabe M, Hubberten H-M, Saito K, Hoefgen R (2010) General regulatory patterns of plant mineral nutrient depletion as revealed by serat quadruple mutants disturbed in cysteine synthesis. Mol Plant 3:438–466

Watanabe M, Hubberten H.-M, Hoefgen R (2012) Plant response to mineral ion availability: transcriptome responses to sulfate, selenium and iron. In: De Kok LJ, Tausz M, Hawkesford MJ, Hoefgen R, McManus MT, Norton RM, Rennenberg H, Saito K, Schnug E, Tabe, L (eds) Sulfur metabolism in plants: mechanisms and application to food security, and responses to climate change. Dordrecht Heidelberg London New York, pp 123–134

Watanabe M, Balazadeh S, Tohge T, Erban A, Giavalisco P, Kopka J, Mueller-Roeber B, Fernie AR, Hoefgen R (2013) Comprehensive dissection of spatio-temporal metabolic shifts in primary, secondary and lipid metabolism during developmental senescence in *Arabidopsis thaliana*. Plant Physiol 162:1290–1310

Wu XY, Kuai BK, Jia JZ, Jing HC (2012) Regulation of leaf senescence and crop genetic improvement. J Integr Plant Biol 54:936–952

Wulff-Zottele C, Gatzke N, Kopka J, Orellana A, Hoefgen R, Fisahn J, Hesse H (2010) Photosynthesis and metabolism interact during acclimation of *Arabidopsis thaliana* to high irradiance and sulphur depletion. Plant Cell Environ 33:1974–1988

Xu G, Fan X, Miller AJ (2012) Plant nitrogen assimilation and use efficiency. Annu Rev Plant Biol 63:153–118

Yamada K, Matsushima R, Nishimura M, Hara-Nishimura I (2001) A slow maturation of a cysteine protease with a granulin domain in the vacuoles of senescing *Arabidopsis* leaves. Plant Physiol 127:1626–1634

Yan JQ, He CX, Wang J, Mao ZH, Holaday SA, Allen RD, Zhang H (2004) Overexpression of the *Arabidopsis* 14-3-3 protein GF14 lambda in cotton leads to a "Stay-Green" phenotype and improves stress tolerance under moderate drought conditions. Plant Cell Physiol 45:1007–1014

Yoshimoto N, Saito K (2012) Molecular and cellular regulation of sulfate transport and assimilation. In: De Kok LJ, Tausz M, Hawkesford MJ, Hoefgen R, McManus MT, Norton RM, Rennenberg H, Saito K, Schnug E, Tabe L (eds) Sulfur metabolism in plants: mechanisms and application to food security, and responses to climate change. Dordrecht Heidelberg London New York, pp 25–33

Zientara-Rytter K, Lukomska J, Moniuszko G, Gwozdecki R, Surowiecki P, Lewandowska M, Liszewska F, Wawrzynska A, Sirko A (2011) Identification and functional analysis of Joka2, a tobacco member of the family of selective autophagy cargo receptors. Autophagy 7:1145–1158

Zörb C, Steinfurth D, Gödde D, Niehaus V, Möhling KH (2012) Metabolite profiling of wheat flag leaf and grains during grain filling phase as affected by sulfur fertilisation. Funct Plant Biol 39:156–166

Zuchi S, Cesco S, Varanini Z, Pinton R, Astolfi S (2009) Sulfur deprivation limits Fe-deficiency responses in tomato plants. Planta 230:85–94

Chapter 9
Nutrient Use and Nutrient Use Efficiency of Crops in a High CO_2 Atmosphere

Sabine Tausz-Posch, Roger Armstrong, and Michael Tausz

Abstract Atmospheric CO_2 concentrations [CO_2] are continually increasing and are predicted to reach ~550 μmol mol^{-1} by 2050, about a 40 % increase from 2013 levels. Such a large increase in one of the key resources for plant growth will have significant effects on all plants, as carbon assimilation and, consequently, growth and yield is stimulated by the so-called 'CO_2 fertilisation effect'. The one sided increase in carbohydrate acquisition leads to changes in the chemical composition of plants: despite decreases in nutrient concentrations in plant tissues, the greater biomass developed by crops under elevated [CO_2] could lead to increased nutrient demand. Nutrient use efficiency in terms of yield divided by available nutrient may improve, but grains or vegetative plant parts have decreased protein and mineral nutrient concentrations, which can diminish market and nutritious value. A number of hypotheses have been proposed to explain the decreases in nutrient concentrations, among them: (1) Dilution by increased biomass, (2) decreased mass flow, (3) changes in root architecture and function, (4) decreased nitrate reduction, and (5) changes in nutrient allocation and remobilisation. In addition, elevated [CO_2] is likely to change soil processes, including nutrient supply. The extent to which some or all of these contribute to changes in crop nutrition and yield quality is currently unknown because most have not been sufficiently tested under relevant field conditions. This chapter gives an overview of the changes in plant nutrition and trade-offs under elevated [CO_2] to point out that current and future efforts towards improved plant nutrient efficiency should explicitly take into consideration rising [CO_2]. In particular, field testing of putative nutrient use efficiency traits and nutrient management strategies should include elevated [CO_2] as a relevant factor in suitable exposure systems such as Free Air CO_2 Enrichment (FACE) technology.

S. Tausz-Posch (✉)
Faculty of Veterinary and Agricultural Sciences, The University of Melbourne, Creswick, VIC 3363, Australia
e-mail: sposch@unimelb.edu.au

R. Armstrong
Department of Environment and Primary Industries, The University of Melbourne, Horsham, VIC 3402, Australia

M. Tausz
Department of Forest and Ecosystem Science, The University of Melbourne, Creswick, VIC 3363, Australia

M.J. Hawkesford et al. (eds.), *Nutrient Use Efficiency in Plants*,
Plant Ecophysiology 10, DOI 10.1007/978-3-319-10635-9_9

Keywords FACE (free air carbon dioxide enrichment) • Carbon dioxide • Climate change • NUE (nitrogen use efficiency) • Metabolism • Nitrogen • Photosynthesis • Nutrient content • Roots • Soils

Introduction

Carbon exchange between large reservoirs in the atmosphere, land and oceans constitutes the complex global carbon cycle. The carbon involved in this cycle is in a state of dynamic equilibrium, with as much as 400 gigatonnes of carbon exchanged each year (Kitchen 2014). Since the industrial revolution in the nineteenth century, emission of anthropogenic greenhouse gases, predominantly in the form of carbon dioxide, has disturbed this equilibrium (Kitchen 2014). In 2012, human activities caused the release of more than 10 gigatonnes additional carbon emissions into the atmosphere, a number that has grown from the past and will grow with future global economic development.

Surpluses of released carbon have led to the increase in atmospheric CO_2 concentrations [CO_2] from a pre-industrial concentration of ~278 μmol mol^{-1} to a current concentration of ~395 μmol mol^{-1}. Atmospheric [CO_2] is predicted to reach ~550 μmol mol^{-1} by 2050 according to the IPCC scenario A1B (Carter et al. 2007). This would mean that every terrestrial plant will be exposed to at least a 40 % greater concentration in one of the key resources for plant growth compared to present conditions. Effects on climate aside, such a large change, especially if it is not matched by similar changes in other plant nutrients, will also have considerable effects on plant production (Ziska 2008). Arguably therefore any crop improvement efforts must account for the direct effects of increasing atmospheric [CO_2] on plant metabolism and crop growth (Ainsworth et al. 2008; Hatfield et al. 2011; Tausz et al. 2013).

Since the 1980s, significant research efforts have been undertaken to understand how plants will perform and grow under atmospheric [CO_2] enrichment. Methodological approaches range from controlled environments such as laboratory growth chambers and glasshouses to closed-top and open-top field chambers and Free Air Carbon dioxide Enrichment (FACE) systems. Early enclosure experiments mark the basic knowledge of our understanding of plant responses to CO_2 enrichment but they also include potential limitations. For example, results from enclosure experiments may include "chamber effects", possibly exaggerating the effects of elevated [CO_2]. Enclosure studies also often include changed radiation conditions, greater than normal growing temperatures, changed microclimate factors such as wind speed or relative humidity and/or restricted root growth due to the use of pots or containers (Ainsworth and Long 2005; Amthor 2001). In contrast, FACE systems allow the effects of CO_2 enrichment on plant metabolism and growth to be studied under natural and fully open air conditions. Although some concerns have been raised about the rapid fluctuations of [CO_2] in FACE systems (Bunce 2012), advantages such as the ability to grow crops in their natural microclimate probably outweigh any such disadvantages. FACE systems might also produce misleading data if they are used in climates and soils atypical for the plants investigated, highlighting the importance to establish FACE experiments in major and representative cropping areas (Amthor 2001; Tausz et al. 2013). There are a number of large-scale (10–20 m diameter plots) FACE facilities in agricultural systems currently in operation globally (Tausz et al. 2013).

Plant Metabolism Changes Under Elevated [CO_2]

CO_2 enrichment affects C3 plants primarily through increases in photosynthesis rate (A) and a reduction in stomatal conductance (g_s); all other effects of CO_2 enrichment on plant metabolism and growth are linked to changes in these processes (Fig. 9.1; Ainsworth and Rogers 2007). Increases in A under CO_2 enrichment occur because of particular properties of the key carbon fixation reaction catalysed by Ribulose-1,5-bisphosphate carboxylase/oxygenase (Rubisco). Firstly, Rubisco has a low affinity for CO_2 as a substrate for carboxylation, which means that the reaction is not saturated at current atmospheric [CO_2]. Consequently, a rise in [CO_2] increases the carboxylation rate in this reaction, resulting in greater A (Drake et al. 1997). Secondly, Rubisco also catalyses the oxygenation of Ribulose-1,5-bisphosphate (RubP) within the photorespiratory pathway. Photorespiration is competitively inhibited by CO_2 and increases in atmospheric [CO_2] will therefore suppress this pathway (Moore et al. 1999).

In contrast to photosynthesis, the mechanisms responsible for the reduction of g_s under CO_2 enrichment remain vague. Ainsworth and Rogers (2007) conducted a meta-analysis of stomatal density responses to CO_2 enrichment and they found that an average 5 % decrease in density was not statistically significant. They concluded that rather than changes in stomatal density, changes in stomatal aperture are responsible for decreased g_s under high [CO_2]. Stomatal aperture is determined

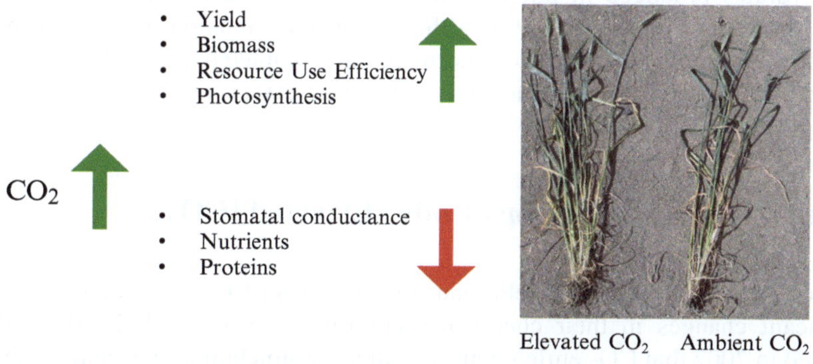

Fig. 9.1 General responses of crops grown under elevated [CO_2]. With rising [CO_2] grain yield, biomass, photosynthesis, nutrient and water use efficiency increase while stomatal conductance, nutrient concentrations and protein concentrations decrease. The picture on the *right* shows wheat (*Triticum aestivum* L.) grown within the Australian Grains Free Air Carbon dioxide Enrichment (*AGFACE*) facility in Horsham, Victoria, Australia, either under an elevated CO_2 concentration of ~550 μmol mol^{-1} (on the *left*) or under an ambient CO_2 concentration of ~395 μmol mol^{-1} (on the *right*). Elevated [CO_2] grown wheat shows a significant increase in tiller number and therefore in biomass and yield

by the turgor pressure in guard cells via ion and organic solute concentrations (Messinger et al. 2006). Rising [CO_2] may be linked to depolarising the membrane potential in guard cells, resulting in stomatal closure (Ainsworth and Rogers 2007; Drake et al. 1997; Messinger et al. 2006).

A meta-analysis on the responses of A and g_s to CO_2 enrichment under FACE conditions confirmed that in C3 crops light-saturated CO_2 uptake rates (A_{sat}) increased on average by ~13 % while stomatal conductance is reduced by ~25 % (Ainsworth and Rogers 2007). Increases in A have stimulating effects on plant growth and productivity (Fig. 9.1; Ziska et al. 2012) and, depending on the plant species investigated, greater A leads to greater biomass accumulation and yield (termed "CO_2 fertilisation effect"). For example, C3 crops respond to high [CO_2] with ~20 % greater biomass and ~17 % greater yield (Ainsworth and Rogers 2007; Tausz-Posch et al. 2012). It is noteworthy that in the same meta-analysis (Ainsworth and Rogers 2007) C4 crops also showed a significant ~11 % increase in A_{sat} in response to CO_2 enrichment. In general, C4 crops are more independent of ambient [CO_2] as they have the ability to concentrate CO_2 in the leaf mesophyll. Their carboxylation reaction is generally close to saturated under current ambient [CO_2]. The positive response of C4 plants to CO_2 enrichment was linked to the improved water status of the crops investigated (Ainsworth and Rogers 2007).

The initial stimulation of C3 photosynthesis may not be sustained when plants are exposed to CO_2 enrichment for long time periods (Kirschbaum 2011; Moore et al. 1999). Plants grown under CO_2 enrichment showed decreased A relative to plants grown under ambient [CO_2] when both were measured at ambient [CO_2]. This down-regulation of photosynthesis under CO_2 enrichment is called "photosynthetic acclimation" (Moore et al. 1999). Photosynthetic acclimation is linked to reduced concentrations of Rubisco in the leaves, and this is hypothesised to result from repression of Rubisco synthesis by accumulating carbohydrates (Drake et al. 1997). If there are insufficient sinks for the increased amounts of carbohydrates produced in photosynthesis, this situation is accentuated (Moore et al. 1999; Stitt and Krapp 1999). Although such acclimation reduces the photosynthetic capacity of plants, it does normally not completely offset the stimulation caused by CO_2 enrichment (Ainsworth and Rogers 2007).

Mineral Nutrition of Crops Under Elevated [CO_2]

With changes in growth and yield, plants grown under CO_2 enrichment also show significant changes in their chemical composition (Erbs et al. 2010). Loladze reported in 2002 that CO_2 enrichment will alter the stoichiometry (relative elemental composition) of plants leading to reduced concentrations of most macro and micro elements in plant tissues. As a consequence there are potential negative consequences for the nutritional value of crops (Loladze 2002). For example, decreased tissue nitrogen (N) concentrations under elevated [CO_2] have been widely reported (Stitt and Krapp 1999). As a large proportion of N in plants is used for protein synthesis, protein concentrations in vegetative parts as well as seeds and grains tend to decrease under elevated [CO_2]. In many crops, particularly cereals such as wheat or rice, concentrations of protein and minerals in grains are important determinants of the nutritional value and functional properties (Erbs et al. 2010; Högy et al. 2009).

The lower tissue N concentrations in conjunction with proportionally increased carbon assimilation and its subsequent incorporation into storage and structural material of the plant results in increased carbon to nitrogen ratio (C/N). This leads to increased nitrogen use efficiency of plants under CO_2 enrichment (Fig. 9.1; Drake et al. 1997). The same principle can be applied for other macro and micronutrients, although less documented than for N. Although overall nutrient concentrations in plant biomass are decreased under CO_2 enrichment, total nutrient removal via grains or plant biomass, for example such as N, can increase under CO_2 enrichment (Conroy and Hocking 1993; Lam et al. 2012b). Such effects will have significant consequences for food quality and the management of our agricultural systems in a future high [CO_2] world.

Nitrogen

N is the ecologically and economically most important mineral nutrient in cropping systems globally, and has thus received intense attention so that knowledge of plant N metabolism is comparably advanced relative to most other nutrients. It is therefore not surprising that the majority of research on interactions of nutrients with high [CO_2] in plants has focused on N. We will therefore discuss major principles of plant nutrition changes under CO_2 enrichment using N as the main example.

Nitrogen Concentrations in Plant Tissues Under CO_2 Enrichment

Dry mass based tissue concentrations of nitrogen (N_m) are generally lower in high [CO_2] grown plants, as confirmed by a number of recent synthesis papers (Ainsworth and Long 2005; Leakey et al. 2009; Stitt and Krapp 1999; Taub and Wang 2008; Wang et al. 2013). For example, leaf N_m decreased by about 13 % across a number of FACE experiments (Ainsworth and Long 2005), or by an average of 9 % for wheat across experiments with different CO_2 exposure systems and CO_2 concentrations (Wang et al. 2013). This decrease is somewhat smaller on a leaf area basis, only about 4 % across a number of FACE experiments (Leakey et al. 2009). The decrease in leaf N is consistently related to a decrease in total leaf protein, specifically, and perhaps exclusively, to a decrease in the amount of Rubisco protein (Leakey et al. 2009). Decreased Rubisco content in leaves is linked to photosynthetic downward acclimation under prolonged exposure to increased [CO_2] and may thus reflect changes in N allocation patterns.

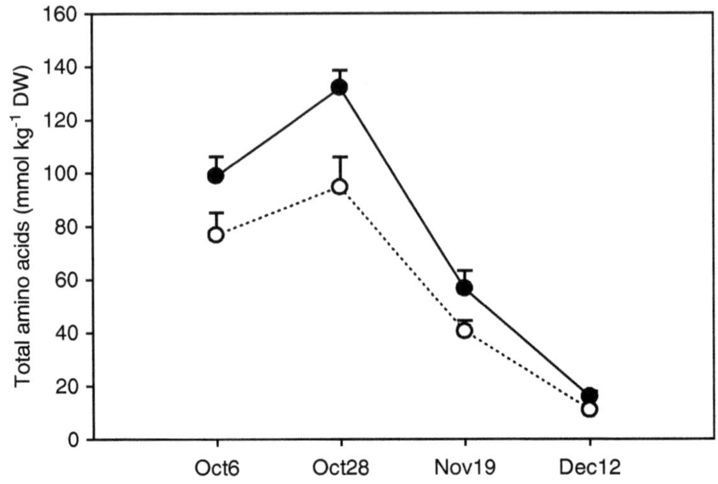

Fig. 9.2 Total free amino acid concentrations in wheat flag leaves from heading (Oct 6) to maturity (Dec 12). Plants were grown either under ambient CO_2 (*closed symbols*) or an elevated CO_2 concentration of ~550 ppm (*open symbols*) within the AGFACE facility, Horsham, Australia. Values are means (standard errors) with four replications in each

For wheat, a decrease in leaf N was also related to decreases in leaf chlorophyll across a number of experiments (by 8 % according to a synthesis by Wang et al. 2013), but such changes in chlorophyll are not consistently found. High [CO_2] can also lead to decreases of leaf nitrate concentrations (e.g. for wheat reported in Hocking and Meyer 1991). There is little information on free amino acids, but recent FACE data indicate about 20–30 % decreases in wheat leaves (Fig. 9.2).

There is quite a range in responses between different experiments, environments or species. Most consistent among such variability is that decreases in leaf N_m are less in legumes than in non-N fixing species, perhaps because legumes can channel excess carbohydrates towards greater nitrogen fixation rates (Rogers et al. 2009).

Decreases in grain protein contents in the final harvest are commonly unfavourable for food or feed quality, and of particular concern in cereals such as wheat, where bread making and marketing quality is largely determined by protein content (Loladze 2002; Taub et al. 2008). Grain N decreased under CO_2 enrichment by between 0 % and 20 % in cereals (wheat, rice, barley; as reviewed by Högy and Fangmeier 2008; Högy et al. 2013; Taub et al. 2008) and by less than 5 % (but still significantly so) in soybean (Taub et al. 2008).

Can Greater N Supply Restore Leaf and Grain N?

It is uncertain to what extent N supply determines the extent to which N_m decreases in leaves or grains under CO_2 enrichment: Some evidence suggested that N_m decreases were most pronounced in N-deficient and less marked or even absent in well fertilised plants (Stitt and Krapp 1999), including in FACE experiments

(Sinclair et al. 2000). This is in line with the suggestion that photosynthetic downward acclimation, which is related to leaf N and (Rubisco) protein, is less pronounced or absent in well-fertilised plants. However, recent synthesis papers suggest that even if high N supply counteracts decreases in N to some extent, N_m remains significantly lower in leaves (Wang et al. 2013) and grains (Taub et al. 2008).

Whilst decreases in tissue concentrations of N are reasonably consistent and well established, results on changes of overall N demand and uptake of the crop per m^2 ground area are variable. N uptake into (above ground) biomass depends on the relative magnitude of biomass increase and decrease in N tissue concentrations. Most analyses suggest that biomass and yield stimulation are relatively greater than the decrease in tissue N concentration, which would lead to greater N demands of bigger crops under CO_2 enrichment (Lam et al. 2012a, b, c) despite decreased N concentrations in tissues. However, others suggest that N uptake remains unchanged (Wang et al. 2013).

It is noteworthy that the plant growth response to elevated [CO_2] in natural ecosystems, which do not receive extra N input, may decrease over time, and this was ascribed to a depletion of N (and other nutrient) reserves (= 'progressive nitrogen limitation' (PNL) hypothesis, e.g. Hungate et al. 2003).

Adjustment of N_{mcrit} Under CO_2 Enrichment

It has been suggested early on that critical tissue concentrations of N (and other mineral nutrients) must be adjusted under high [CO_2] (Conroy and Hocking 1993; Fig. 9.3). Tissue concentrations of nutrients such as N are often used as a diagnostic tool to assess a plant's nutritional status. Critical tissue (such as leaf) concentration of N (N_{mcrit}) would indicate sufficient N supply. N_{mcrit} is the value of N_m that is needed for optimum growth, assuming otherwise non-limiting conditions. At leaf N greater than N_{mcrit} growth will not increase further (Fig. 9.3). Conroy and Hocking (1993) suggested that critical concentrations need adjustment towards lower values under high [CO_2] (Fig. 9.3). As such results were derived from pot experiments, field evaluation in cropping systems will be important.

Does CO_2 Enrichment Change N Response of Plants?

Similarly to other essential nutrients, insufficient N supply to crops can limit the [CO_2]-driven growth and yield enhancement; N-limited crops sometimes do not show a significant effect of CO_2 enrichment on biomass, or the relative yield and biomass increase is less than in well fertilised crops (Stitt and Krapp 1999). As a consequence, the N response curve may change under elevated [CO_2]. N response curves (Fig. 9.4) are derived from fertiliser experiments (fertiliser added at different

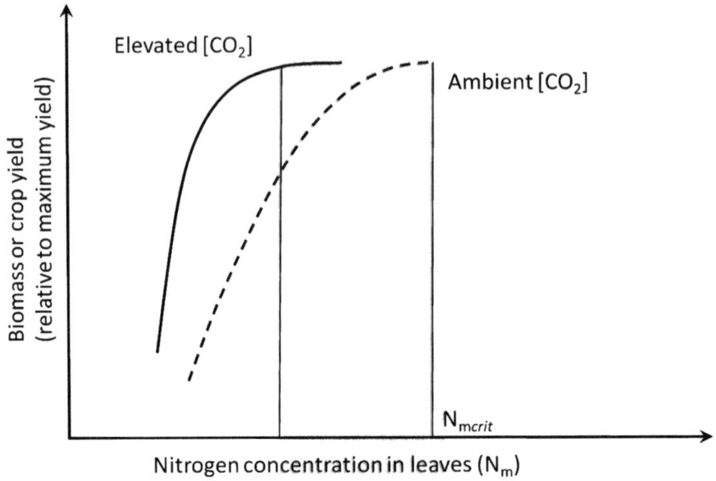

Fig. 9.3 Scheme illustrating the determination of critical N concentration (N_{mcrit}) in leaves and proposed changes under elevated [CO_2]. Concrete results reported in Conroy and Hocking (1993)

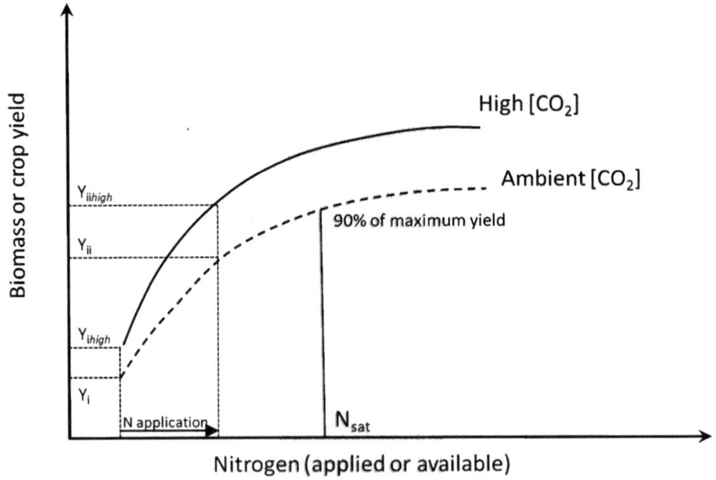

Fig. 9.4 Hypothetic nitrogen response curves (Mitscherlich curve*s*) of crops grown under ambient and high [CO_2]. N_{sat} indicates sufficient N supply to achieve 90 % of maximum growth or yield. Y_i indicates yield without additional (fertiliser) N, Y_{ii} denotes yield after a certain fertiliser application (*arrow*). It is not resolved whether and under which conditions N_{sat} changes under elevated [CO_2]. Agronomic efficiency of fertiliser application can be calculated as the difference between yields with and without fertiliser (Y_{ii}-Y_i) divided by the fertiliser applied (Table 9.1). In this hypothetic system, agronomic efficiency of the fertiliser applied (*arrow*) would be greater under elevated than ambient [CO_2], because the yield difference is greater at the higher N application

rates), and are an important tool to assess N limitations and requirements of crops (Hawkesford 2011). Figure 9.4 illustrates the important questions: (1) Is the amount of N required for maximum growth different between high and ambient [CO_2]? (2) Is the N response (either in % yield increase per kg N or in kg yield increment per kg N) greater under high [CO_2] than under ambient [CO_2]? Work on potted wheat under relatively high CO_2 enrichment concluded that the total amount of N required for maximum growth remains unchanged, and N-responsiveness both in absolute (g biomass per g added N) and relative terms (in % biomass per g added N) was greater under high [CO_2] (Conroy and Hocking 1993). Whilst this is in line with the notion of generally greater resource use efficiency of plants under CO_2 enrichment, there are few, if any, reports of field data (such as full N response curves under free air conditions).

Nitrogen Use Efficiencies Under CO_2 Enrichment

Changes in tissue concentrations in conjunction with biomass and yield changes lead to changes in Nitrogen Use Efficiency (NUE) in their various definitions and its components (Good et al. 2004; Hawkesford 2011; Table 9.1). A frequently used definition for NUE is the ratio of biomass accumulation or yield over the amount of available N, and according to this definition NUE is closely related to yield (Hawkesford 2011). As yield increases under CO_2 enrichment, so will by definition NUE of the crop, provided available N remains unchanged. NUE as defined above can be divided into Nitrogen Uptake Efficiency (NUpE, N taken up into the biomass as a proportion of total available N) and Nitrogen Utilisation Efficiency (NUtE, the ratio of yield or biomass over the N taken up (Hawkesford 2011). Depending on the relative magnitude of changes in N uptake, growth and yield, and N concentrations in biomass, the different components of NUE may change (Table 9.1 for an overview).

On a physiological level, changes in Photosynthetic Nitrogen Use Efficiency (PNUE) are the result of lower concentrations of leaf N in conjunction with greater assimilation rates under CO_2 enrichment leading per definition to greater PNUE, where PNUE is defined as carbon assimilation rate (expressed in μmol C fixed per m^2 leaf area and s) per unit leaf N (expressed as g N per m^2 leaf area; N_{area}). A review of FACE experiments confirmed an increase in PNUE, and this increase was mainly due to the large increase in assimilation rate, and to a minor extent to a decrease in N_{area} (Leakey et al. 2009).

An important aspect of NUE in agricultural systems is the proportion of N applied to crops (as fertiliser) that is utilised by the target crop. A significant proportion of the applied N can either remain in the soil (although potentially available to subsequent crops) or can be lost from the soil-plant systems by several processes, including denitrification, volatilisation and leaching. For example, in Australian wheat production systems actual fertiliser N recovered in the wheat plant itself ranged from approximately 20–60 % (Chen et al. 2008). Recent

Table 9.1 Effects of elevated [CO_2] on different aspects of plant nutrient (Nut) metabolism and plant nutrient (Nut) use efficiencies. Nutrient use efficiency definitions after (Hawkesford 2011)

Symbol	Parameter	Explanation	Major trends under elevated [CO_2]
Nut_m	Dry mass based tissue concentration	g or mol nutrient per g tissue dry weight; e. g. mg g^{-1} or μmol g^{-1}	Decrease in vegetative biomass and grains, most consistent for N but common also for many other nutrients
Nut_{area}	Leaf area based nutrient content	g or mol nutrient per m^2 leaf area	Usually decreases, perhaps to a lesser extent than Nut_m
NutUp	Nutrient uptake (into biomass), also nutrient demand	g per m^2 plot surface area	Commonly increases if biomass stimulation is greater than Nut_m reduction, or remains unchanged
NutYield	Nutrient yield	g in harvested fraction per plot surface area	Commonly increases due to yield stimulation
NutUE	Nutrient use efficiency	Yield per available nutrient	Increases due to increase in yield
NutUpE	Nutrient uptake efficiency	Amount of nutrient taken up into biomass per available nutrient	Commonly increases due to greater increase in biomass than decrease in Nut_m; may remain unchanged
NutUtE	Nutrient utilisation efficiency	Yield per amount of nutrient in biomass	Increase or no change
NutHI	Nutrient harvest index or nutrient partitioning quotient	Fraction of total nutrient in (above ground) biomass that ends up in harvest	Decreases for N with ample N supply; may remain unchanged (e. g. at low N supply)
PNutUE	Photosynthetic nutrient use efficiency	Carbon assimilation rate per amount of nutrient in leaf	Increases for nitrogen (and other nutrients) mainly due to stimulation of assimilation
FUE	Fertiliser use efficiency	Amount of nutrient taken up per amount of nutrient applied	Appears unchanged for N, but little information available
NutAE	Agronomic nutrient efficiency	Yield increase per nutrient applied with fertiliser	Possibly increases for N, but only few field experiments use more than 2 nutrient application rates

experimentation using ^{15}N labelled granular N (urea) fertiliser in dryland grain systems in southern Australia under FACE found that although the biomass, plant N content and the amount of N absorbed from the soil by wheat was increased, elevated [CO_2] did not affect the proportion of fertiliser N taken up by the plant (Lam et al. 2012a). However, in order to meet the greater overall demand for N by the crop, the authors suggested that greater amounts of N (as fertiliser) would be required in future climates. Interestingly, elevated [CO_2] did increase the rate of N_2O emissions, a potent greenhouse gas, from these wheat systems. This could contribute to radiative forcing in the atmosphere, although the absolute amount of N lost from the system (as N_2O) would be negligible in terms of N budget (Lam et al. 2013).

Other Mineral Nutrients

Whilst effects of CO_2 enrichment on plant N are relatively consistent and well documented, there is little consensus about most other mineral nutrients. It is important to recognise that different mineral nutrients are not independent of each other: they can share or co-regulate common uptake and metabolic pathways, and are regulated in an interdependent manner by plant growth or assimilate availability. We have more detailed knowledge on interactions between P and N or S and N (Hawkesford and De Kok 2007; White and Hammond 2008), but interactions apply to other mineral nutrients as well. Although not investigated in great detail under high [CO_2], it is safe to assume that such interactions play an important role in modifying the CO_2 response of selected mineral elements and their metabolism.

Recent reviews mostly suggest that elevated [CO_2] also leads to decreased tissue concentrations of mineral nutrients other than N, but results are inconsistent: For example, one recent synthesis paper found significant decreases in 11 (if N is excluded) elements in elevated [CO_2] grown plant tissues, including Mg, K, P, S, Fe, Ca, Mn, and Zn (McGrath and Lobell 2013). In apparent contrast, Duval et al. (2012) found only a few significant changes in leaf mineral concentrations (excluding N) in crops (excluding legumes) and N-fixers: in that analysis, among all investigated mineral nutrients, only leaf Mg concentrations were significantly decreased under CO_2 enrichment in both functional groups. In legume leaves, B and Fe decreased as well, whereas Mn even increased significantly. However, results for grasses were different again, and perhaps most surprisingly, this study found no significant effects of elevated [CO_2] on grain nutrients (Duval et al. 2012). In several other studies, which focused on the effects on grains rather than vegetative plant parts, there seems to be overall agreement that the elements N, S, Mg, Ca, Zn, and Na are significantly reduced under CO_2 enrichment while contrasting results are reported for K, Mn, P and Fe (Fangmeier et al. 1999; Fernando et al. 2012; Högy and Fangmeier 2008; Högy et al. 2013; Manderscheid et al. 1995).

Some of these discrepancies may be due to different nutrient supply conditions, which is important for some micronutrients. Apparently contradictory results may also point out some in principle problems with many synthesis papers, where the analysed data set may be biased by particular exposure or growing systems, or factors analysed independently (such as, for example, crop type) may be strongly confounded by other factors. For example, it is common that data from different crop types or plant functional groups come from separate groups of experiments (for example, different large scale FACE experiments), and are therefore potentially confounded by factors such as climate or soil. Further, environmental growing conditions can mediate nutrient responses to elevated [CO_2]: in one FACE study, variations in rainfall and/or temperature had a significant effect on the response on macro and micro mineral concentrations in wheat grains to CO_2 enrichment (Fernando et al. 2012).

The interactions between elevated [CO_2] and mineral nutrition may be markedly different for different nutrients: in contrast to N, where critical nutrient concentrations are reduced (Fig. 9.2), critical concentrations of P generally remain the same or are increased when C3 plants are grown under elevated [CO_2] (Ghannoum et al. 2007). Increases in critical P concentrations of C3 plants grown under elevated [CO_2] have been attributed to changes in competition for P. As photosynthetic carbon fixation is preferred over the photorespiratory cycle, more P is needed for the energy carrier ATP (Ghannoum et al. 2007). Similar to N, several studies have noted that growth (and yield) response to high [CO_2] depends on P nutrition. The relative importance of this interaction between P and high [CO_2] however varies with species. Lam et al. (2012c) found that high [CO_2] increases biomass of two pulse species, chick pea (*Cicer arietinum*) by 18–64 % and field pea (*Pisum sativum*) by 24–57 %, as well as the pasture legume barrel medic (*Medicago truncatulata*), and that this effect was greater when P supply was non-limiting.

Proposed Mechanisms for Decreased Nutrient Concentrations in Elevated [CO_2]

The mechanisms responsible for and controlling decreased nutrient concentrations under elevated [CO_2] are not completely understood, and there are a number of hypotheses under consideration. Most of these hypotheses are connected to studies conducted on N but, depending on the nutrient, similar principles may apply for other nutrients. An overview is given in Fig. 9.5.

Dilution by Increased Biomass and Carbohydrate Production

CO_2 enrichment increases C fixation and dry matter accumulation in plants. Increased dry matter is mostly derived from greater carbohydrate (C, H and O) accumulation and concentrations of other macro and micro nutrients will decrease in biomass if their uptake is not increased (Taub and Wang 2008). For example, Poorter et al. (1997) investigated the chemical composition and construction costs of leaves of 27 wild and agricultural species at ambient and elevated [CO_2]. They found that the strongest response of plants to CO_2 enrichment in respect to their chemical composition is the increase in concentrations of total non-structural carbohydrates (TNC) resulting from stimulated photosynthesis under high [CO_2]. The next strongest responses to CO_2 enrichment were decreases in protein and mineral concentrations leading the authors to conclude that a dilution effect caused by accumulating TNC significantly contributes to decreasing nutrient concentrations under CO_2 enrichment. Similar conclusions were made by Taub and Wang (2008), who used graphical vector analyses to study biomass dilution effects under CO_2 enrichment.

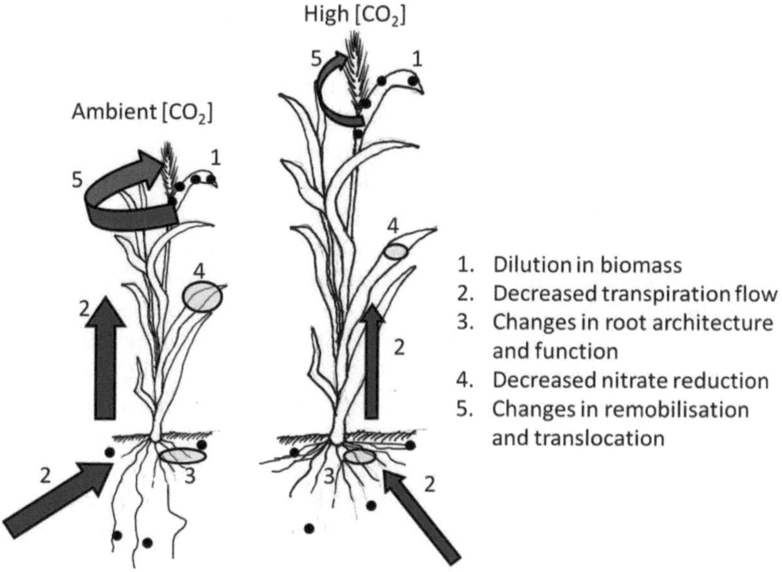

Fig. 9.5 Schema of plant based mechanisms that potentially contribute to decreases in mineral nutrient (symbolized by *black circles*) concentrations in crops grown under elevated [CO_2]. *1* Dilution in increased biomass by increased carbohydrate supply. *2* Decreased mass flow due to decreased stomatal conductance. *3* Changes in root architecture and function. More root mass in top soil may improve access to nutrients, but this may come at a cost of access to deeper layers. Uptake physiology may change. *4* Decreased rate of nitrate (and possibly sulfate) reduction. *5* Adverse changes in remobilisation from leaves and translocation to grains

Biomass dilution, however, does not entirely account for the total decrease in nutrient concentrations. For example, when Poorter et al. (1997) cross-checked decreasing nutrient concentrations for their direct dependence on increased TNC, the minerals and proteins expressed on a TNC-free biomass basis remained significantly reduced, even if less so than on a total dry weight basis. This suggests that processes other than dilution by TNC contribute to decreased nutrient concentrations under CO_2 enrichment. Simple dilution by greater (structural) biomass production remains a possibility, but if biomass dilution is exclusively responsible for decreasing nutrient concentrations under CO_2 enrichment then all nutrients would decrease equally in concentration. However, as reported earlier, decreases in macro- and micronutrients can vary greatly among each other and across studies (between 0.7 and 19.5 % or 3.7 and 18.3 %) (Fangmeier et al. 1999; Fernando et al. 2012; Högy and Fangmeier 2008; Högy et al. 2009; Manderscheid et al. 1995), and there seems to be no general relationship with growth stimulation by elevated [CO_2]. It has been repeatedly suggested that factors relating to nutrient uptake efficiency and metabolism are involved in decreased nutrient concentrations under high [CO_2]. It has to be acknowledged, however, that an exact quantification of possible biomass dilution effects within experiments, particularly under free air growth conditions, is still missing.

Reduced Mass Flow

Transpiration is the evaporation of water into the atmosphere from the leaves and stems of plants. It is a passive process mainly governed by the vapour pressure deficit of the atmosphere and the soil moisture content, but mediated by the variable resistance of the stomata. CO_2 enrichment decreases stomatal conductance leading to decreased transpiration in plants (Ainsworth and Rogers 2007). Transpiration has several functions, and one of its functions is to drive the mass flow of nutrients in the soil to the rhizosphere and root surfaces (Cramer et al. 2009). Mass flow driven by transpiration contributes significantly to the delivery of nutrients to the plant, and then the translocation within the plant. For example, Barber (1995) reported that mass flow contributed by more than 70 % for N, S, Mg and Ca to the nutrient concentrations in *Zea mays* with the remaining percentages taken up by diffusion and interception of roots. The effectiveness of mass flow depends on factors that influence transpiration. Decreased transpiration rates under CO_2 enrichment will reduce the mass flow of nutrients in the rhizosphere and hence decrease nutrient availability for plant uptake, as well as decrease the delivery rate of nutrients to the above ground plant parts. It was hypothesised early that this decreases mineral nutrition of plants (Conroy and Hocking 1993; McGrath and Lobell 2013).

Evidence that reduced mass flow is partially responsible for decreases in nutrient concentrations can be provided by comparing the extent to which individual nutrients decrease under high [CO_2]. For example, the largest decrease in concentration was found for mobile nutrients such as N, Mg or Ca that are supplied to the rhizosphere by mass flow. Conversely, the concentration of less mobile nutrients, e.g. P, and those that are mainly dependent on physiological uptake processes was decreased to a lesser extent by CO_2 enrichment (Högy and Fangmeier 2008; Högy et al. 2013; Taub and Wang 2008). A first conclusion might be that biomass dilution leads to a decrease in concentration of all nutrients while concentrations of mobile nutrients are further decreased via decreased transpiration mass flow. A recent analysis aimed at testing the mass flow hypothesis on synthesised literature data had to resort to pairing mass flow and nutrient uptake data from different studies, because "No published studies that simultaneously examined mass flow and nutrient concentration for plants grown in elevated [CO_2] were available" (McGrath and Lobell 2013), thus pointing out a major gap in experimental verification.

Root Architecture and Function

Roots are the first plant organs receiving nutrients from the soil and root morphology and architectural traits such as length, depth, branching and curving determine a plant's access to nutrients (and other resources such as water) in the soil. Despite their importance to plant growth, most studies on impacts of climate change variables such as elevated [CO_2] have focused on above ground traits rather than roots

(Benlloch-Gonzalez et al. 2014). The basic development of a root system is given by its genetic predisposition but conditions such as, for example, soil nutrient availability or soil moisture content will greatly affect root growth as soon as growth commences (Rich and Watt 2013).

All C in roots comes from the re-allocation of photosynthates from photosynthesising above ground biomass with up to 50 % of all the C fixed being transported below ground during early development (Gregory et al. 1996). Considering that C3 crops increase their light–saturated CO_2 uptake rates on average by ~13 % when grown under CO_2 enrichment (Ainsworth and Rogers 2007), the amount of fixed C allocated to roots increases in plants growing under high CO_2 and this can lead to changes in root architectural traits (Pritchard and Rogers 2000). For example, high [CO_2] significantly increases the root biomass and roots often undergo the greatest relative dry weight gain among all plant organs (recently reviewed by Madhu and Hatfield 2013). CO_2 enrichment also changes the vertical distribution of roots in a way that more roots accumulate in the top soil layers as compared to the deeper layers in the soil. This leads to a shallower root system with greater root density in the top layers (Pritchard and Rogers 2000; Pritchard et al. 2006; VanVuuren et al. 1997). Although this may increase access to nutrients which are generally concentrated in topsoils (e.g. Ho et al. 2005), it may also have undesirable consequences where rooting access to subsoil water is important (Lilley and Kirkegaard 2011).

Changes in root architectural traits will affect the spatial patterns of soil nutrient exploitation. A dense root system with many lateral branches and root hairs is considered to be better for nutrient acquisition as compared to plants with a sparse root system because of their greater root surface (Kong et al. 2013). Total nutrient uptake should therefore benefit from increased root length and mass under CO_2 enrichment. However, it has been reported that root physiological activity often is reduced in crops grown under CO_2 enrichment (Fitter 1996; Bassirirad et al. 1996), with potential repercussions on specific uptake capacity of roots. Uptake systems for nutrients such as N in the form of nitrate and ammonium (Miller and Chapman 2011), or S in the form of sulphate (Hawkesford and De Kok 2006), are now well characterised and the specific transport systems are subject to physiological regulation. Whilst it is likely that changes in the carbon to nutrient balance in plants affect such regulation, it is unknown how growth under elevated [CO_2] affects nutrient transporters and their regulation.

Decreased Nitrate Reduction

Another hypothesis regarding the decrease in (specifically) plant N nutrition was recently put forward by Bloom et al. (2010). Their paper showed a significant decrease of nitrate reduction under elevated [CO_2] in wheat. The hypothesis was that the decrease in photorespiration affected nitrate assimilation whereby a number of mechanisms could contribute (Bloom et al. 2010). Yong et al. (2007) reported photosynthetic acclimation to CO_2 enrichment in rice plants with excessive N supply, but not in those with lower N supply. It was speculated that C and N reduction compete for electrons from the photosynthetic light reaction, and that this leads to an inhibition in nitrate assimilation. Bloom et al. (2010) also suggested that the inhibition of nitrate assimilation by elevated [CO_2] might affect nitrate uptake rates as a follow-on effect. If impaired nitrate reduction is a reason for declining tissue N concentrations then meeting N needs through ammonium NH_4^+ could alleviate the problem. It has been suggested that ammonium-based fertilisers or the use of nitrification inhibitors (to avoid conversion of ammonia into nitrate in the soil) could be considered, provided NH_4^+ toxicity and detrimental effects on soils can be managed (Bloom 2009). A recent study seems to corroborate this hypothesis because plants supplied with NH_4^+ as an N source showed better responses to elevated [CO_2] than those supplied with NO_3^- (Carlisle et al. 2012). However, this approach has not yet been tested under realistic field conditions. A similar principle could also apply to S reduction, but whether S reduction rates are affected by elevated [CO_2] has not been tested.

Remobilisation of Nutrients

The remobilisation patterns of nutrients from leaves to developing seeds have particular significance for grains of cereal crops (Gregersen 2011). For example, Palta and Fillery (1993) investigated N remobilisation patterns in wheat and they found that N acquisition from the soil stopped by anthesis and that ~69 % of the N allocated to the spike was derived from remobilisation from the vegetative plant parts. Remobilisation strategies are particularly relevant for low rainfall cropping systems where crops are water-limited during the grain filling period and therefore rely heavily on the nutrients already stored in vegetative plant parts pre-anthesis, with flag leaves representing the strongest source of nutrients (Palta and Fillery 1993). Nutrient remobilisation is closely linked to leaf senescence processes and for optimum remobilisation exact timing of processes is crucial (Gregersen 2011; Yang and Zhang 2006).

Nutrient and carbohydrate remobilisation patterns are already of interest in the quest for more nutrient efficient crops (Gregersen 2011), but little is known about whether changes in remobilisation patterns contribute to the decrease in nutrient concentrations, particularly in grains, under CO_2 enrichment. In a study on barley it was proposed that developing grains have a greater sink capacity for N, as they receive ample supplies of carbohydrates. N reserves in the leaves may become depleted faster, leading to accelerated senescence of flag leaves (Fangmeier et al. 2000). One recent study on canola in a high rainfall system found that elevated [CO_2] significantly decreased N remobilisation to seeds (Franzaring et al. 2012): The NHI (see definition in Table 9.1) decreased by as much as 65 % under ample N supply. It seems that changes in nutrient remobilisation can be significant, but field studies with elevated [CO_2] under soil drying conditions, for example in low rainfall systems, where the remobilisation of nutrients is relatively most important (Palta et al. 1994; Yang et al. 2000), are not yet available.

Interactions of Soils with Elevated Atmospheric [CO_2]

In many agricultural systems, production is limited either by nutrients or other factors such as water. As a consequence, plant growth is not restricted by C availability (Poorter and Perez-Soba 2001) until these other limitations are overcome. Nutrient use efficiency (NutUE) reflects the ability of plants to access nutrients from soil (and fertilisers) as well as the dynamics of the nutrient once it is absorbed.

In contrast to the rapid growth in knowledge of how high [CO_2] alters the above ground physiological responses of plants, much less is known about potential interactions between high [CO_2], different soil processes and plant responses. For example, soil available N levels have been hypothesised to gradually decline under high [CO_2]. This so-called progressive nitrogen limitation (PNL, Luo et al. 2004), will result in available soil N levels becoming increasingly limiting as N and C are sequestered in plant biomass and soil organic matter. As C/N ratios of plant residues increase (as indicated in previous sections), soil N mineralisation rates will decrease (Prior et al. 2008). Consequently, an understanding of how high [CO_2] will affect NutUE of plants must account for the influence of different soil properties on both the capacity to supply nutrients to plants as well as potentially influence the ability of the plant to access available nutrients (and other critical resources such as water) in the soil.

Strong interactions can exist between different mineral nutrients (especially N) at both a physiological level (see above) as well as in a broader agronomic context. For example, the old weathered soils that underpin many Australian grain production systems are very deficient in P (Donald 1964) and McDonald (1989) noted the importance of N and P interactions in controlling grain yield in cereal crops. Only a relatively small proportion of P in the soil is in a chemical form that is immediately 'available' to plants. Consequently, farmers generally manage soil P deficiencies to plants by applying P fertilisers. However, a relative large proportion – up to 95 % (McBeath et al. 2012) – of the fertiliser P becomes unavailable for the crops. For example, P can be strongly adsorbed onto soil particles, precipitated as insoluble forms (normally associated with iron, aluminium or calcium) or immobilised by soil biota. Considerable research has been dedicated to investigating the potential of plants (at both an intra and inter species level) to access these 'unavailable' forms of P as a strategy to improve NutUE (e.g. Armstrong et al. 1993; Osborne and Rengel 2002; Wang et al. 2010). Although high [CO_2] can actually temporarily increase P immobilisation (as organic P) in the rhizosphere of crops (Jin et al. 2013), there is no evidence that high [CO_2] alters the ability of plants to acquire P from different sparingly soluble sources of P (Hocking and Barret 2003; Jin et al. 2014).

Differences in NutUE between plants depend on the ability of roots to access nutrients in the soil and fertiliser; especially for relatively immobile nutrients such as P, most nutrient acquisition is via root interception (Barber 1995). In many environments, soils can exhibit a range of physicochemical properties such as high aluminium, sodicity, salinity and high boron that can significantly restrict root growth, and therefore uptake of soil nutrients and water (Adcock et al. 2007). It has been hypothesised that high [CO_2] may ameliorate the negative effect of some of these constraints such as high salinity (Poorter and Perez-Soba 2001), supported by the finding that high [CO_2] generally increases assimilate allocation to below ground processes including root growth and exudation (Lynch and St. Clair 2004). However, Munns et al. (1999) found that whereas there was a positive interaction between [CO_2] and salinity at low CO_2 levels, there was no CO_2 fertilisation effect at high [CO_2].

Outlook: Traits and Practices to Improve Crop Nutrition and Quality Under High [CO_2]

Given the extent of nutrient deficiencies globally and the financial (e.g. steadily increasing fertiliser prices) and environmental (e.g. eutrophication of water ways) cost of managing nutrients in agricultural systems it is not surprising that considerable research has been devoted to improving NutUE in crops especially that of N and P in both low and high input systems (Wiesler et al. 2001; Hawkesford 2011). This research has encompassed genetic solutions (Hirel et al. 2007), improved fertiliser forms (e.g. Chen et al. 2008; McLaughlin et al. 2011) as well as systems approaches such as better predictions of crop nutrient requirements through improved soil testing and predictions of crop nutrient demand often via the use of computer simulation models (e.g. Carberry et al. 2002; Moeller et al. 2009). As shown in this chapter, elevated [CO_2] shifts the relative resource availability towards photosynthetically fixed C and therefore affects the trade-offs that are key to optimising plant traits and crop management in practice (Sadras and Calderini 2009). Recently, the argument that crop improvement should explicitly take into account elevated [CO_2] has gained some traction, particularly as it was shown that traits selected by breeders over the past 100 years were not necessarily beneficial under increased [CO_2] (Ainsworth et al. 2008; Tausz et al. 2013; Ziska et al. 2012). Similarly, management practices should be evaluated under elevated [CO_2], as plant growth changes in amount and timing, and this will affect nutrient demand, uptake and utilisation.

Most aspects of plant NutUE will improve as a consequence of the CO_2 fertilisation effect (Table 9.1), but the overall increased nutrient demand of crops accumulating more biomass ensures that the quest for optimum plant nutrition and NutUE remains of paramount importance. Elevated [CO_2] will only accentuate the "yield quality conundrum", e.g. the fact that mineral nutrients or protein concentrations decrease as crop yields increase (Hawkesford 2011). The main challenge will be how to supply sufficient N to maintain grain protein concentrations under high [CO_2], but the issues are similar for other nutrients, as their global supplies are becoming limited (such as P), they play important role in grain quality (such as S), or micronutrients such as Fe and Zn that are crucial to human health. Genetic (breeding or biotechnological) and management improvements targeting the quality yield conundrum particularly under elevated [CO_2] are therefore a high priority in agricultural research.

For the successful development of new improved plant varieties a number of pre-requisites must be met: (i) there must be genetic variability available (ii) there must be an ability to readily identify the genes associated with this phenotypic variability and (iii) the trait must be readily inheritable (Foulkes et al. 2009). Although genetic variation for nutrient efficiency has been identified in many arable crops (e.g. Svečnjak and Rengel 2006 in canola), there are relatively few reports of the development of new varieties with greater NutUE. There are also arguments that rather than directly selecting for traits such as improved NutUE, breeding for yield improvement alone (which many plant breeders indicate is their number one priority) has also contributed to increased N harvest index and increased N uptake (Sadras and Lawson 2013). However, as highlighted in this chapter, there may be a number of opportunities to identify specific target traits, for example employing the developing knowledge on the mechanisms underlying the decline of plant nutrients under elevated [CO_2].

Because a key challenge in developing new varieties is managing the complexity of interactions between traits and environmental conditions, elevated [CO_2] must be taken into consideration in realistic field settings. FACE experiments can play a key role in assessing breeding traits, but also evaluating nutrient management under the future climate conditions (including drought). Current FACE experiments mostly target [CO_2] expected for 2040–2050 (550 μmol mol^{-1} air), which is timely for selection efforts considering that the turnaround time from trait identification to a new variety can be 10–20 years (Chapman et al. 2012). Whilst changes in crop management practices, such as new fertiliser products, application rates and timing strategies, may require less lead time than the development of new crop varieties, they will be most effectively applied and tested in the right combination with the right crop varieties under the relevant environmental conditions. Dealing with below ground processes, especially those associated with roots and soils, poses particular difficulties, and adds complexity to the system. This complexity cannot be avoided, as shown by the importance of roots and soil processes in the interaction of high [CO_2] with plant nutrition. The use of 3-D functional simulation modelling (Dunbabin et al. 2013) offers considerable promise, especially when combined with new technology such as x-ray computer tomography that allows better imaging of root growth within soils. More flexible exposure systems that combine the advantages of a FACE system (free air without enclosure effects) with the use of large soil cores (such as the "SoilFACE" system as part of the AGFACE facilities; Butterly et al. 2012) or lysimeters and rhizotron access also offer opportunities to investigate relevant plant nutrition processes in the field.

In summary, it will be important to consider increasing [CO_2] in further improvements of crop nutrition and NutUE, either by genetic or managements adaptations. Facilities that enable crop growth under elevated [CO_2] will therefore play an increasing role in such efforts, alongside technologies that control other key factors such as water supply or temperature regimes.

Acknowledgements The Australian Grains Free Air CO_2 Enrichment (AGFACE) facility and related experiments are jointly run by The University of Melbourne and the Victorian State Department of Environment and Primary Industries (DEPI) with funding from the Grains Research and Development Corporation (GRDC), the Australian Government, and the Australian Research Council (ARC). Data in Fig. 9.2 were derived from collaboration of STP with Malcolm Hawkesford, Rothamsted Research, UK.

References

Adcock D, McNeill AM, McDonald GK, Armstrong RD (2007) Subsoil constraints to crop production on neutral and alkaline soils in south-eastern Australia: a review of current knowledge and management strategies. Aust J Exp Agr 47:1245–1261

Ainsworth EA, Long SP (2005) What have we learned from 15 years of free-air CO_2 enrichment (FACE)? A meta-analytic review of the responses of photosynthesis, canopy. New Phytol 165:351–371

Ainsworth EA, Rogers A (2007) The response of photosynthesis and stomatal conductance to rising CO_2: mechanisms and environmental interactions. Plant Cell Environ 30:258–270

Ainsworth EA, Beier C, Calfapietra C, Ceulemans R, Durand-Tardif M, Farquhar GD, Godbold DL, Hendrey GR, Hickler T, Kaduk J, Karnosky DF, Kimball BA, Koerner C, Koornneef M, Lafarge T, Leakey ADB, Lewin KF, Long SP, Manderscheid R, McNeil DL, Mies TA, Miglietta F, Morgan JA, Nagy J, Norby RJ, Norton RM, Percy KE, Rogers A, Soussana JF, Stitt M, Weigel HJ, White JW (2008) Next generation of elevated CO_2 experiments with crops: a critical investment for feeding the future world. Plant Cell Environ 31:1317–1324

Amthor JS (2001) Effects of atmospheric CO_2 concentration on wheat yield: review of results from experiments using various approaches to control CO_2 concentration. Field Crop Res 73:1–34

Armstrong RD, Helyar KR, Prangnell R (1993) Direct assessment of mineral phosphorus availability to tropical crops using ^{32}P labelled compounds. Plant Soil 150:278–287

Barber SA (1995) Soil nutrient bioavailability: a mechanistic approach. Wiley, New York

Bassirirad H, Tissue DT, Reynolds JF, Chapin FS (1996) Response of *Eriophorum vaginatum* to CO_2 enrichment at different soil temperatures: effects on growth, root respiration and PO_4^{3-} uptake kinetics. New Phytol 133:423–430

Benlloch-Gonzalez M, Berger J, Bramley H, Rebetzke G, Paltra JA (2014) The plasticity of the growth and proliferation of wheat root system under elevated CO_2. Plant Soil 374:963–976

Bloom AJ (2009) As carbon dioxide rises, food quality will decline without careful nitrogen management. Calif Agr 63:67–72

Bloom AJ, Burger M, Rubio-Asensio JS, Cousins AB (2010) Carbon dioxide enrichment inhibits nitrate assimilation in wheat and *Arabidopsis*. Science 328:899–903

Bunce JA (2012) Responses of cotton and wheat photosynthesis and growth to cyclic variation in carbon dioxide concentration. Photosynthetica 50:395–400

Butterly C, Armstrong R, Chen D, Mathers N, Tang C (2012) Effect of elevated CO_2 and N level on growth of wheat and field pea. In: Yunusa I (ed) Capturing opportunities and overcoming obstacles in Australian agronomy. Proceedings of the 16th Australian agronomy conference 2012, Armidale

Carberry PS, Probert ME, Dimes JP, Keating BA, McCown RL (2002) Role of modelling in improving nutrient efficiency in cropping systems. Plant Soil 245:193–203

Carlisle E, Myers S, Raboy V, Bloom A (2012) The effects of inorganic nitrogen form and CO_2 concentration on wheat yield and nutrient accumulation and distribution. Frontiers Plant Sci 3:195

Carter TR, Jones RN, Lu X, Bhadwal S, Conde C, Mearns LO, O'Neill BC, Rounsevell MDA, Zurek MB (2007) New assessment methods and the characterisation of future conditions.

In: Parry ML, Canziani OF, Palutikof JP, van der Linden PJ, Hanson CE (eds) Contribution of working group II to the fourth assessment report of the intergovernmental panel on climate change. Cambridge University Press, Cambridge, pp 133–171

Chapman SC, Chakraborty S, Dreccer MF, Howden SM (2012) Plant adaptation to climate change-opportunities and priorities in breeding. Crop Pasture Sci 63:251–268

Chen D, Suter HC, Islam A, Edis R, Freney JR, Walker CN (2008) Prospects of improving efficiency of fertiliser nitrogen in Australian agriculture: a review of enhanced efficiency fertilisers. Aust J Soil Res 46:289–301

Conroy J, Hocking P (1993) Nitrogen nutrition of C-3 plants at elevated atmospheric CO_2 concentrations. Physiol Plantarum 89:570–576

Cramer MD, Hawkins HJ, Verboom GA (2009) The importance of nutritional regulation of plant water flux. Oecologia 161:15–24

Donald CM (1964) Phosphorus in Australian agriculture. J Aust Inst Agric Sci 6:75–105

Drake BG, GonzalezMeler MA, Long SP (1997) More efficient plants: a consequence of rising atmospheric CO_2? Annu Rev Plant Phys 48:609–639

Dunbabin VM, Postma JA, Schnepf A, Pagès L, Javaux M, Wu L, Leitner D, Chen YL, Rengel Z, Diggle AJ (2013) Modelling root–soil interactions using three–dimensional models of root growth, architecture and function. Plant Soil 372:93–124

Duval BD, Blankinship JC, Dijkstra P, Hungate BA (2012) CO_2 effects on plant nutrient concentration depend on plant functional group and available nitrogen: a meta-analysis. Plant Ecol 213:505–521

Erbs M, Manderscheid R, Jansen G, Seddig S, Pacholski A, Weigel HJ (2010) Effects of free-air CO_2 enrichment and nitrogen supply on grain quality parameters and elemental composition of wheat and barley grown in a crop rotation. Agr Ecosyst Environ 136:59–68

Fangmeier A, De Temmerman L, Mortensen L, Kemp K, Burke J, Mitchell R, van Oijen M, Weigel HJ (1999) Effects on nutrients and on grain quality in spring wheat crops grown under elevated CO_2 concentrations and stress conditions in the European, multiple-site experiment 'ESPACE-wheat'. Eur J Agron 10:215–229

Fangmeier A, Chrost B, Högy P, Krupinska K (2000) CO_2 enrichment enhances flag leaf senescence in barley due to greater grain nitrogen sink capacity. Environ Exp Bot 44:151–164

Fernando N, Panozzo J, Tausz M, Norton RM, Fitzgerald GJ, Myers S, Walker C, Stangoulis J, Seneweera S (2012) Wheat grain quality under increasing atmospheric CO_2 concentrations in a semi-arid cropping system. J Cereal Sci 56:684–690

Fitter AH (1996) Characteristics and functions of root systems. In: Waisel Y, Eshel A, Kafka U (eds) Plant roots: the hidden half. Marcel Dekker, New York, pp 1–20

Foulkes MJ, Reynolds MP, Sylvester-Bradley R (2009) Genetic improvement of grain crops: yield potential. In: Sadras VO, Calderini DF (eds) Crop physiology applications for genentic improvement and agronomy. Academic, Amsterdam, pp 355–386

Franzaring J, Gensheimer G, Weller S, Schmid I, Fangmeier A (2012) Allocation and remobilisation of nitrogen in spring oilseed rape (*Brassica napus* L. cv. Mozart) as affected by N supply and elevated CO_2. Environ Exp Bot 83:12–22

Ghannoum O, Searson MJ, Conroy JP (2007) Nutrient and water demands under global climate change. In: Newton PCD, Carran RA, Edwards GR, Niklaus PA (eds) Agroecosystems in a changing climate. Taylor & Francis, Boca Raton, pp 53–83

Good AG, Shrawat AK, Muench DG (2004) Can less yield more? Is reducing nutrient input into the environment compatible with maintaining crop production? Trends Plant Sci 9:597–605

Gregersen PL (2011) Senescence and nutrient remobilisation in crop plants. In: Hawkesford MJ, Barraclough P (eds) The molecular and physiological basis of nutrient use efficiency in crops. Wiley, Hoboken, pp 83–102

Gregory PJ, Palta JA, Batts GR (1996) Root systems and root:mass ratio – carbon allocation under current and projected atmospheric conditions in arable crops. Plant Soil 187:221–228

Hatfield JL, Boote KJ, Kimball BA, Ziska LH, Izaurralde RC, Ort D, Thomson AM, Wolfe D (2011) Climate impacts on agriculture: implications for crop production. Agron J 103:351–370

Hawkesford MJ (2011) An overview of nutrient use efficiency and strategies for crop improvement. In: Hawkesford MJ, Barraclough P (eds) The molecular and physiological basis of nutrient use efficiency in crops. Wiley, Hoboken, pp 5–19
Hawkesford MJ, De Kok LJ (2006) Managing sulphur metabolism in plants. Plant Cell Environ 29:382–395
Hawkesford MJ, De Kok LJ (2007) Sulfur in plants: an ecological perspective. Springer, Dordrecht
Hirel B, Le Gouis J, Ney B, Gallais A (2007) The challenge of improving nitrogen use efficiency in crop plants: towards a more central role for genetic variability and quantitative genetics within integrated approaches. J Exp Bot 58:2369–2387
Ho MD, Rosas JC, Brown KM, Lynch JP (2005) Root architectural tradeoffs for water and phosphorus acquisition. Funct Plant Biol 32:737–748
Hocking P, Barret D (2003) Elevated atmospheric CO_2 does not increase the capacity of plants to access fixed soil phosphorus from an oxisol. In: Second international symposium on P dynamics in the soil-plant continuum, Uniprint, Perth, pp 100–101
Hocking PJ, Meyer CP (1991) Carbon-dioxide enrichment decreases critical nitrate and nitrogen concentrations in wheat. J Plant Nutr 14:571–584
Högy P, Fangmeier A (2008) Effects of elevated atmospheric CO_2 on grain quality of wheat. J Cereal Sci 48:580–591
Högy P, Wieser H, Kohler P, Schwadorf K, Breuer J, Franzaring J, Muntifering R, Fangmeier A (2009) Effects of elevated CO_2 on grain yield and quality of wheat: results from a 3-year free-air CO_2 enrichment experiment. Plant Biol 11:60–69
Högy P, Brunnbauer M, Koehler P, Schwadorf K, Breuer J, Franzaring J, Zhunusbayeva D, Fangmeier A (2013) Grain quality characteristics of spring wheat (*Triticum aestivum*) as affected by free-air CO_2 enrichment. Environ Exp Bot 88:11–18
Hungate BA, Dukes JS, Shaw MR, Luo Y, Field CB (2003) Nitrogen and climate change. Science 302:1512–1513
Jin J, Tang C, Armstrong R, Butterly C, Sale P (2013) Elevated CO_2 temporally enhances phosphorus immobilization in the rhizosphere of wheat and chickpea. Plant Soil 368:315–328
Jin J, Tang C, Hogarth TW, Armstrong R, Sale P (2014) Nitrogen form but not elevated CO_2 alters plant phosphorus acquisition from sparingly soluble phosphorus sources. Plant Soil 374:109–119
Kirschbaum MUF (2011) Does enhanced photosynthesis enhance growth? Lessons learned from CO_2 enrichment studies. Plant Physiol 155:117–124
Kitchen LD (2014) Turning knowledge into action. In: Dudonis C (ed) Global climate change – turning knowledge into action. Pearson Education Inc, NJ, pp 341–342
Kong LG, Wang FH, Lopez-Bellido L, Garcia-Mina JM, Si JS (2013) Agronomic improvements through the genetic and physiological regulation of nitrogen uptake in wheat (*Triticum aestivum* L.). Plant Biotech Rep 7:129–139
Lam SK, Chen DL, Norton R, Armstrong R, Mosier AR (2012a) Nitrogen dynamics in grain crop and legume pasture systems under elevated atmospheric carbon dioxide concentration: a meta-analysis. Glob Change Biol 18:2853–2859
Lam SK, Chen D, Norton R, Armstrong R (2012b) Nitrogen demand and the recovery of ^{15}N-labelled fertilizer in wheat grown under elevated carbon dioxide in southern Australia. Nutr Cycl Agroecosys 92:133–144
Lam SK, Chen D, Norton R, Armstrong R (2012c) Does phosphorus stimulate the effect of elevated [CO_2] on growth and symbiotic nitrogen fixation of grain and pasture legumes? Crop Past Sci 63:53–62
Lam SK, Chen D, Norton R, Armstrong R, Mosier AR (2013) Influence of elevated atmospheric carbon dioxide and supplementary irrigation on greenhouse gas emissions from a spring wheat crop in southern Australia. J Agric Sci 151:201–208
Leakey ADB, Ainsworth EA, Bernacchi CJ, Rogers A, Long SP, Ort DR (2009) Elevated CO_2 effects on plant carbon, nitrogen, and water relations: six important lessons from FACE. J Exp Bot 60:2859–2876

Lilley JM, Kirkegaard JA (2011) Benefits of increased soil exploration by wheat roots. Field Crop Res 122:118–130

Loladze I (2002) Rising atmospheric CO_2 and human nutrition: toward globally imbalanced plant stoichiometry? Trends Ecol Evol 17:457–461

Luo Y, Su B, Currie WS, Dukes JS, Finzi A, Hartwig U, Hungate B, McMurtrie RE, Oren R, Parton WJ, Pataki DE, Shaw RM, Zak DR, Field CB (2004) Progressive nitrogen limitation of ecosystem response to rising atmospheric carbon dioxide. Bioscience 54:731–739

Lynch JP, St. Clair SB (2004) Mineral stress: the missing link in understanding how global climate change will affect plants in real world soils. Field Crop Res 90:101–115

Madhu M, Hatfield JL (2013) Dynamics of plant root growth under increased atmospheric carbon dioxide. Agron J 105:657–669

Manderscheid R, Bender J, Jäger HJ, Weigel HJ (1995) Effects of season long CO_2 enrichment on cereals. 2. Nutrient concentrations and grain quality. Agr Ecosyst Environ 54:175–185

McBeath TM, McLaughlin MJ, Kirby JK, Armstrong RD (2012) The effect of soil water status on fertiliser, topsoil and subsoil phosphorus utilisation by wheat. Plant Soil 358:337–348

McDonald GK (1989) The contribution of nitrogen fertiliser to the nitrogennutrition of rainfed wheat crops in Australia: a review. Aust J Exp Agr 29:455–481

McGrath JM, Lobell DB (2013) Reduction of transpiration and altered nutrient allocation contribute to nutrient decline of crops grown in elevated CO2 concentrations. Plant Cell Environ 36:697–705

McLaughlin MJ, McBeath TM, Smernik R, Stacey SP, Ajiboye B, Guppy C (2011) The chemical nature of P accumulation in agricultural soils – implications for fertiliser management and design: an Australian perspective. Plant Soil 349:69–87

Messinger SM, Buckley TN, Mott KA (2006) Evidence for involvement of photosynthetic processes in the stomatal response to CO_2. Plant Physiol 140:771–778

Miller AJ, Chapman N (2011) Transporters involved in nitrogen uptake and movement. In: Hawkesford MJ, Barraclough P (eds) The molecular and physiological basis of nutrient use efficiency in crops. Wiley-Blackwell, Chichester, pp 193–210

Moeller C, Asseng S, Berger J, Milroy SP (2009) Plant soil available water at sowing in Mediterranean environments – is it a useful criterion to aid nitrogen fertiliser and sowing decisions? Field Crop Res 114:127–136

Moore BD, Cheng SH, Sims D, Seemann JR (1999) The biochemical and molecular basis for photosynthetic acclimation to elevated atmospheric CO_2. Plant Cell Environ 22:567–582

Munns R, Cramer GR, Ball MC (1999) Interactions between rising CO_2, soil salinity and plant growth. In: Luo Y, Mooney HA (eds) Carbon dioxide and environmental stress. Academic, San Diego, pp 139–167

Osborne LD, Rengel Z (2002) Genotypic differences in wheat for uptake and utilisation of P from iron phosphate. Aust J Agr Res 53:837–844

Palta JA, Fillery IR (1993) Nitrogen accumulation and remobilization in wheat of ^{15}N-urea applied to a duplex soil at seeding. Aust J Exp Agr 33:233–238

Palta JA, Kobata T, Turner NC, Fillery IR (1994) Remobilization of carbon and nitrogen in wheat as influenced by postanthesis water deficits. Crop Sci 34:118–124

Poorter H, Perez-Soba M (2001) The growth response of plants to elevated CO_2 under non-optimal environmental conditions. Oecologia 129:1–20

Poorter H, VanBerkel Y, Baxter R, Den Hertog J, Dijkstra P, Gifford RM, Griffin KL, Roumet C, Roy J, Wong SC (1997) The effect of elevated CO_2 on the chemical composition and construction costs of leaves of 27 C-3 species. Plant Cell Environ 20:472–482

Prior SA, Torbert HA, Runion GB, Rogers HH, Kimball BA (2008) Free-air CO_2 enrichment of sorghum: soil carbon and nitrogen dynamics. J Environ Qual 37:753–758

Pritchard SG, Rogers HH (2000) Spatial and temporal deployment of crop roots in CO_2-enriched environments. New Phytol 147:55–71

Pritchard SG, Prior SA, Rogers HH, Davis MA, Runion GB, Popham TW (2006) Effects of elevated atmospheric CO_2 on root dynamics and productivity of sorghum grown under conventional and conservation agricultural management practices. Agr Ecosyst Environ 113:175–183

Rich SM, Watt M (2013) Soil conditions and cereal root system architecture: review and considerations for linking Darwin and Weaver. J Exp Bot 64:1193–1208
Rogers A, Ainsworth EA, Leakey ADB (2009) Will elevated carbon dioxide concentration amplify the benefits of nitrogen fixation in legumes? Plant Physiol 151:1009–1016
Sadras VO, Calderini DF (2009) Crop physiology: applications for genetic improvement and agronomy. Academic Press, Elsevier, Burlington
Sadras VO, Lawson C (2013) Nitrogen and water-use efficiency of Australian wheat varieties released between 1958 and 2007. Eur J Agron 46:34–41
Sinclair TR, Pinter PJ, Kimball BA, Adamsen FJ, LaMorte RL, Wall GW, Hunsaker DJ, Adam N, Brooks TJ, García RL, Thompson T, Leavitt S, Matthias A (2000) Leaf nitrogen concentration of wheat subjected to elevated CO_2 and either water or N deficits. Agr Ecosyst Environ 79:53–60
Stitt M, Krapp A (1999) The interaction between elevated carbon dioxide and nitrogen nutrition: the physiological and molecular background. Plant Cell Environ 22:583–621
Svečnjak Z, Rengel Z (2006) Canola cultivars differ in nitrogen utilisation efficiency at vegetative stage. Field Crop Res 97:221–226
Taub DR, Wang X (2008) Why are nitrogen concentrations in plant tissues lower under elevated CO_2? A critical examination of the hypotheses. J Integr Biol 50:1365–1374
Taub DR, Miller B, Allen H (2008) Effects of elevated CO_2 on the protein concentration of food crops: a meta-analysis. Global Chang Biol 14:565–575
Tausz M, Tausz-Posch S, Norton RM, Fitzgerald GJ, Nicolas ME, Seneweera S (2013) Understanding crop physiology to select breeding targets and improve crop management under increasing atmospheric CO_2 concentrations. Env Exp Bot 88:71–80
Tausz-Posch S, Seneweera S, Norton RM, Fitzgerald GJ, Tausz M (2012) Can a wheat cultivar with high transpiration efficiency maintain its yield advantage over a near-isogenic cultivar under elevated CO_2? Field Crop Res 133:160–166
Van Vuuren MMI, Robinson D, Fitter AH, Chasalow SD, Williamson L, Raven JA (1997) Effects of elevated atmospheric CO_2 and soil water availability on root biomass, root length, and N, P and K uptake by wheat. New Phytol 135:455–465
Wang X, Tang C, Guppy CN, Sale PWG (2010) Cotton, wheat and white lupin differ in phosphorus acquisition from sparingly soluble sources. Env Exp Bot 69:267–272
Wang L, Feng Z, Schjoerring JK (2013) Effects of elevated atmospheric CO_2 on physiology and yield of wheat (*Triticum aestivum* L.): a meta-analytic test of current hypotheses. Agr Ecosyst Environ 178:57–63
White PJ, Hammond JP (2008) The ecophysiology of plant-phosphorus interactions. Springer, Dordrecht
Wiesler F, Behrens T, Horst WJ (2001) The role of nitrogen-efficient cultivars in sustainable agriculture. Sci World 1(S2):61–69
Yang JC, Zhang JH (2006) Grain filling of cereals under soil drying. New Phytol 169:223–236
Yang JC, Zhang JH, Huang ZL, Zhu QS, Wang L (2000) Remobilization of carbon reserves is improved by controlled soil-drying during grain filling of wheat. Crop Sci 40:1645–1655
Yong ZH, Chen GY, Zhang DY, Chen Y, Chen J, Zhu JG, Xu DQ (2007) Is photosynthetic acclimation to free-air CO_2 enrichment (FACE) related to a strong competition for the assimilatory power between carbon assimilation and nitrogen assimilation in rice leaf? Photosynthetica 45:85–91
Ziska LH (2008) Rising atmospheric carbon dioxide and plant biology: the overlooked paradigm. DNA Cell Biol 27:165–172
Ziska LH, Bunce JA, Shimono H, Gealy DR, Baker JT, Newton PCD, Reynolds MP, Jagadish KSV, Zhu CW, Howden M, Wilson LT (2012) Food security and climate change: on the potential to adapt global crop production by active selection to rising atmospheric carbon dioxide. Proc R Soc B Biol Sci 279:4097–4105

Chapter 10
Monitoring Plant Nutritional Status

Moez Maghrebi, Fabio Francesco Nocito, and Gian Attilio Sacchi

Abstract Methods and techniques effective in achieving yield objectives, optimizing the use of resources and preventing environmental contamination are defined as agronomic Best Management Practices (BMPs). Considering fertilisations BMPs pursue the aims to match mineral nutrient supply with crop requirements, minimizing their losses from the field. If spatial and temporal information about crop needs were available, precision fertilisation approaches could be planned in order to increase fertiliser use efficiency and to improve some economic and environmental aspects related to the crop systems. The state of the art and the research perspectives on the fine tuning of optical devices allowing proximal or remote sensing at sub-field scale of crops traits related to plant nutritional status, as well as on the exploitation of the gene fusion concept in developing transgenic bioindicators monitoring the nutritional status of the plants are here reviewed and discussed. Concerning the latter aspect particular attention is paid to the development of synthetic promoters conferring to the biondicator nutrient specificity and also able to target the expression of the associated reporter genes in organs in which their signals should be early and easily detectable.

Keywords Agronomy • Nutrient monitoring • Fertilizers • Precision agriculture • Critical concentrations • Bioindicators

Introduction

Mineral nutrient availability is one of the most important factors determining yield in agriculture. Conventional farming requires a continuous and large supply of fertilisers to the soil in order to replace the nutrients removed with plant harvest. In the predicted scenario of a rising demand for food and energy for the expected world population of about nine billion in 2050 (Godfray et al. 2010) a dramatic increase in the use of fertilisers in the crop systems is foreseen. Since fertiliser production and distribution have a high demand for energy, in the last decade the

M. Maghrebi • F.F. Nocito • G.A. Sacchi (✉)
Department of Agricultural and Environmental Sciences – Production, Landscape, Agroenergy, Università degli Studi di Milano, via Celoria 2, 20133 Milano, Italy
e-mail: gianattilio.sacchi@unimi.it

M.J. Hawkesford et al. (eds.), *Nutrient Use Efficiency in Plants*,
Plant Ecophysiology 10, DOI 10.1007/978-3-319-10635-9_10

price of fertilisers, although fluctuating, have been burgeoning and it is highly improbable that this trend will change in the next few years (http://faostat3.fao.org/home/index.html).

Only a fraction of the nutrient provided to the soil as fertiliser is taken up from the crops. This fraction, expressed as a percentage, is defined as the crop *Apparent Recovery* (*AR*; Craswell and Godwin 1984):

$$AR = 100(\mathrm{N}_F - \mathrm{N}_{nF})/F$$

where N_F and N_{nF} are the total amount of the nutrient absorbed by the crop, if fertilised or not, respectively, and F is the amount of the nutrient added to the soil with the fertiliser. For the main crops quite low average values of *AR* are reported: about 35 % for N (Raun and Johnson 1999), 10–30 % for P (Malhi et al. 2002) and seldom higher than 50 % for K (Rengel and Damon 2008). The low capacity of crops in removing the nutritional elements added to the soil has significant environmental implications and reflects limits in the management of the fertilisation practices, the existence of constraints due to both chemical-physical and microbiological soil properties and plant intrinsic biological limits. *AR* does not consider yield traits. On the contrary, the *Agronomic Efficiency* (*AE*) of the fertiliser, defined by the ratio $\mathrm{d}Y/\mathrm{d}F$, where $\mathrm{d}Y$ is the infinitesimal yield (Y) and $\mathrm{d}F$ is the infinitesimal increase in the amount of the nutrient in the soil (F) due to the fertiliser application, considers such traits. *AE* can in turn be expressed considering two components: the crop *Removal Efficiency* (*RE*), defined as the ratio $\mathrm{d}N_F/dF$, where $\mathrm{d}N_F$ is the infinitesimal incremental amount of the nutrient taken up by plant after the fertilisation, and the *Physiologic Efficiency* (*PE*), defined as the ratio $\mathrm{d}Y/\mathrm{d}N_F$:

$$AE = RE \times PE = \mathrm{d}N_F/\mathrm{d}F \times \mathrm{d}Y/\mathrm{d}N_F = \mathrm{d}Y/\mathrm{d}F$$

In the field, both *RE* and *PE,* and then *AE,* depend on the interaction between genetic and environmental factors. In other chapters of this book the molecular and genetic aspects determining *AE* for some essential elements are extensively reviewed. Here, in a perspective of precision farming (Pierce and Nowak 1999), some strategies to improve *AE* by optimising dN_F throughout the use of plant-based sensor systems are presented and discussed.

Fertiliser Best Management Practices

Definitions

In order to limit the intrinsic risks of diffuse pollution due to intensive agriculture, both local and supranational authorities are committed to the fine tuning of methods or techniques found to be the most effective and practical means in achieving yield objective optimisation and preventing contaminations of soils, water resources and air. As a whole, these recommended measures are defined as agronomic Best

Management Practices (BMPs) and include: the choice of variety, planting date, row spacing, seeding rates, integrated pest management, weed control, disease control, and nutrient management. Focusing on the input into the field of inorganic nutrients, BMPs are considered the management practices that foster the effective and responsible use of fertiliser matching nutrients supply with crop requirements and minimise their losses from the fields.

Toward a Fertiliser Precision Management

Fertiliser BMPs include the identification of the: (a) the *right product* by matching the fertiliser characteristics to the crop site specific needs and soil properties; (b) the *right time* by synchronising the presence of the nutrient with the moment of crop maximum demand and uptake capacity; (c) the *best rate* by matching the amount of fertiliser input to crop needs in order to avoid over-input leading to nutrient leaching and other losses to the environment, as well as starvation conditions; (d) the *right place* by making sure the presence of the nutrients where plants can efficiently take up them.

Although in a field a relatively high spatial variability in the crop requirements of a specific nutrient could exist, fertilisers are uniformly applied, to avoid yield gaps and considering the less fertile portion of soil, where the crops have the maximum demand. The amounts of nutrient provided in excess with the fertiliser can be absorbed by the crops without resulting in any benefit in term of yield or leached towards the underground water becoming a concern for the quality of the environment.

If spatial and temporal information about crop needs of nutrients were available site- and time-specific inputs of the nutrient could be planned, resulting in a precision fertilisation approach, and in a win-win option to increase fertiliser nutrient efficiency and to improve the economic and environmental sustainability of the crop systems. In other words, the optimisation of dN_F term, in the equation defining *AE*, requires detailed information for decision support systems, allowing farmers to adopt the minimal nutrient input for maximal return, according to a Fertiliser Best Management Practices (FBMPs) approach.

Evaluating Plant Nutritional Status

Soil and Leaf Analyses

Within a field the spatial variability of the soil chemical-physical and biological characteristics, including the amount of bioavailable forms of the mineral nutrients essential for crops, can be pronounced. Mapping this variability and plant nutritional status at high temporal and spatial resolution represents the first step towards the setup of site-specific FBMPs. In the last years, several sensor-based techniques to assess parameters indicative of the nutritional status of soil-plant systems have been proposed to replace, or support traditionally used physical measures and chemical analysis. Results on soil properties obtained from electromagnetic induction sensors along with those derived from the use of ground conductivity meters and radiometers analysing canopy reflectance, appear to provide data which can be used to target N-fertilisation to specific field conditions (Adamchuck et al. 2011). However, soil and plant chemical analyses are still widely used in developing methods for the evaluation of the nutritional status of the crops. In particular, the evaluation of elemental concentrations in plant tissues can be helpful in diagnosing nutrient deficiency. This strategy is currently used to assess nutrient availability and guide fertility programmes for many fruit tree crops. Nevertheless, the usefulness of this approach in order to develop FBMPs for herbaceous crops is rather debatable, since the concentration of a specific element may change among the leaves of a single plant, and may also change over time within a single leaf (Barker and Pilbeam 2007). Thus, plant sampling, in term of timing and tissues to choose, is the most critical step (Kalra 1998). Moreover, elemental analysis detects only severe and long-term deficiency since a plant's initial response to nutrient limitation is to activate mechanisms aimed at maintaining the ionic homeostasis of their cells (Schatchtman and Shin 2007; Gojon et al. 2009). Finally, conferring a diagnostic value to the concentration of a single element could be misleading since complex cross-talk connections between the regulatory mechanisms controlling the ionic homeostasis in plants exist (Rouached et al. 2010).

Nutrient-Critical Concentration and Dilution Curve

The *critical concentration of a nutrient* (n_c) stands for the concentration of the nutrient in the shoots above which, in the absence of other growth limiting factors, the plant is sufficiently supplied with the nutrient to achieve its maximum potential yield. In other words, when the n_c is achieved and maintained, further supplies of the nutrient will not influence the growth of the plants and, in the absence of any sort of demand-driven negative-regulation of its uptake, it could be uselessly accumulated in the plant tissues. For some nutrients it is possible to define the so-called *toxicological value*, which indicates the concentration above which further nutrient accumulation induces damage on cell metabolism and structure. All the concentrations of the nutrient between its critical and toxicological values define the so-called *luxury range* (Fig. 10.1), which depends on the chemical properties of the element and on its biochemical roles. The fine-tuning of the application of fertiliser to maintain the concentration of the mineral nutrients in the plant tissues as close as possible to their critical values represents a FBMP approach.

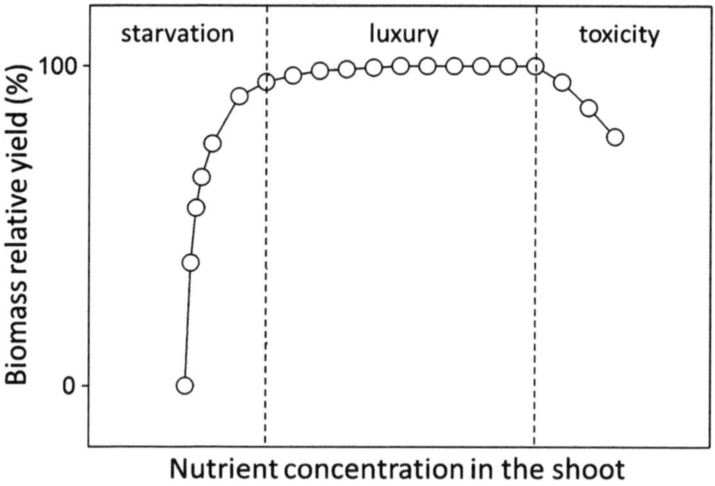

Fig. 10.1 Relationship between the concentration of a nutrient in the shoot and the biomass relative yield

The critical concentration of a nutrient is not a constant value since it depends on genotype, environmental conditions, and the developmental stage of the plants (Lemaire and Gastal 2009; Greenwood et al. 1986). For instance, in the above-ground tissue of cereals the value of n_c (as % on the DW basis) changes during the plant developmental stages according to the dilution function:

$$n_c = aW^{-b}$$

where W is the maximum above-ground biomass in a specific stage of the plant cycle, a represents the concentration of n_c in the shoot when the crop mass is 1 Mg DM ha^{-1}, and b represents the dilution coefficient.

The dilution curve of a specific cultivar is obtained by plotting the n_c values, at a given developmental stage and field situation determined by a set a fertilisation experiments versus the accumulated plant shoot biomass (Fig. 10.2). Once the dilution curve for a specific nutrient is known, it is possible to evaluate the nutritional status of the crop by evaluating its nutritional index (*NI*) defined as the ratio between the actual nutrient concentration in the shoots and the corresponding value of n_c. Indeed, if *NI* is lower than 1 the crop is in a suboptimal nutritional status for a given nutrient and needs to be fertilised; if the *NI* is higher than 1 the crop is in the luxury range.

The actual feasibility of such an approach for a sustainable management of N fertilisation is conditioned by the limited availability of easy and low cost methods to rapidly estimate the value of W in the field and, mainly by the actual concentrations of the nutrient in the shoots over a representational cropping area. The value of W can be extrapolated by the crop leaf area index (LAI) currently evaluated through remote sensing approaches (Zheng and Moskal 2009). The evaluation of the actual

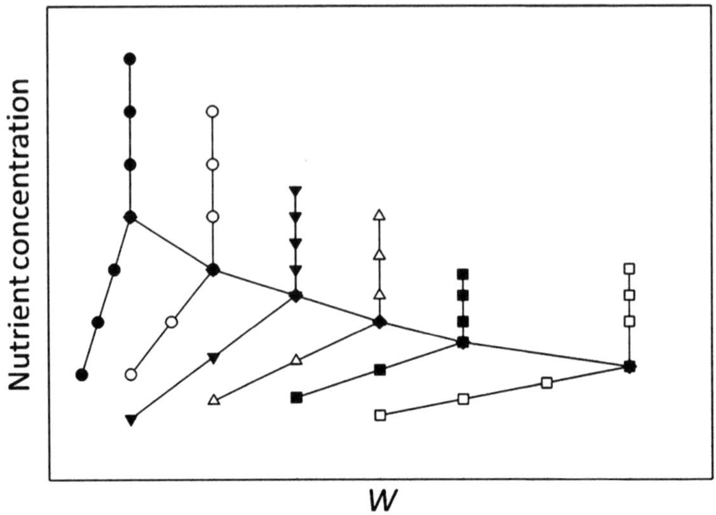

Fig. 10.2 Dilution curve for a generic essential nutrient as defined by six n_c values determined for six plant developmental stages

concentration of the nutrient, without adopting the traditional chemical analyses, is the most difficult challenge to overcome.

Since nitrogen is the nutritional element that most often affects crop production and the current world use of N fertilisers is approximately 90 million metric tons (with an estimated cost of about $50 billion), it is reasonable that several research efforts have been focused on the fine tuning of non-invasive methods for the determination of N levels in shoots throughout the entire growth cycle of crops as a guide to N-FBMPs (for an exhaustive review see Samborski et al. 2009). Assuming nitrogen as an example, in the next paragraph we briefly summarize the advances on the non-destructive approaches developed or under investigation for monitoring the nutritional status of a crop.

Non-destructive Monitoring of Crop Nutritional Status: The Example of Nitrogen

Nitrogen availability affects chlorophyll content in leaves (Schlemmer et al. 2005) and as a consequence the level of this pigment is considered a good indicator of the nitrogen nutritional status of a crop (Samborski et al. 2009 and references therein). Instruments analysing the spectral properties of leaf tissue to estimate their chlorophyll content (optical chlorophyll meters) have been developed to evaluate the need for agricultural N applications. Due to the pigment's light absorption properties (in the visible wavelength range), the higher the chlorophyll content, the higher the reflectance of the leaf (in the 525–680 nm range) and consequently, the higher the amount of red light absorbed. Combining light absorbance measures at 660 nm and near-infrared (NIR) light transmittance at 940 nm, which in turn depend on leaf moisture content and thickness, a good estimation of chlorophyll per unit area has been obtained in the major crop leaves. It has been proved that the chlorophyll meter can detect the early signs of N stress not yet detectable by visual analysis using a leaf colour chart (Debaeke et al. 2006).

Recently, it has been proposed a hand-held instrument (Dualex®), exploiting chlorophyll as an internal sensor of photons, enables the user to contemporaneously assess the level of both photosynthetic pigments in the mesophyll and flavonoids in the epidermis of the leaf. Briefly, comparing the amount of chlorophyll fluorescence emitted under UV excitation (λ_{380}) with that emitted under visible light (λ_{660}) whether absorbed or not, the instrument is able to evaluate the level of flavonoids absorbing in the UV range. Contemporaneously, by comparison of the light transmittance at λ_{720}, in the range of chlorophyll absorption, and at λ_{840}, in the range influenced by leaf structural properties but not by chlorophyll, a reliable evaluation of the levels of the photosynthetic active pigment is obtained. Dualex® thus allows measurement of a Nitrogen Balance Index (NBI®), which indicates the ratios of both chlorophyll and flavonoids units and as a result is related to leaf N content (Cartelat et al. 2005) since leaf flavonoids can be considered an indicator of N availability. Indeed, in N-starved plants the concentration of carbon-based secondary metabolites increase (Hamilton et al. 2001) and in particular, due to the enhanced expression of specific transcription factors involved in controlling their biosynthetic pathway, those of anthocyanin and flavonols (Lea et al. 2007). Several experimental evaluations suggest that in the case of wheat and corn Dualex® seems to furnish more reliable information about the N status of the plants with respect to other hand-held optical systems (Tremblay et al. 2012).

Since leaf N status influences the quantum yield of PSII electron transport and then the chlorophyll fluorescence parameters (Lu and Zhang 2000), canopy fluorescence quenching analyses could be considered suitable for sensing crop N status (Tremblay et al. 2012). In particular, the recent introduction of a hand-held fluorimeter (Multiplex®) equipped with LEDs generating four wavelengths (λ_{375}, λ_{450}, λ_{530}, λ_{630}) and detectors monitoring fluorescence at three wavelengths (λ_{447} or λ_{590} if the excitation at λ_{450} is used or not, respectively, λ_{665} and λ_{735}) seems to be quite promising for the in-season assessment of crop N status (Tremblay et al. 2012 and references therein). Combining different excitation and emission bands the instrument provides independent parameters related to chlorophyll, flavonoids and N content of the plants (Tremblay et al. 2012).

The devices described above determine optical parameters for individual leaves or, in the case of Multiplex®, at a typical distance of a few centimeters thus monitoring circular canopy surfaces of not more than 10 cm in diameter. Consequently, they are not particularly suitable in evaluating the N status of a crop at field scale. Sensors, analysing canopy reflectance properties and thus its N status and needs are also available (Erdle et al. 2011). They are classified as passive (Yara N-Sensor®/Field Scan and FiledSpec® Portable Spectroradiometer) or active (GreenSeeker® and Crop Circle™) non-contact sensors depending on the sunlight reflected by the canopy or on their own specific light sources in the visible (650 or 590 nm) and NIR (770 or 880 nm) range, respectively. Spectral data collected by these devices allow the calculation of the so-called normalised differences vegetation index (NDVI) according to the formula:

$$NDVI = (NIR - Vis)/(NIR + Vis)$$

where *NIR* and *Vis* stand for the spectral reflectance measurements acquired in the *NIR* or visible (red) regions, respectively. The *NDVI* value is about 0.5 when the vegetation chlorophyll content and thus, in the absence of any other stress factors, plant N status is optimal; conversely in sub-optimal conditions the value of *NDVI* is much lower.

Several examples of the use of these portable proximal sensors (which can also be mounted on tractors) in the fine-tuning of variable-rate technology for site-specific N fertilisation exist (Solari et al. 2008; Diacono et al. 2013). The possibilities to easily and efficiently translate the information on N crop status, obtained by the hand-held or proxy sensor approaches described as above, in site- and time-specific recommendations for FBMPs can be invalidated by a plethora of biotic and/or biotic stressors, including a non-optimal availability of nutrients other than N, which influence the chlorophyll content of the leaves. Therefore, the parameters and the vegetation index obtained are usually validated by setting up standardisation procedures providing for plots of the same cultivar in the same environment at different N availability. In this way genetic, environmental and agronomical factors can be eliminated as potential sources of error and making the data obtained by the sensor-based approaches more reliable (Samborski et al. 2009; Diacono et al. 2013).

Hyperspectral radiometers providing contemporaneous reflectance measurements over a relatively narrow wavebands (<10 nm), should make it possible to identify specific regions of the spectrum which could be used to develop new indices, highly sensitive to plant N status and unaffected by other exogenous factors (Hansen and Schjoerring 2003). Indeed, an increasing number of studies suggest that field as well as airborne or spaceborne hyperspectral canopy radiometric data can be useful for estimating plant nitrogen concentration in cultivated or natural environments (Ollinger et al. 2008, Stroppiana et al. 2009), although recently some criticisms about the remote sensing of leaf tissue constituents by hyperspectral data have been raised (Knyazikhin et al. 2013).

Leaf chlorophyll concentration is also an indirect diagnostic symptom for N status of the crop. However, it is important to take into account that reduced chlorophyll biosynthesis is a relatively late response to N starvation which only becomes evident after the plant has initiated other molecular and physiological responses for maintaining N homeostasis (Schatchtman and Shin 2007; Gojon et al. 2009).

Unfortunately, non-destructive reliable monitoring approaches comparable with those above described for N have not been developed for the other mineral nutrients whose availability affects crop yield (in particular P, K and S). Thus, for these nutrients the chance to adopt FBMPs is limited to the classic chemical evaluation of plant tissues and soils.

Plant Bioindicator for Nutritional Status

Bioindication and Biomonitoring

The development of quick and inexpensive methods to determine changes in nutrient bioavailability is required in order to monitor soil nutrient dynamics for better fertiliser management for a variety of crops in different environmental conditions. Developing bioassays based on the use of specific plant sentinels or bioindicators, may represent a reliable and efficient strategy to obtain quick, accurate and low-cost information about nutrient availability changes in a given crop system. Thus, the use of these modern biotechnologies could allow the non-destructive analysis of plants under field conditions. Development of these kinds of tools represents a new and challenging area of research.

Plants respond to nutrient supply or shortage through a complex of physiological, morphological, and developmental responses, which are under the control of several gene pathways. Microarray technology is a convenient tool for rapid analysis of plant gene expression patterns under a variety of environmental and nutritional conditions. Genome-wide microarray analyses showed extensive changes in the expression of several genes involved in primary and secondary metabolism, nutrient transport, protein synthesis, regulation of gene expression and cellular growth processes (Maruyama-Nakashita et al. 2003; Wang et al. 2003; Bi et al. 2007; Li et al. 2010; Kant et al. 2010; Ma et al. 2012). Such studies not only improved our general understanding on plant responses to nutrient availability but also provided a reliable data from which to develop new molecular strategies for real-time monitoring of plant nutritional status.

Recently, Yang et al. (2011) used multiple whole genome microarray experiments to identify gene expression biomarkers capable of assessing plant responses under limiting and sufficient nitrogen conditions. Using logistic regression statistical approaches, they identified a common set of genes in maize whose expression profiles quantitatively assessed the extent of plant stress under different nitrogen conditions. Interestingly, such a biomarker gene set is independent of maize genotype, tissue type, developmental stage, and environment (including plants grown under controlled conditions and in the field), and thus has the potential to be used as an agronomic tool for real-time monitoring and to optimise nitrogen fertiliser usage.

The Gene Fusion Concept Enables to Define a New Class of Transgenic Bioindicators

The existence of gene pools, which specifically respond to the nutritional status of the plant, has introduced a new class of bioindicators, based on the concept of gene fusion (Fig. 10.3). A generic nutrient-responsive gene is formally considered as

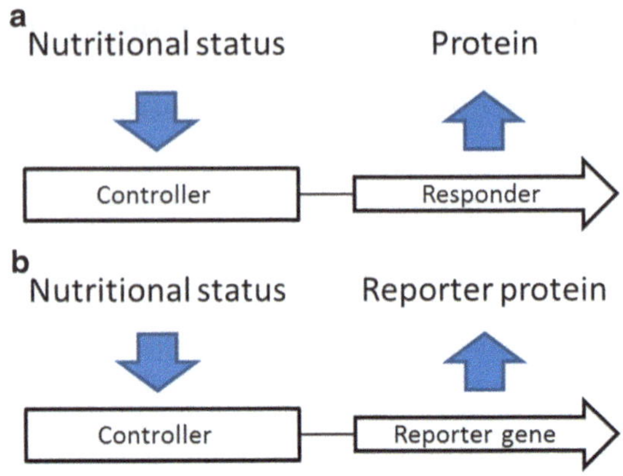

Fig. 10.3 Gene fusion is a DNA construction in which the coding sequence from one gene (reporter) is transcribed and/or translated under the direction of the controlling sequence of another gene (controller). (**a**) Gene structure. (**b**) Gene fusion

consisting of two parts: the promoter or controller that senses the nutritional status of the plant and directs the synthesis of a new product from the second component, the responder. By replacing the original sequence of the responder gene with a new and easily studied gene, called a reporter gene, it should be possible to obtain valuable information about the activity of the promoter. Such a molecular manipulation should provide information about the nutritional status of the plant by simply measuring the activity of the reporter protein.

Plant biologists to study how a particular gene is controlled when measurement of the gene product is too difficult have extensively used the gene fusion concept. The elective tool used in this type of studies is the GUS gene fusion system, which uses *uidA* from *E. coli* as a reporter gene. This gene encodes a β-glucuronidase able to hydrolyse a wide range of β-D-glucuronide substrates producing coloured or other compounds in amounts proportional to enzyme activity (Jefferson 1989). Assays for testing the activity of GUS in genetically modified plants are carried out on plant material or plant extracts under laboratory conditions. A variety of glucuronides is commercially available and can be used for fluorometric, spectrophotometric, luminometric and histochemical GUS analyses, qualitative as well as quantitative (Gallagher 1992).

Unfortunately, reporter systems based on the activity of β-glucuronidase (GUS) have severe intrinsic limitations that preclude their application under field conditions since they require the enzyme (GUS) and its substrate be brought together to produce the hydrolytic products necessary for the analysis. Likewise, luciferin–luciferase imaging systems have been used in plants, but their application in the field is hampered by the low level of light emission and the need for sensitive photon-counting cameras to detect signal (de Ruijter et al. 2003). Thus, new tools to perform non-destructive analysis are essential to develop specific sentinel plants to be directly used under field conditions.

In a pioneering paper, Jefferson (1993) listed some criteria useful to develop new reporter systems suitable for agricultural molecular biology. Some of the key criteria, such an in vivo reporter system are that it should be: (i) non-destructive; (ii) non-disruptive to avoid physiological alteration of crop performance; (iii) useful and functional in most crop species; (iv) inexpensive and capable of being used everywhere; (v) simply to detect with little or no instrumentation; (vi) easy to use under field conditions.

Naturally fluorescent proteins could offer a valuable alternative to the use of GUS reporter systems since, in contrast to GUS, the detection of their expression does not require the addition of a substrate. Green fluorescent protein (GFP), a spontaneously fluorescent protein, was initially isolated from the luminescent marine jellyfish (*Aequorea victoria*). GFP emits a highly and stable bright green fluorescence after absorbing blue light (Tsien 1998). The wild type *Aequorea* protein has a major excitation peak at 395 nm which is about three times higher in amplitude than a minor peak at 475 nm. In normal solution, excitation at 395 nm gives emissions peaking at 508 nm, whereas excitation at 475 nm gives a maximum at 503 nm (Heim et al. 1994). Since its discovery GFP from *Aequorea victoria* has become a frequently used tool in plant biology. The first studies on transgenic plants expressing wild type GFP proved the usefulness of this protein as an in vivo and real-time visible marker and encouraged researchers to modify it in order to obtain new variants that could be more effectively synthesised in plant cells and macroscopically detectable at the whole plant level (Stewart 2001). One of these modified versions of GFP is the mGFP5er variant that produces a stable protein targeted to the endoplasmic reticulum as the result of the addition of a N-terminal *Arabidopsis* basic chitinase fusion and a C-terminal HDEL fusion (Haseloff et al. 1997). The coding sequence contains three mutations that enhance the folding of mGFP5er at higher temperatures and allows excitation of the protein using either ultraviolet (395 nm) or blue (473 nm) light (Siemering et al. 1996). In addition, new fluorescence colours have been created through mutagenesis of the natural protein giving longer excitation and emission wavelengths and to enhance the fluorescence brightness. The new colours range from blue and cyan (EBFP and ECFP) to yellow (EYFP); such new proteins have excitation/emission peaks at 383/474, 434/472 and 514/527 nm, respectively (Spiess et al. 2005; Mena et al. 2006).

Fluorescent proteins have been largely used as visual genetic labels at the whole plant, tissue and cell levels, since they offer a fast and easy-to-use non-destructive tool with which the efficiency and timing of gene expression can be evaluated. Detection and quantification of fluorescence at the whole-plant level normally requires the use of complex and expensive laboratory instruments (scanning laser and fluorescence imaging systems). Portable instruments, such as fibre optic probe fluorometers, have recently been designed to assess GFP fluorescence under field conditions (Harper and Stewart 2000; Millwood et al. 2003). However, because bioindicators need to be disseminated over a wide area, a successful field application of these plants also requires a cost-effective remote monitoring system providing real-time information about the nutritional status of a whole crop system. Recently Adams et al. (2011) proposed an alternative method for crop monitoring in which sentinel plants and sensing units are deployed in tandem at specific locations. Ideally, such a system integrates biological and sensory technologies with communication technologies to provide a practical field-deployable telemetry system.

The gene fusion concept could be used to measure complex phenomena, even in the absence of mechanistic knowledge of how that phenomenon works (Jefferson 1993). This technology is completely general and could be exploited to develop transgenic bioindicators providing signals whose intensity is proportional to the concentration of a given analyte in growing environment (i.e. mineral nutrients, pollutants, water, etc.) or to the intensity of a biotic or abiotic stress that plants could experience during their growth. Potential targeted traits to be monitored are only limited by the availability of specific promoters (or controllers) driving the reporter expression under a specific condition.

Recently these technologies have been applied in plants to develop model transgenic bioindicators of the nutritional status to be used for laboratory purposes. To date, reporter gene activity has been used to assess the phosphate, sulfate and magnesium status in *Arabidopsis* and also to detect the level of nickel in the growing medium (Hammond et al. 2003; Krizek et al. 2003; Maruyama-Nakashita et al. 2006; Kamiya et al. 2012). In all these studies GUS, GFP and LUC have been successfully used as reporter genes to indicate nutritional status under the control of promoter sequences indirectly identified by microarray analyses.

Hammond and co-workers (2003) first proposed the creation of an *Arabidopsis* transgenic bioindicator, able to monitor plant phosphorous status. They fused GUS with the promoter of the phosphate starvation responsive gene *SQD1* (a gene involved in the synthesis of sulfolipids), obtaining an *Arabidopsis* transgenic line in which GUS activity increased following P starvation. Interestingly, the reporter responses to P withdrawal were much more rapid and quantitative than phenotypic observations, showing this approach is particularly suitable for developing efficient systems for monitoring plant P status. More recently Kamiya et al. (2012) used a similar approach to establish a novel monitoring system for magnesium in plants. In particular they obtained an *Arabidopsis* transgenic line that expressed luciferase (LUC) under the control of the Mg deficiency-inducible *CAX3* promoter. The transgenic lines showed a clear response under low Mg conditions and the degree of luminescence reflected the accumulation of endogenous *CAX3* mRNA. However, *CAX3* induction does not seem to be specific to low Mg, since the levels of other ions (Ca^{2+} and Na^{+}) or P starvation may influence transcription (Shigaki and Hirschi 2000).

Notwithstanding some limitations *Arabidopsis* 'smart' plants could also be used as tools in basic research aimed at isolating novel mutants disrupted in nutrient homeostasis or identifying plants with enhanced nutrient use efficiency. For instance, the key transcription factor, SLIM1, regulating the sulfur assimilatory pathway has recently been identified by screening *Arabidopsis* mutants carrying a fluorescent reporter gene under the control of the sulfur limitation-responsive promoter of the SULTR1;2 sulfate transporter (Maruyama-Nakashita et al. 2006). Using this approach it is possible to identify all the potential genes involved in controlling the expression of SULTR1;2 under sulfur shortage, since in this condition the relative mutants will display altered fluorescence emissions as compared to the wild type bioindicator.

Strategies to Enhance the Specificity of Bioindicators

In the future, the use of smart plant technology in crops would provide rapid bioassay methods to obtain valuable information about nutrient availability in the soil solution and/or the nutritional status of the plants allowing efficient temporal and special application of fertilisers and the development of decision-making systems for precision farming. To date the exploitation of these technologies is limited, not only by the lack of telemetry systems suitable for plant monitoring across a large area, but also by the lack of precise information to design a transformation-cassette that would enable the nutrient-specific control of reporter activity. Thus, the choice of a core promoter to confer specific transgene expression, represent the major challenge we have to face in order to develop the next generation of bioindicators.

A typical plant promoter consists of CAAT and TATA boxes for recognition of DNA-dependent RNA polymerase, several-tens of bp upstream of the transcription initiation site (Yoshida and Shinmyo 2000). Specific DNA sequences, called *cis*-elements, generally upstream of the core promoter, drive the cell- or organ-specific expression of the downstream gene under certain environmental conditions. Specific factors, called *trans*-factors (or transcription factors), bind to the *cis*-elements affecting RNA polymerase activity. Generally, multiple-*cis*-elements and *trans*-factors work together to induce the full regulation of gene expression, since gene expression is generally under the control of several factors (Yoshida and Shinmyo 2000; Venter 2007).

In recent years, a wide range of different promoters have been characterised and extensively used for regulating the expression of transgenes in plant cells (Venter 2007). In several cases, the *cis*-elements that are necessary for transcriptional regulation and the *trans*-factors that interact with these elements have been identified. From these studies has emerged a complex picture in which DNA sequence *cis*-elements that are important for regulation are scattered over thousands of base pairs, and these elements interact with *trans*-factors that can be either ubiquitous or highly restricted in their distribution. In this way diverse expression patterns may be achieved through combinations of a limited number of regulatory elements and *trans*-acting factors. The knowledge of these combinatorial mechanisms should allow the generation different transcription patterns by 'cut and pasting' the components in different ways.

Analysis of the cauliflower mosaic virus (CaMV) 35S promoter has contributed to the understanding of transcriptional regulatory mechanisms and has allowed the design of inducible transgene expression cassettes. The −343 to −46 upstream region relative to the site of initiation of transcription (+1) of the promoter is responsible for the strength of transcription. Two regions, −343 to −208 and −208 to −90, are responsible for transcriptional activation, and the −90 to −46 region plays an accessory role by further increasing the transcriptional activity (Odell et al. 1985; Fang et al. 1989). Artificial promoters are generally constructed by a combinatorial design of different promoter elements, with the minimal core

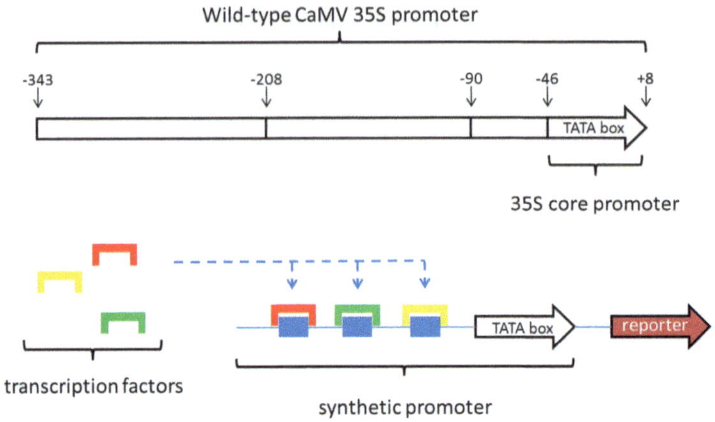

Fig. 10.4 Schematic representation of a synthetic promoter useful for bioindication purposes. The core region of the CaMV 35S promoter is fused with a combinatorial engineering of cis-elements (*blue boxes*) which, following the interaction with specific transcription factors, drives the reporter expression under particular conditions

DNA fragment (−46 to +8 bp) of the CaMV 35S promoter as the main component (Fig. 10.4). The core-promoter region contains a TATA-box necessary for recruiting RNA polymerase II and the orchestrated assembly of general transcription factors to form the pre-initiation complex (Novina and Roy 1996). The CaMV 35S core-promoter is ideal for transcription initiation and has been used in several synthetic plant promoters in which combinatorial engineering of *cis*-element have been introduced upstream of the core-promoter sequence.

The use of synthetic promoters allowing for targeted inducibility of a reporter gene is of considerable interest to develop engineering strategies aimed at creating plant bioindicators for real-time monitoring of nutritional status. For these purposes promoter sequence domains or *cis*-elements conferring nutrient- and organ-specificity should be combined in order to target the reporter expression in organs (shoot and leaves) in which signals should be easily detectable.

Many different plant promoters have been described as able to restrict gene expression to particular cells, tissues or organs. The *GaMYB2* promoter is cotton fibre- and *Arabidopsis* trichome-specific, and can drive gene expression specifically in glandular cells (head cells) of glandular trichomes in transgenic tobacco (Shangguan et al. 2008). Some *cis*-elements regulating tissue-specific gene expression have also been identified. For instance, mesophyll expression module 1 (*Mem1*), a 41 bp fragment of the *ppcA1* promoter, directs mesophyll-specific expression. The tetranucleotide sequence, CACT has been identified as a key component of *Mem1* by evolutionary and functional studies (Gowik et al 2004). More recently, Ye et al. (2012) identified a rice green tissue-specific expression gene, *DX1*, and described two novel tissue-specific *cis*-elements (GSE1 and GSE2) within the *DX1* promoter. In particular, GSE1 acted as a positive regulator in all green tissues, whereas GSE2 acted as a positive regulator only in sheath and stem tissues.

Obviously, nutrient-specific *cis*-elements are equally as important for reporter expression as tissue-specific *cis*-elements. Nutrient-inducible plant promoters contain multiple *cis*-acting elements, only some of which may specifically contribute to nutrient inducibility. A number of potential nutrient responsive *cis*-elements have recently been identified in the promoter of several nutrient responsive genes and have been indicated as key regulatory factors of gene expression under different nutritional conditions.

Sulfur-responsive elements (SUREs) have been identified in the promoter regions of the *Arabidopsis NIT3* nitrilase and β-subunit β-conglycinin gene from soybean (Awazuhara et al. 2002; Kutz et al. 2002), although no consensus sequences have been shown yet. However, an interesting study on *Arabidopsis* sulfate transporter *SULTR1;1* promoter demonstrates that a 5 bp sequence is essential to promote sulfur response of SULTR1;1 (Maruyama-Nakashita et al. 2005). Such a sequence also appears in the promoter regions of many sulfur-responsive genes, suggesting its involvement in the transcriptional control of a gene set required for adaptation to sulfur-limiting conditions. Deletion analysis of the barley *IDS2* (iron deficiency-specific clone no. 2) gene promoter allowed the identification of two *cis*-acting elements, iron-deficiency-responsive element 1 and 2 (IDE1 and IDE2), which synergistically induced iron-specific expression in tobacco roots. Finally, comparative analyses of several nitrite reductase gene promoters from various higher plants have recently allowed identification of a conserved sequence motif as nitrate-responsive *cis*-element (Konishi and Yanagisawa 2010).

What Do Bioindicators Sense? A Key Problem

Modification of promoter architecture necessary for manipulating gene reporter activity requires accurate studies of the regulatory network involved in controlling gene expression under different nutritional conditions. Unfortunately, for the most part, these aspects are still largely unknown preventing the optimal design of a synthetic nutrient-inducible promoter, particularly in cases where a mineral nutrient undergoes complex assimilatory metabolisms (i.e. nitrate or sulfate) or interacts with other nutrients. In all these cases the specific question to be answered is: what do synthetic nutrient-specific promoters sense?

For example, the transcriptional regulatory mechanisms involved in sulfate uptake and assimilation reasonably result from direct sensing of the plant nutritional status rather than from the composition of the external soil solution (Lappartient and Touraine 1997; Lappartient et al. 1999). This control involves an inter-organ signaling mechanism in which key intermediates of the sulfate assimilatory pathway may act as negative or positive signals in modulating the expression of the sulfur-responsive genes. Adequate levels of sulfur compounds would repress gene expression through a negative feedback loop preventing excessive sulfate uptake and reduction; vice versa a contraction of the intermediates along the assimilatory pathway would unrepress gene transcription allowing sulfate to enter the pathway. A second regulatory loop, involving OAS as a key intermediate, would act in promoting gene unrepression when nitrogen and carbon supply exceeds sulfur availability within the cells (Hawkesford 2000). In this context the need to dissect the molecular mechanisms involved in the nutritional signal perception and transduction is evident since, in several cases, the relationships existing between gene expression and the levels of the signal-intermediates are not always clear. Further research is needed to associate single gene expressions to a specific nutritional signal or sulfur-nutritional status.

Genome-wide expression analyses have revealed that nitrate supply induces changes in the expression of several genes, not only those involved in nitrate reduction and assimilation. Such behaviour is likely both due to the direct effects of nitrate itself and indirect effects caused by changes in nitrogen metabolite content or nitrogen nutritional status. In fact, nitrate is thought to act as a signal molecule influencing the expression of a number of genes, since their expression is rapidly induced by nitrate even in mutants severely compromised for nitrate reductase activity (Wang et al. 2004). In addition, it has been shown that nitrate-inducible expression NADH/nitrate reductase mRNA in maize roots, scutella and leaves also occurs in the presence of inhibitors of protein synthesis, suggesting that the signal transduction system mediating this response is constitutively expressed in plant cells, independently of the presence or the absence of nitrate in the growing medium (Price et al. 2004). Results of these studies clearly shows that dissection analyses of the signal transduction pathways controlling gene expression under different nitrogen supply should provide important information to define smart plants able to sense the cellular level of nitrite or the general nitrogen nutritional status of a crop system.

Conclusions

Growing varieties with enhanced efficiency and/or modifying the environment in which the crop is grown can increase the nutrient use efficiency of a crop production system. The selection of varieties with improved nutrient use efficiency is a more generic approach, which necessarily requires deep knowledge of the genetic variation and inheritance of these traits. On the other hand, the improvement of the agricultural practices aimed at sustaining the nutrient needs of a crop may provide more immediate advantages in terms of cost and environmental quality.

Although much progress has been made in improving fertilisation practices there remain considerable uncertainties about the persistence of nutrients in the soil and their actual availability to the plants. The development of quick and inexpensive methods to determine changes in nutrient bioavailability in the soil or the nutritional status of the plants are desirable for better fertiliser management for different crops in a variety of environmental conditions. It could be particularly important not only for areas of intensive agriculture but also for agriculture in developing countries where accessibility to fertilisers could be a problem.

(continued)

The development of optical devices allowing the remote monitoring at sub-field scale of traits specifically linked to molecular, biochemical or physiological consequences of scarcity or excess, of specific essential elements other than nitrogen, is surely a challenge for future research. Following on from the *spectranomic* approach developed by Asner and Martin (2008) as the application of metabolomics for the assessment of biochemical traits at a landscape scale, Brunetti et al. (2013) suggest an airborne hyperspectral analysis whereby changes in absorbance and reflectance patterns of specific chemical compounds within leaves could further improve detection of biochemical phenotypes across a relatively large scale. A combination of these approaches could define exciting new frontiers for the fine-tuning of continuous and non-destructive methods of monitoring plant nutritional status in the field. In the next years, whereby multidisciplinary studies, efforts have to be planned to actually translate in field the knowledge accumulated that seem to promise interesting deliverables.

References

Adamchuck VI, Viscarra Rossel RA, Sudduth KA, Lammers PS (2011) Sensor fusion for precision agriculture. In: Thomas C (ed) Sensors fusion-foundation and applications. InTech Publishers, Croatia, pp 27–40

Adams JP, Topsakal E, Yuceer C (2011) Fluorescing phytosensors: a speculated next step for environmental monitoring. Plant Mol Biol Biotechnol 2:16–25

Asner GP, Martin RE (2008) Airborne spectranomics: mapping canopy chemical and taxonomic diversity in tropical forests. Front Ecol Environ 7:269–278

Awazuhara M, Kim H, Goto DB, Matsui A, Hayashi H, Chino M, Kim SG, Naito S, Fujiwara T (2002) A 235-bp region from a nutritionally regulated soybean seed-specific gene promoter can confer its sulfur and nitrogen response to a constitutive promoter in aerial tissues of *Arabidopsis thaliana*. Plant Sci 163:75–82

Barker AV, Pilbeam DJ (2007) In: Baker AV, Pilbeam DJ (eds) Handbook of plant nutrition. CRC Publisher, Boca Raton, pp 3–18

Bi YM, Wang RL, Zhu T, Rothstein SJ (2007) Global transcription profiling reveals differential responses to chronic nitrogen stress and putative nitrogen regulatory components in *Arabidopsis*. BMC Genomics 8:281

Brunetti C, George RM, Tattini M, Field K, Davey MP (2013) Metabolomics in plant environmental physiology. J Exp Bot 64:4011–4020

Cartelat A, Cerovic ZG, Goulas Y, Meyer S, Lelarge C, Prioul JL, Barbottien A, Jeuffroy MH, Gate P, Agati G, Moja I (2005) Optically assessed contents of leaf polyphenolics and chlorophyll as indicators of nitrogen deficiency in wheat (*Triticum aestivum* L.). Field Crop Res 91:35–49

Craswell ET, Godwin DC (1984) The efficiency of nitrogen fertilizers applied to cereals grown in different climates. In: Tinker PB, Luachli A (eds) Advances in plant nutrition, vol 1. Praeger Publisher, New York, pp 1–55

de Ruijter NCA, Verhees J, van Leeuwen W, van der Krol AR (2003) Evaluation and comparison of the GUS, LUC, and GFP reporter system for gene expression studies in plants. Plant Biol 5:103–115

Debaeke P, Rouet P, Justes E (2006) Relationship between the normalized SPAD index and the nitrogen nutrition index: application to durum wheat. J Plant Nutr 29:75–92
Diacono M, Rubino P, Montemurro F (2013) Precision N management of wheat. A review. Agron Sustain Dev 33:219–241
Erdle K, Mistele B, Schmidhalter U (2011) Comparison of active and passive spectral sensors in discriminating biomass parameters and nitrogen status in wheat cultivars. Field Crop Res 124:74–84
Fang RX, Nagy F, Sivasubramaniam S, Chua NH (1989) Multiple *cis* regulatory elements for maximal expression of the cauliflower mosaic virus 35S promoter in transgenic plants. Plant Cell 1:141–150
Gallagher SR (1992) GUS protocols: using the GUS gene as a reporter of gene expression. Academic, San Diego, pp 1–221
Godfray HCJ, Beddington JR, Crute IR, Haddad L, Lawrence D, Muir JF, Pretty J, Robinson S, Thomas SM, Toulmin C (2010) Food security: the challenge of feeding 9 billion people. Science 327:812–818
Gojon A, Nacry P, Davidian JC (2009) Root uptake regulation: a central process for NPS homeostasis in plants. Curr Opin Plant Biol 12:328–338
Gowik U, Burscheidt J, Akyildiz M, Schlue U, Koczor M, Streubel M, Westhoff P (2004) Cis-regulatory elements for mesophyll-specific gene expression in the C_4 plant *Flaveria trinervia*, the promoter of the C_4 phosphoenolpyruvate carboxylase gene. Plant Cell 16:1077–1090
Greenwood DJ, Neeteson JJ, Draycott A (1986) Quantitative relationships for the dependence of growth rate of arable crops on their nitrogen content, dry weight and aerial environment. Plant Soil 91:281–301
Hamilton J, Zangerl A, DeLucia E, Berenbaum M (2001) The carbon-nutrient balance hypothesis: its rise and fall. Ecol Lett 4:86–95
Hammond JP, Bennett MJ, Bowen HC, Broadley MR, Eastwood DC, May ST, Rahn C, Swarup R, Woolaway KE, White PJ (2003) Changes in gene expression in *Arabidopsis* shoots during phosphate starvation and the potential for developing smart plants. Plant Physiol 132:578–596
Hansen PM, Schjoerring JK (2003) Reflectance measurement of canopy biomass and nitrogen status in wheat crops using normalized difference vegetation indices and partial least squares regression. Remote Sens Environ 86:542–553
Harper BK, Stewart CN (2000) Patterns of green fluorescent protein expression in transgenic plants. Plant Mol Biol Report 18:141a–141i
Haseloff J, Siemering KR, Prasher DC, Hodge S (1997) Removal of a cryptic intron and subcellular localization of green fluorescent protein are required to mark transgenic *Arabidopsis* plants brightly. Proc Natl Acad Sci U S A 94:2122–2127
Hawkesford MJ (2000) Plant responses to sulphur deficiency and the genetic manipulation of sulphate transporters to improve S-utilization efficiency. J Exp Bot 51:131–138
Heim R, Prasher DC, Tsien RY (1994) Wavelength mutations and posttranslational autoxidation of green fluorescent protein. Proc Natl Acad Sci U S A 91:12501–12504
Jefferson RA (1989) The GUS reporter gene system. Nature 342:837–838
Jefferson RA (1993) Beyond model systems: new strategies, methods, and mechanisms for agricultural research. Ann N Y Acad Sci 700:53–73
Kalra YP (1998) Handbook of reference methods for plant analysis. CRC Press, Boca Raton
Kamiya T, Yamagami M, Hirai MY, Fujiwara T (2012) Establishment of an in planta magnesium monitoring system using CAX3 promoter-luciferase in *Arabidopsis*. J Exp Bot 63:355–363
Kant S, Bi YM, Rothstein SJ (2010) Understanding plant response to nitrogen limitation for the improvement of crop nitrogen use efficiency. J Exp Bot 62:1499–1509
Knyazikhin Y, Schull MA, Stenberg P, Mõttus M, Rautiainen M, Yang Y, Marshak A, Carmona PL, Kaufmann RK, Lewis P, Disney MI, Vanderbilt V, Davis AB, Baret F, Jacquemoud S, Lyapustin A, Myneni RB (2013) Hyperspectral remote sensing of foliar nitrogen content. Proc Natl Acad Sci U S A 110:E185–E192

Konishi M, Yanagisawa S (2010) Identification of a nitrate-responsive *cis*-element in the *Arabidopsis NIR1* promoter defines the presence of multiple *cis*-regulatory elements for nitrogen response. Plant J 63:269–282

Krizek BA, Prost V, Joshi RM, Stoming T, Glenn TC (2003) Developing transgenic *Arabidopsis* plants to be metal-specific bioindicators. Environ Toxicol Chem 22:175–181

Kutz A, Muller A, Hennig P, Kaiser WM, Piotrowski M, Weiler EW (2002) A role for nitrilase 3 in the regulation of root morphology in sulphur-starving *Arabidopsis thaliana*. Plant J 30:95–106

Lappartient AG, Touraine B (1997) Glutathione-mediated regulation of ATP sulfurylase activity, SO_4^{2-} uptake, and oxidative stress response in intact canola roots. Plant Physiol 114:177–183

Lappartient AG, Vidmar JJ, Leustek T, Glass AMD, Touraine B (1999) Inter-organ signalling in plants: regulation of ATP sulfurylase and sulfate transporter genes expression in roots mediated by phloem-translocated compound. Plant J 18:89–95

Lea US, Slimestad R, Smedvig P, Lillo C (2007) Nitrogen deficiency enhances expression of specific MYB and bHLH transcription factors and accumulation of end products in the flavonoids pathway. Planta 225:1245–1253

Lemaire G, Gastal F (2009) Quantifying crop responses to nitrogen deficiency and avenues to improve nitrogen use efficiency. In: Sadras VO, Calderini DF (eds) Crop physiology: applications for genetic improvement and agronomy. Elsevier Publisher, Adelaide, pp 171–211

Li L, Liu C, Lian X (2010) Gene expression profiles in rice roots under low phosphorus stress. Plant Mol Biol 72:423–432

Lu C, Zhang J (2000) Photosynthetic CO_2 assimilation, chlorophyll fluorescence and photoinhibition as affected by nitrogen deficiency in maize plants. Plant Sci 151:135–143

Ma TL, Wu WH, Wang Y (2012) Transcriptome analysis of rice root responses to potassium deficiency. BMC Plant Biol 12:161

Malhi BS, Haderlein LK, Pauly DG, Johnston AM (2002) Improving fertilizer phosphorus use efficiency. Better Crop 86:8–9

Maruyama-Nakashita A, Inoue E, Watanabe-Takahashi A, Yamaya T, Takahashi H (2003) Transcriptome profiling of sulfur-responsive genes in *Arabidopsis* reveals global effects of sulfur nutrition on multiple metabolic pathways. Plant Physiol 132:597–605

Maruyama-Nakashita A, Nakamura Y, Watanabe-Takahashi A, Inoue E, Yamaya T, Takahashi H (2005) Identification of a novel *cis*-acting element conferring sulfur deficiency response in *Arabidopsis* roots. Plant J 42:305–314

Maruyama-Nakashita A, Nakamura Y, Tohge T, Saito K, Takahashi H (2006) *Arabidopsis* SLIM1 is a central transcriptional regulator of plant sulfur response and metabolism. Plant Cell 18:3235–3251

Mena MA, Treynor TP, Mayo SL, Daugherty PS (2006) Blue fluorescent proteins with enhanced brightness and photostability from a structurally targeted library. Nat Biotechnol 24:1569–1571

Millwood RJ, Halfhill MD, Harkins D, Russotti R, Stewart CN (2003) Instrumentation and methodology for quantifying GFP fluorescence in intact plant organs. Biotechniques 34:638–643

Novina CD, Roy AL (1996) Core promoters and transcriptional control. Trends Genet 12:351–355

Odell JT, Nagy F, Chua NH (1985) Identification of DNA sequences required for activity of the cauliflower mosaic virus 35S promoter. Nature 313:810–812

Ollinger SV, Richardson AD, Martin ME, Hollinger DY, Frolking SE, Reich PB, Plourd LC, Katul GG, Munger JW, Oren R, Smith ML, Paw UKT, Bolstad PV, Cook BD, Day MC, Martin TA, Monson RK, Schmid HP (2008) Canopy nitrogen, carbon assimilation, and albedo in temperate and boreal forests: functional relations and potential climate feedbacks. Proc Natl Acad Sci U S A 105:19336–19341

Pierce FJ, Nowak P (1999) Aspects of precision agriculture. In: Sparks DL (ed) Advances in agronomy. Academic Press, New York, pp 1–85

Price J, Laxmi A, St Martin SK, Jang JC (2004) Global transcription profiling reveals multiple sugar signal transduction mechanisms in *Arabidopsis*. Plant Cell 16:2128–2150

Raun WR, Johnson GV (1999) Improving nitrogen use efficiency for cereal production. Agron J 91:357–363

Rengel Z, Damon PM (2008) Crops and genotypes differ in efficiency of potassium uptake and use. Physiol Plant 133:624–636

Rouached H, Secco D, Arpa BA (2010) Regulation of ion homeostasis in plants: current approaches and future challenges. Plant Sign Behav 5:501–502

Samborski SM, Tremblay N, Fallon E (2009) Strategies to make use of plant sensors-based diagnostic information for nitrogen recommendations. Agron J 101:800–816

Schatchtman DP, Shin R (2007) Nutrient sensing and signalling: NPKS. Annu Rev Plant Biol 58:47–69

Schlemmer MR, Francis DD, Shanahan JF, Schepers JS (2005) Remotely measuring chlorophyll content in corn leaves with differing nitrogen levels and relative water content. Agron J 97:106–112

Shangguan XX, Xu B, Yu ZX, Wang LJ, Chen XY (2008) Promoter of a cotton fiber MYB gene functional in trichomes of *Arabidopsis* and glandular trichomes of tobacco. J Exp Bot 59:3533–3542

Shigaki T, Hirschi K (2000) Characterization of CAX-like genes in plants: implications for functional diversity. Gene 257:291–298

Siemering KR, Golbik R, Sever R, Haseloff J (1996) Mutations that suppress the thermosensitivity of green fluorescent protein. Curr Biol 6:1653–1663

Solari F, Shanahan JF, Ferguson R, Schepers JS, Gitelson A (2008) Active sensor reflectance measurements of corn nitrogen status and yield potential. Agron J 100:571–579

Spiess E, Bestvater F, Heckel-Pompey A, Toth K, Hacker M, Stobrawa G, Feurer T, Wotzlaw C, Berchner-Pfannschmidt U, Porwol T, Acker H (2005) Two-photon excitation and emission spectra of the green fluorescent protein variants ECFP, EGFP and EYFP. J Microsc 217:200–204

Stewart CN (2001) The utility of green fluorescent protein in transgenic plants. Plant Cell Rep 20:376–382

Stroppiana D, Boschetti M, Brivio PA, Bocchi S (2009) Plant nitrogen concentration in paddy rice from field canopy hyperspectral radiometry. Field Crop Res 111:119–129

Tremblay N, Wang Z, Cerovic ZC (2012) Sensing crop nitrogen status with fluorescence indicators. A review. Agron Sustain Dev 32:451–464

Tsien RY (1998) The green fluorescent protein. Annu Rev Biochem 67:509–544

Venter M (2007) Synthetic promoters: genetic control through *cis* engineering. Trends Plant Sci 12:118–124

Wang R, Okamoto M, Xing X, Crawford NM (2003) Microarray analysis of the nitrate response in *Arabidopsis* roots and shoots reveals over 1,000 rapidly responding genes and new linkages to glucose, trehalose-6-phosphate, iron, and sulfate metabolism. Plant Physiol 132:556–567

Wang R, Tischner R, Gutiérrez RA, Hoffman M, Xing X, Chen M, Coruzzi G, Crawford NM (2004) Genomic analysis of the nitrate response using a nitrate reductase-null mutant of *Arabidopsis*. Plant Physiol 136:2512–2522

Yang XS, Wu J, Ziegler TE, Yang X, Zayed A, Rajani MS, Zhou D, Basra AS, Schachtman DP, Peng M, Armstrong CL, Caldo RA, Morrell JA, Lacy M, Staub JM (2011) Gene expression biomarkers provide sensitive indicators of in planta nitrogen status in maize. Plant Physiol 157:1841–1852

Ye R, Zhou F, Lin Y (2012) Two novel positive *cis*-regulatory elements involved in green tissue-specific promoter activity in rice (*Oryza sativa* L ssp.). Plant Cell Rep 31:1159–1172

Yoshida K, Shinmyo A (2000) Transgene expression systems in plant, a natural bioreactor. J Biosci Bioeng 90:353–362

Zheng G, Moskal LM (2009) Retrieving leaf area index (LAI) using remote sensing: theories, methods and sensors. Sensors 9:2719–2745

Index

M.J. Hawkesford et al. (eds.), *Nutrient Use Efficiency in Plants*, Plant Ecophysiology 10, DOI 10.1007/978-3-319-10635-9

MIX
Papier aus verantwortungsvollen Quellen
Paper from responsible sources
FSC® C105338

If you have any concerns about our products,
you can contact us on
ProductSafety@springernature.com

In case Publisher is established outside the EU,
the EU authorized representative is:
Springer Nature Customer Service Center GmbH
Europaplatz 3, 69115 Heidelberg, Germany

Printed by Libri Plureos GmbH
in Hamburg, Germany